T0251306

TV White Space Spectrum Technologies

Regulations, Standards, and Applications

TV White Space Spectrum Technologies

Regulations, Standards, and Applications

Edited by

Rashid A. Saeed and Stephen J. Shellhammer

CRC Press
Taylor & Francis Group
Boca Raton London New York

CRC Press is an imprint of the
Taylor & Francis Group, an **informa** business
AN AUERBACH BOOK

CRC Press
Taylor & Francis Group
6000 Broken Sound Parkway NW, Suite 300
Boca Raton, FL 33487-2742

© 2012 by Taylor & Francis Group, LLC
CRC Press is an imprint of Taylor & Francis Group, an Informa business

No claim to original U.S. Government works

International Standard Book Number: 978-1-4398-4879-1 (Hardback)

Visit the Taylor & Francis Web site at
http://www.taylorandfrancis.com

and the CRC Press Web site at
http://www.crcpress.com

Contents

About the Editors

Rashid A. Saeed received his BSc in Electronics Engineering from Sudan University of Science and Technology (SUST) and his PhD in Communication Engineering from Universiti Putra Malaysia (UPM). He served as a senior researcher at MIMOS Berhad and then at Telekom Malaysia R&D, where he was awarded the Platinum Badge for Outstanding Research Achievement Award. Since 2010, he has been an assistant professor of electrical engineering at Universiti Islam Antarabangsa Malaysia.

Rashid has published and is responsible for over 70 research papers, tutorials, talks, and book chapters on the topic of UWB, cognitive radio, and radio resources management. He was awarded two US patents and has filed for eight more. Rashid is a certified WiMAX engineer (RF and core network) and is a Six Sigma–certified Black Belt, based on DMAIC++ from Motorola University. He is one of the contributors of IEEE-WCET wireless certification in its earlier stages, and is a senior member of the IEEE, IEM Malaysia, and Sigma Xi.

Stephen J. Shellhammer leads a cognitive radio project within the Qualcomm Corporate Research and Development Department. He is currently the chair of the IEEE 802.19 working group on wireless coexistence, leading a project on TV white space coexistence. He was also the technical lead on spectrum sensing within the IEEE 802.22 working group. He is currently a member of the IEEE 802 executive committee and was also the chair of the IEEE 802.15.2 task group on wireless coexistence. Before joining Qualcomm, he was the Director of the Advanced Development Department at Symbol Technologies, and later worked at Intel in its wireless local area network division.

Stephen has a BS in Physics from the University of California, San Diego; an MSEE from San Jose State University; and a PhD in Electrical Engineering from the University of California, Santa Barbara. He was an adjunct professor at SUNY Stony Brook, where he taught graduate courses in electrical engineering. He is a senior member of the IEEE.

Contributors

Yohannes D. Alemseged
Smart Wireless Laboratory
Wireless Network Research Institute
National Institute of Information
   and Communications Technology
   (NICT)
Yokosuka, Japan

Borhanuddin M. Ali
Engineering Department
Universiti Putra Malaysia
Serdang, Malaysia

Tuncer Baykas
Smart Wireless Laboratory
Wireless Network Research Institute
National Institute of Information
   and Communications Technology
   (NICT)
Yokosuka, Japan

Danijela Čabrić
Department of Electrical Engineering
University of California, Los Angeles
Los Angeles, California

Mark Cummings
Kennesaw State University
Atlanta, Georgia
and
Orchestral Networks
Silicon Valley, California

Randy L. Ekl
Motorola Solutions, Inc.
Schaumburg, Illinois

Stanislav Filin
Smart Wireless Laboratory
Wireless Network Research Institute
National Institute of Information
   and Communications Technology
   (NICT)
Yokosuka, Japan

Benoît Pierre Freyens
University of Canberra
Canberra, Australia

Mario Gerla
University of California, Los Angeles
Los Angeles, California

Pål Grønsund
Department of Informatics
University of Oslo
Oslo, Norway

David P. Gurney
Motorola Solutions, Inc.
Schaumburg, Illinois

Hiroshi Harada
Smart Wireless Laboratory
Wireless Network Research Institute
National Institute of Information
    and Communications Technology
    (NICT)
Yokosuka, Japan

Jukka Henriksson
Fairspectrum Ltd
Espoo, Finland

Wendong Hu
Corallia, Inc.
San José, California

Mika Kasslin
Nokia Research Center
Helsinki, Finland

Heikki Kokkinen
Fairspectrum Ltd
Helsinki, Finland

Zhou Lan
Smart Wireless Laboratory
Wireless Network Research Institute
National Institute of Information
    and Communications Technology
    (NICT)
Yokosuka, Japan

Mark Loney
Australian Communications & Media
    Authority
Canberra, Australia

Paulo Marques
Instituto de Telecomunicações – Pólo
    de Aveiro
Aveiro, Portugal

Preston Marshall
Department of Electrical Engineering
University of Southern California
Los Angeles, California

Rania A. Mokhtar
Engineering Faculty
Sudan University of Science and
    Technology (SUST)
Khartoum, Sudan

Markus Muck
Infineon Technologies
Neubiberg, Germany

Joseph W. Mwangoka
Instituto de Telecomunicações – Pólo
    de Aveiro
Aveiro, Portugal

Bruce D. Oberlies
Motorola Solutions, Inc.
Schamburg, Illinois

Richard Paine
Marysville, Washington

Jihoon Park
Department of Electrical Engineering
University of California, Los Angeles
Los Angeles, Califonia

Przemysław Pawełczak
Department of Electrical Engineering
University of California, Los Angeles
Los Angeles, Califonia

Gregory J. Pottie
University of California, Los Angeles
Los Angeles, California

M. Azizur Rahman
Smart Wireless Laboratory
Wireless Network Research Institute
National Institute of Information
   and Communications Technology
   (NICT)
Yokosuka, Japan

Alex Reznik
InterDigital
King of Prussian, Pennsylvania

Jonathan Rodriguez
Instituto de Telecomunicações – Pólo
   de Aveiro
Aveiro, Portugal

Paivi Ruuska
Nokia Research Center
Helsinki, Finland

Ahmed K. Sadek
Qualcomm
San Diego, California

Cong Shen
Qualcomm
San Diego, California

Chin-Sean Sum
Smart Wireless Laboratory
Wireless Network Research Institute

National Institute of Information
   and Communications Technology
   (NICT)
Yokosuka, Japan

Chen Sun
Smart Wireless Laboratory
Wireless Network Research Institute
National Institute of Information
   and Communications Technology
   (NICT)
Yokosuka, Japan

Ha Nguyen Tran
Smart Wireless Laboratory
Wireless Network Research Institute
National Institute of Information
   and Communications Technology
   (NICT)
Yokosuka, Japan

Gabriel Porto Villardi
Smart Wireless Laboratory
Wireless Network Research Institute
National Institute of Information
   and Communications Technology
   (NICT)
Yokosuka, Japan

Jianfeng Wang
Philips Research North America
Briarcliff Manor, New York

Junyi Wang
Smart Wireless Laboratory
Wireless Network Research Institute
National Institute of Information
   and Communications Technology
   (NICT)
Yokosuka, Japan

Risto Wichman
School of Electrical Engineering
Aalto University
Espoo, Finland

Wenyi Zhang
Department of Electronic Engineering
and Information Science
University of Science and Technology
of China
Hefei, China

Introduction

Rashid A. Saeed and Stephen J. Shellhammer

Overview

The need for additional spectrum to provide wireless services has been growing at an accelerating pace. Smartphone use has exploded, and meeting the demand for wireless broadband services is a challenge. One of the main issues is the scarcity of spectrum to support these wireless services. Cognitive radio (CR) is a promising new technology for enabling access to additional spectrum that is so badly needed. A number of conferences focusing on cognitive radio have emerged in the past few years, and there has also been an increase in the number of publications focusing on cognitive radio, including a number of recent cognitive radio books published [1–14]. However, up till now there has not been a book published with a specific focus on cognitive radio in the TV white space (TVWS). This volume covers a broad range of topics on the TV white space wireless networks.

TVWS—Broadcast Television to Wireless Broadband

TV white space is the portion of the TV bands that is unused by licensed services. There are many TV channels in very high frequency (VHF) and ultrahigh frequency (UHF) TV bands. In some geographic regions the utilization of these TV channels is high, while in other regions many of the channels are unused. Regulatory agencies, like the Federal Communication Commission (FCC) in the United States, have been developing regulations to allow wireless networks to obtain access to these unused channels while ensuring that these wireless networks not cause harmful interference to the licensed services in the TV bands. The set of unused TV channels, referred to as TV white space, can change based on geographic location and on time.

Since the TV white space spectrum, in both the VHF and UHF band, is below 1GHz it will provide much better RF propagation than the systems deployed in the Industrial, Scientific and Medical (ISM) bands, allowing for a reliable, cost-effective, and better coverage in rural areas and metropolitan applications, such as intelligent transportation, emergency and public safety, and smart grid.

Why TVWS is Unique

Operation in the TV white space is conditioned on the devices not causing harmful interference to the licensed services in the TV bands. Cognitive radio technology is proposed in TV white space to ensure that the TV white space devices do not cause harmful interference to the incumbent services operating in the occupied TV channels.

There are several cognitive radio technologies that have been developed to ensure this protection of licensed services. The first cognitive radio technology is actually a combination of two technologies: geo-location positioning and a database of incumbent systems. Geo-location positioning is the ability of a TV white space device to know its location in terms of latitude and longitude. An example of a geo-location positioning technology is a global positioning system (GPS) receiver. The TV white space database is a database which stores information about all the licensed services in the TV bands. For example, the database stores the location of a TV broadcast station, its channel of operation, and its transmit power. From this information it can calculate a geographic region in which TV receivers can receive the TV broadcast signal and are to be protected from harmful interference. The TV white space database provides information to the TV white space devices as to which channels they can utilize in their known location. In some cases, like for fixed outdoor systems that are professionally installed, the GPS system may not be required, because the fixed system which has been professionally installed is at a registered location, which has been certified by the professional installer.

The second cognitive radio technology used to protect incumbent systems, referred to in the literature as *spectrum sensing,* is a technology embedded in the TV white space device that makes measurements of the radio frequency (RF) TV channels. From those RF measurements. it determines which channels are occupied by incumbent systems and which are unoccupied and are hence TV white space.

In the first case the cognitive radio technology is split between the TV white space device (geo-location) and an external entity (TV white space database). In the second case the cognitive radio technology (spectrum sensing) is embedded entirely within the TV white space device. Given that it is becoming difficult for regulators to free up spectrum that is fully cleared of incumbent systems, it is likely that these cognitive radio technologies and other cognitive radio technologies will be used widely in the future to allow commercial wireless systems to share spectrum with incumbent systems. The TV white space represents the first example of what will likely become a popular regulatory structure in the future.

Book Outline

This book contains contributions by many coauthors from the wireless industry, standards developers, and academia in TV white space, dynamic spectrum access, and cognitive radio fields. It provides an overview of many topics relevant to TV white space. The book is organized in four sections. **Section I** focuses on regulations, spectrum policies, channelization and pre-system requirements. This must be the starting point because, without regulatory rules that permit wireless operation in the TV white space, no networks or devices can be built or deployed in the band. Chapters on regulation and policy cover wide regions from around the world, from the United States to Europe, Australia, Japan, and Singapore. **Section II** focuses on TV white space standards efforts in different standard-developing organizations (SDOs), with emphasis on the IEEE 802.22 wireless regional area network (WRAN) standard. **Section III** addresses coexistence of the TV white space devices with licensed services and also coexistence of the different networks of TV white space devices. The chapters in this section describe coexistence techniques between all potential standards, technologies, devices, and service providers. In this section there are two chapters on spectrum sensing looking to strengthen and enrich sensing research, as the FCC did not consider sensing mature enough to be adopted as mandatory requirements in their regulations. This part also presents database as a reliable scheme for coexistence. **Section IV** discusses some other important aspects in TV white space, including spectrum allocation, use cases, and security. The four sections are discussed below in detail.

Section I

Chapter 1 provides a summary of TV white space regulations around the world. There is a detailed study of the United States Federal Communication Commission (FCC) rules because the United States was the first country to develop such regulations and because the rules evolved over many years. The chapter begins with a history of the regulations beginning with the Spectrum Policy Task Force in 2002, then moving on to the proposed regulations which resulted in significant industry involvement and feedback to the FCC, and ending with a summary of the final rules that were issued in 2010. After going over the FCC regulations, the chapter summarizes the ongoing development of TV white space regulations in other regions of the globe.

In Europe the Conference on Postal and Telecommunications Administrations (CEPT) has begun studying different possible TV white space regulations. The European Communication Committee (ECC), within CEPT, has established a project called SE43 on cognitive radio systems in the TV white space. A summary of that SE43 technical report is provided in Chapter 1. The European rules share similarities with those in the United States, but are different in a number of ways. In addition, the final regulations must be provided by the individual European member states, which could each be a little different. The Office of Communication (Ofcom) in the United Kingdom has been developing regulations. It is not uncommon for the United Kingdom to be able to develop regulations before other European member states, because being an island eliminates the border issues that other member states must address. The chapter also provides a summary of the regulatory activities in Canada, where there has been a focus on rural broadband systems to provide service to rural areas that may be currently unserved or underserved. The chapter concludes by discussing the activities in several Asian countries, including Japan, Korea, and Singapore.

Chapter 2 takes a look at how establishing TV white space regulations now will potentially impact future regulations. The authors from Australia analyze these issues in their country, but those same issues could equally well apply to other countries. The authors bring up the issue of a future Digital Dividend 2, in which some of the current TV broadcast spectrum is reallocated to licensed wireless telecommunications. If unlicensed TV white space devices are deployed today in the unused TV channels, what will be the effect on future licensed telecommunication systems? These TV white space devices will be designed to share spectrum with broadcast TV systems, but will they be able to adapt to Digital Dividend 2 when licensed telecommunication systems are deployed in the band? The authors make comparisons of the wireless microphone systems that have been allowed to operate in unused TV channels for many years and the newer TV white space devices that will be permitted to operate in those unused channels. The chapter provides background on TV white space concepts as well as an overview of Australian spectrum regulations. The authors discuss the impact of TV white space devices on the potential Digital Dividend 2, in which a portion of the current TV broadcast spectrum is reallocated for cellular telecommunication networks.

Chapter 3 discusses the impact of the TV white space regulations on the system requirements of the TV white space devices and networks (i.e., maximum antenna heights used in the system, operation of a database, spectrum masks, maximum transmitted power, etc.). This chapter describes the TV white space database requirement that must be met to protect the incumbent systems in the TV bands. A summary of the device requirements is also provided.

Chapter 4 provides an analysis of the quantity of TV white space in the United States. One of the questions that comes up when considering use of the TV white space is "how much spectrum is there?" The number of available channels depends on which type of devices one is considering. One can set up a fixed network consisting entirely of fixed TV white space devices. Similarly, one can set up a portable network consisting entirely of portable TV white space devices. And finally, one can set up a network consisting of a combination of both fixed and portable devices. The number of TV white space channels available depends on which type of network one is considering. This chapter analyzes the FCC regulations and applies them to databases of standard-power and low-power TV broadcast stations.

Chapter 5 evaluates the potential for TV white space operation in Europe. It provides background material on digital terrestrial TV in Europe, describing the digital video broadcasting project for terrestrial reception (DVB-T) deployment in Europe. The chapter gives an overview of the CEPT ECC SE43 project on cognitive radio in the TV white space, as well as an overview of the European Telecommunications Standards Institute (ETSI) project on reconfigurable radio systems. Then the authors develop a case study of TV white space in Finland, calculating the number of potential TV white space channels for different regions in Finland. The chapter ends by applying a model of innovation growth to the TV white space and makes predictions about the level of TV white space usage in the future.

Section II

Chapter 6 covers standards for physical (PHY) layer and medium access control (MAC) layers, operating in the TV white space. The first standard considered is the ECMA-392 standard for wireless networking of personal/portable devices in the TV white space. This standard was developed specifically for the TV white space and supports both master–slave and peer-to-peer topologies. The MAC and PHY layers have been designed to support wireless delivery of high-definition video within the home. The standard supports both geo-location/database and spectrum sensing for protection of incumbent systems. Another standard that specifically targets the TV white space is IEEE 802.22, which was recently published by the IEEE. This is a standard for wireless regional area networks (WRANs). The standard was originally derived from IEEE 802.16e, but has been enhanced to support TV white space operation. The PHY layer has been modified to provide support for longer-range operation. In addition to the traditional data and management planes in wireless network architecture, IEEE 802.22 has added a cognitive plane for the various cognitive radio technologies. The standard supports geo-location and database access for protection of incumbent systems. The standard also provides support for spectrum sensing and includes an informative annex on various spectrum sensing techniques that were evaluated by the IEEE 802.22 working group. Finally, the standard provides several techniques for neighboring WRAN cells to coexist. The next standard project that is considered IEEE 802.11af, which is an amendment to the popular IEEE 802.11 wireless local area network standard, often referred to by its market name Wi-Fi. This project began in 2010 and is still under development. This amendment makes enhancements to the MAC and PHY layer for operation in the TV white space. The PHY layer is an orthogonal frequency domain multiplex (OFDM) PHY design with support for multiple bandwidths, which are based on earlier amendments. The PHY layer is expected to be a clock-scaled version of an earlier OFDM PHY to fit in the TV white space channels, which is a narrow bandwidth that the PHY layers used in the ISM bands. The final MAC/PHY that is considered is an ad hoc committee within the IEEE Standards Coordinating Committee 41 (SCC41) which is looking into starting a new MAC/PHY project for the TV white space.

Chapter 7 discusses TVWS coexistence and dynamic spectrum access (DSA) standardization activities at upper layers. This includes IEEE 802.19, IEEE 1900.4, and ETSI. The IEEE 802.19 project is developing a standard that will utilize the location awareness of the TV white space devices to improve coexistence between different wireless networks, even if they are designed according to different standards, while IEEE 1900.4 WG is developing standards specifying management architectures enabling distributed decision making for optimized radio resource usage in heterogeneous wireless access networks. ESTI activities are quite broad and cover CR and software defined radio (SDR) architectures and related application programming interfaces (APIs).

Chapter 8 discusses system-level analysis of OFDMA-based networks in TV white spaces with emphasis on IEEE 802.22 as a case study. The chapter develops a theoretical framework for throughput evaluation in opportunistic spectrum access (OSA) systems. The chapter also describes NS-2 simulation and implementation for IEEE 802.22 based on the OS-OFDMA scheme. Results show the impact of temporal and/or spatial wireless microphone operation on IEEE 802.22 performance.

Chapter 9 presents spectrum sharing in inter-network using an IEEE 802.22 scenario. The chapter discusses a distributed, message-based, cooperative, and real-time spectrum sharing concept called on-demand spectrum contention (ODSC) that is used in IEEE 802.22. The ODSC protocol is described, and beacon period framing (BPF) protocol that enables reliable, efficient, and scalable inter-network communications is also discussed. Other interesting inter-network discussions like dedicated or in-band radio frequency issues and backhaul-based or over-the-air inter-network communications are highlighted as well.

Section III

Chapter 10 presents various spectrum sensing aspects and issues related to TV white space. Special attention has been paid to sensing in ongoing TV white space standards including, for example, IEEE 802.22 and the IEEE 802.11af. Throughout the chapter, many TVWS scenarios were studied and elaborated. The chapter shows that sensing is crucial for devices that do not register in any geo-location database, and how spectrum sensing can be used by service providers to improve their systems.

Chapter 11 explores energy detection-based distributed spectrum sensing (DSS), where three distributed sensing schemes are discussed: cooperative sensing, collaborative sensing, and selective sensing. In distributed sensing, multiple sensors make RF measurements and share their local measurement results and then, by combining measurement results, form multiple independent sensors from which an improved global spectrum sensing result can be obtained. The chapter provides a theoretical analysis based on analytical models providing results for the probability of detection in Rayleigh fading channel, showing the improvement in sensing performance, possible by using distributed spectrum sensing.

Chapter 12 presents protection of incumbent systems in the TV white space by using a white space database, as well as spectrum sensing. As the coexistence issues with the incumbent are described, the chapter discusses many of the aspects that must be considered in a TV white space database. These aspects include cooperative databases, synchronization between databases, access security, and authentication. Since in some cases the database may not be accessible or updated with current incumbent information, this may lead to reduced frequency agility and spectrum utilization efficiency. The chapter introduces a new system architecture, which is based on both spectrum sensing and database access.

Section IV

Chapter 13 describes acquisition principles for acquiring spectrum that lead to efficient TVWS spectrum allocation. The chapter reviews the roles of a number of elements that may lead to reliable and efficient allocation of the TV white spaces. The chapter begins with a discussion of the various approaches to spectrum regulation: command and control, spectrum commons, and market-based approach. Among the approaches to spectrum policy that is considered in this chapter is having an intermediary operating between the TV white space networks and the incumbent systems. The chapter provides a number of motivations for the need of intermediaries to facilitate the functioning of the TVWS systems from a market perspective and presents cognition as an enabling technology. The chapter then provides a summary of many of the factors that must be considered in the protection of the incumbent systems. The realization of many of these factors will ultimately be embedded in the TV white space database. The chapter ends with a description of several methods of making spectrum available for TV white space devices. These methods can be divided into device-centric approaches like spectrum sensing, and network-centric approaches like database access.

Chapter 14, which is a follow-up from Chapter 13 by the same authors, continues discussing elements of efficient TV white space allocation with a focus on business models. Two business models are considered: TV White Space Information Provision and Wireless Service Provision in the TV white space. In the first model, the spectrum context broker sells or provides reliable spectrum context to potential TV white space devices. The model combines all the roles related to spectrum context collection, processing and dissemination. In the second model, an operator acquires reliable spectrum context information from the broker and provides wireless services to the end user.

Chapter 15 provides an overview of several application use cases in TV white space. Many of the use cases are similar to wireless use cases in other frequency bands. However, TV white space has its unique characteristics that impact the types of use cases that are a good fit for this spectrum. One of the characteristics of the TV white space spectrum is its low frequency, which leads to long-range operation. Another characteristic is the fact that it is a license-exempt frequency band, which makes it possible to deploy devices and networks without having to purchase the rights to the spectrum access. Some of the potential use cases introduced in this chapter include large area connectivity, utility grid networks, transportation and logistics, mobile connectivity, high speed vehicle broadband access, office and home networks, and emergency and public safety.

Chapter 16 explains the security and privacy issues that may threaten TV white space applications. The chapter characterizes security and privacy in to three pairs of antagonistic parameters: the first one is associated with all wireless communications, while the other two are new and unique to TVWS technology. The first pair of parameters is related to access to the system and deals with denial of access when it should be granted, and granting spectrum access when it should not be, thereby causing interference to an incumbent system. The second pair of parameters is related to protection of incumbent systems and is concerned with providing access when it should be provided, and not granting spectrum access when it should be granted, resulting in a device or a network of devices unable to utilize spectrum to which they should have access. The third pair of parameters concerns accountability and privacy. The chapter also discusses a set of privacy threats that are inherent with the TVWS database server. The authors also believe that the privacy and security solution space is complicated by the political interactions of large commercial players (e.g., TV broadcasters and wireless MIC vendors) with significant financial considerations.

References

1. Joseph Mitola III, *Cognitive Radio Architecture: The Engineering Foundations of Radio XML,* Wiley-Interscience, 2006.
2. Hector C. Weinstock, *Focus on Cognitive Radio Technology,* Nova Science Pub Inc, 2006.
3. Hüseyin Arslan, *Cognitive Radio, Software Defined Radio, and Adaptive Wireless Systems,* Springer, 2007.
4. Qusay Mahmoud, *Cognitive Networks: Towards Self-Aware Networks,* Wiley-Interscience, 2007.
5. Yang Xiao and Fei Hu, *Cognitive Radio Networks,* Auerbach Publications, 2008.
6. Bruce A. Fette, *Cognitive Radio Technology,* Academic Press, Second Edition , 2009.
7. Alexander M. Wyglinski, Maziar Nekovee, and Thomas Hou, *Cognitive Radio Communications and Networks: Principles and Practice,* Academic Press, 2009.
8. Ekram Hossain, Dusit Niyato, and Zhu Han, *Dynamic Spectrum Access and Management in Cognitive Radio Networks,* Cambridge University Press, 2009.
9. Lars Berlemann and Stefan Mangold, *Cognitive Radio and Dynamic Spectrum Access,* Wiley, 2009.
10. Kwang-Cheng Chen and Ramjee Prasad, *Cognitive Radio Networks,* Wiley, 2009.
11. Linda E. Doyle, *Essentials of Cognitive Radio,* Cambridge University Press, 2009.
12. Yan Zhang, Jun Zheng and Hsiao-Hwa Chen, *Cognitive Radio Networks: Architectures, Protocols, and Standards,* CRC Press, 2010.

13. Peyton D. Mcguire and Harry M. Estrada, *Cognitive Radio: Terminology, Technology and Techniques (Media and Communications-Technologies, Policies and Challenges)*, Nova Science Pub Inc, 2010.
14. K. J. Ray Liu and Beibei Wang, *Cognitive Radio Networking and Security: A Game-Theoretic View*, Cambridge University Press, 2010.

REGULATIONS
AND PROFILES

Chapter 1

TV White Space Regulations

Stephen J. Shellhammer, Cong Shen,
Ahmed K. Sadek, and Wenyi Zhang

Contents

1.1 Introduction

This chapter provides an overview of the regulations regarding operation in the TV white space (TVWS). The chapter begins with a short overview of the Federal Communications Commission (FCC) spectrum policy task force report that later led to the FCC regulations. That is followed by a summary and analysis of the United States FCC report and order (R&O) [1], which the FCC followed up with a second memorandum opinion and order [2] addressing a number of petitions for reconsideration they received on the original report and order. This includes an explanation of the distinction of fixed and portable TV white space devices and the requirements on both classes of device. The transmit power and spectral mask limits will be summarized. Rules for geo-location, database access, and spectrum sensing will be included. Though the FCC has not specified how it will test these devices, the chapter will include a description of how the FCC tested the prototypes that were submitted to the FCC prior to the FCC issuing the R&O. This overview of the prototype testing gives insight into how the FCC may test TV white space devices when they are submitted to the FCC for certification.

We then explore TV white space regulations under development in other areas including Canada, Europe, the United Kingdom, and Asia. These rules are still under development, and the summary will be based on proposed rules as provided by the regulatory agency in those countries.

1.2 FCC Spectrum Policy Task Force

The Spectrum Policy Task Force was established in June 2002 to assist the Commission in identifying and evaluating changes in spectrum policy that will increase the public benefits derived from the use of the radio spectrum. According to the FCC web site,

The Task Force was charged to:

■ Provide specific recommendations to the Commission for ways in which to evolve the current "command and control" approach to spectrum policy into a more integrated, market-oriented approach that provides greater regulatory certainty, while minimizing regulatory intervention.

■ Assist the Commission in addressing ubiquitous spectrum issues, including, interference protection, spectral efficiency, effective public safety communications, and implications of international spectrum policies.

In November 2002 the Task Force issued its report to the Commission [3].

The FCC was faced with an ever increasing need for additional spectrum to support new advanced wireless technologies. It was felt that the spectrum policy at that time was not keeping pace with the ever-increasing demand for additional spectrum.

The task force began by issuing a public notice soliciting industry comment on the FCC policy and recommendations for methods of improving the spectrum policy. The questions raised in the public notice were divided into five categories: market-oriented allocation and assignment policies, interference protection, spectral efficiency, public safety communications, and international issues. Over 200 submissions were received from industry.

Let us look at some of the key points of the executive summary of the spectrum policy task force report [1].

> Advances in technology create the potential for systems to use spectrum more intensively and to be much more tolerant of interference than in the past.

When many of the FCC policies were first set, analog communication was the norm. Since then advances have been made, and virtually all new wireless devices use digital communications. In addition, it is very common to use various forms of error-correcting codes to improve robustness. This typically includes some form of forward error correction and also automatic repeat request to allow retransmissions. As a result, modern wireless communication systems do not require the signal-to-noise ratio (SNR) of legacy system. And hence they are more tolerant of interference than the older analog system.

> In many bands, spectrum access is a more significant problem than physical scarcity of spectrum, in large part due to legacy command-and-control regulation that limits the ability of potential spectrum users to obtain such access.

The report is indicating that in some bands the spectrum is not used fully. They point out that one of the reasons for the inefficiency is the way the FCC regulates the spectrum.

> To increase opportunities for technologically innovative and economically efficient spectrum use, spectrum policy must evolve towards more flexible and market-oriented regulatory models.

The report is recommending that the FCC spectrum policy needs to be modernized so as to allow for more innovative spectrum policies. One approach is to set up the policies to be less rigid and to include the necessary flexibility to adapt to the needs of the market. Changes in FCC policy occur over many years, while the market operates at a much more rapid pace. By adding in flexibility, the policy would be more supportive of rapid changes in the market.

> Such models must be based on clear definitions of the rights and responsibilities of both licensed and unlicensed spectrum users, particularly with respect to interference and interference protection.
> No single regulatory model should be applied to all spectrum: the Commission should pursue a balanced spectrum policy that includes both the granting of exclusive spectrum usage rights through market-based mechanisms and creating open access to spectrum commons, with command-and-control regulation used in limited circumstances.

The task force is recognizing that no single policy can be used in all cases. The report highlights the need for both licensed spectrum and unlicensed spectrum. Licensed spectrum provides exclusive access to a given band of spectrum. Unlicensed spectrum is shared by a number of wireless technologies, none of which have exclusive access to the band of spectrum.

> The Commission should seek to implement these policies in both newly allocated bands and in spectrum that is already occupied, but in the latter case, appropriate transitional mechanisms should be employed to avoid degradation of existing services and uses.

Any changes to the FCC spectrum policy must protect the rights of the owners of spectrum and ensure that new entrants do not cause harmful interference to the systems currently using the spectrum. This is a major part of the TV white space regulations that began with this report. Here the task force is recommending that the Commission implement these new spectrum policies not only in newly allocated spectrum, but also in bands that are already allocated and in use. Of course, if these new policies are implemented in frequency bands currently in use, the policies must ensure protection of the current users. This is the case with the TV white space, where the new devices must protect the current TV broadcasts.

Next the task force makes several recommendations on spectrum use:

> Preliminary data and general observations indicate that many portions of the radio spectrum are not in use for significant periods of time,

and that spectrum use of these "white spaces" (both temporal and geographic) can be increased significantly.

Here the task force introduces the concept of spectrum "white spaces," which are portions of the spectrum that are not in use in a given geographic location, or are not used for a period of time. The TV white space is a good example of this white space.

The task force recommends:

> As a result, it is important to evolve from current spectrum policies, which reflect an environment made up of a limited number of types of operations, to policies that reflect the increasingly dynamic and innovative nature of spectrum use.

The TV white space policy relies on dynamic spectrum access technology where a device or network's access to the spectrum is dynamic, in that what channels are used varies with both location and time.

The report describes three models for spectrum rights: exclusive-use model, commons model, and the command-and-control model.

In the exclusive-use model the licensee of the frequency band has the exclusive rights to use and transfer of that frequency band in a given geographic region. Not only can the licensee use the band, the licensee can transfer that band to another party through sale or sublease. This is a market-driven model of spectrum rights.

In a commons model, devices can operate and share the frequency band with other unlicensed devices. These devices may be governed by technical standards or etiquettes, but they do not have exclusive access and are not afforded protection from interference from other devices operating in the same band. An example of such a model is the 2.4 GHz frequency band used by Wi-Fi, Bluetooth, and cordless telephones.

The command-and control model is the current model used by the Commission in the majority of the frequency bands, in which the allowable spectrum uses are limited based on regulatory judgments.

The task force recommends increasing the use of both the exclusive-rights model and the commons model. One of the specific recommendations given in the report is that the Commission takes steps to make "white space" in the broadcast bands available for other uses. This led to the TV white space notice of proposed rulemaking in 2004.

1.3 FCC Notice of Inquiry

After the release of the spectrum policy task force report, the FCC issued a Notice of Inquiry (NOI) [4] in December 2002, requesting comments from the industry on unlicensed use of unused portions of the TV broadcast spectrum.

The NOI describes how in many locations there are unused TV channels, and now with advances in technology, it is possible for wireless devices to sense the spectrum for unused TV channels, and that it is also possible for these devices to be "location aware" and able to use the location knowledge to determine what channels are available. The NOI points out which channels are available and the number of channels available change with location and with time, so these new wireless technologies need to be smart devices that are capable of reliably identifying the available channels.

The NOI asked for industry comment on a number of questions on how the rules should be structured. Here is a sampling of some of the key questions asked in the NOI.

> What restrictions, if any, should be placed on the applications or numbers of unlicensed devices that would be permitted in the TV broadcast bands, and why would such restrictions be needed? For example, should applications be limited to fixed uses?

The FCC ultimately considered both fixed devices and personal/portable devices.

> How would new unlicensed devices affect the ability of broadcasters to provide ancillary services such as data after the digital transition?

Wireless microphones are ancillary services used in the TV bands which could be affected by new wireless devices in the TV bands.

> What power and/or field strength limits are necessary for unlicensed transmitters within the TV bands to prevent interference to TV reception? Could unlicensed devices operate in TV bands with a power greater than the 1 watt maximum permitted for Part 15 devices in the ISM bands or power greater than the general Part 15 limit?

Clearly one of the most important issues is the transmit power limit on these devices. The higher the power the higher the possibility of causing interference. Typically, unlicensed devices in other frequency bands (e.g., 2.4 GHz) are limited to 1 watt (30 dBm), so that was a good starting point for consideration in this frequency band.

> What separation distances or D/U ratios should be established between unlicensed devices and the service of analog, digital, Class A, and low-power TV and TV translator stations? What assumptions should be used to determine these protection criteria? Should TV stations be protected only within their grade B or noise limited service contours, or

should unlicensed devices be required to protect TV reception from interference regardless of the received TV signal strength? Is protection necessary only for co-channel and adjacent channel stations? What special requirements, if any, are necessary to protect TV reception in areas where a station's signal is weak? Would minimum performance standards for receivers facilitate the sharing of TV spectrum with unlicensed devices?

In order to protect the TV receivers, the wireless devices must be far enough away from the TV receivers so that the path loss is sufficient to ensure that the interfering signal is sufficiently weak. How to measure this distance is a critical issue. Besides asking about physical distance, the question refers to an FCC term "D/U ratio." This is the desired to undesired ratio. An equivalent term would be signal-to-interference ratio (SIR), which is measured in dB. Another term is the "grade B or noise-limited service contour" which is a physical contour around a TV broadcast station which is calculated using FCC broadcast models to predict the signal power as a function of distance and transmitter parameters. This gets addressed in detail in the FCC report and order.

What technical requirements are necessary to protect other operations in the TV bands, including the PLMRS and CMRS* in the areas where they operate on TV channels and low power auxiliary stations such as wireless microphones and wireless assist video devices? Could technical requirements be developed that would allow unlicensed devices to co-exist with new licensed services on former TV channels 52–69? Should unlicensed transmitters be required to protect unlicensed medical telemetry transmitters operating on TV channels 7–46 from interference?

PLMRS is public land mobile radio service, and CMRS is commercial mobile radio service. These two services are permitted to use some TV channels, between channels 14 and 20, in 13 specific metropolitan areas. Channels 52–69 represent the 700 MHz digital dividend licensed spectrum that has been auctioned for cellular use. The R&O does not allow these new wireless devices to operate in those channels, because they will be used for licensed cellular services.

Should any antenna requirements be imposed? Can technologies such as "smart antennas," which automatically change their directivity as necessary, assist unlicensed devices in sharing the TV bands? Should unlicensed devices be required to use an integrated transmitting antenna and be prevented from using external amplifiers and antennas?

---

* PLMRS: Public Land Mobile Radio Service; CMRS: Commercial Mobile Radio Service.

Clearly antenna gain will affect the level of interference caused by the wireless devices. The R&O does include rules regarding both the transmit antenna and the sensing antenna.

> What are the specific capabilities that an unlicensed transmitter should have to successfully share spectrum with licensed operations in the TV broadcast band without interference? Are there transmission protocols that could enable efficient sharing of spectrum?

Given the advances in technology, modern wireless technologies incorporate new protocols to share spectrum between different devices. The FCC is asking for comment on what type of protocols could be used in the TV band.

> Could GPS or other location techniques be incorporated into an unlicensed device so it could determine its precise location and identify licensed users in its vicinity by accessing a database? Would such an approach be reliable, and could it be combined with other methods to prevent interference to licensed services? What specific methods could be used to protect low-power auxiliary stations such as wireless microphones that are not listed in a database?

One of the methods of protecting legacy services is location awareness. This question asks about requirements on the location system and how this can be combined with a database of the legacy services. In addition, some wireless microphone services may not be included in the database, due to unplanned use, so there is interest in how these devices could be protected.

The FCC received many comments on these questions and the other questions in the NOI. After reviewing those comments in detail, it issued a notice of proposed rulemaking providing possible rules and soliciting comments on the proposed rules.

1.4 FCC Notice of Proposed Rulemaking

In 2004 the FCC issued a notice of proposed rulemaking (NPRM) which provides possible rules for TV white space. The FCC solicited questions on the proposed rules. The NPRM also includes a list of the 76 parties that filed comments on the previous NOI. The FCC considered these comments in drafting the NPRM.

The proposed rules would require that these unlicensed devices include "smart radio" technology to identify unused TV channels.

The FCC recommended three different methods for identification of these unused TV channels.

- Allow existing television and/or radio stations to transmit information on TV channel availability directly to an unlicensed device.
- Employ geo-location technologies such as the Global Positioning Satellite (GPS) system.
- Employ spectrum sensing techniques that would determine if the signals of authorized TV stations are present in an area.

The NPRM provides some background for the need and rationale for the proposed rules. In Appendix B of the NPRM the proposed rules are provided.

The NPRM introduces the idea of two classes of unlicensed devices in this band: fixed and personal/portable. An example of fixed devices is wireless Internet service provider (WISP) base stations. Examples of personal/portable devices are wireless local area network (WLAN) access points and clients.

We will provide a brief overview of the proposed rules.

The unlicensed devices would operate in one or more contiguous TV channels. The maximum transmit power for fixed devices would be 1 watt (30 dBm), and for portable devices 100 mw (20 dBm). An antenna gain of up to 6 dBi could be used without changing the transmit power. If the antenna gain is higher than 6 dBi, then the transmit power would need to be reduced by the amount in dB that the antenna gain exceeds 6 dBi.

A fixed device either must include a GPS system and access to a database to identify unused channels, or the device must be professionally installed and the installer must ensure that at this location the device operates on an unused TV channel.

A personal/portable device must be able to receive a signal from either an unlicensed device or a TV or radio station indicating to the personal/portable device which channels are unused. The personal/portable device must only operate on an unused TV channel.

The NPRM provides the definitions of the protected contours which must be protected. These contours are defined in terms of electromagnetic field strength measured in microvolts/meter in the dB domain, which is indicated by the notation "dBu."

The NPRM also provides the required desired-to-undesired ratio (D/U) that must be maintained to protect the TV receivers within the protected contours.

After the release of the NPRM the FCC evaluated several sets of prototypes before issuing the report and order. Section 1.5 provides an overview of these tests.

1.5 FCC Prototype Testing

Prior to issuing the report and order the FCC ran two series of tests on TV white space prototypes. The results of the test on the first set of prototypes were issued in July 2007 [5].

In the 2007 test the FCC received prototypes from two companies, which are indicated in the report as Prototype A and Prototype B. Both prototypes have spectrum sensing capability, while only Prototype A has transmit capability.

One of the primary questions that FCC was attempting to answer is whether it is feasible to sense for incumbent signals at lower levels as low as −116 dBm, and therefore many of the tests were on the spectrum sensing capability of these prototypes.

Prototype A was tested in both the lab and in the field. Prototype B was only tested in the lab because it was not built for field testing.

We will provide a brief summary of the test results. More detail can be found in [5].

In the laboratory test, Prototype A was able to reliably sense ATSC* signals at power levels of −95 dBm or higher. It was not able to reliably sense signals as low as −116 dBm.

In the laboratory tests Prototype B was able to reliably sense ATSC signals at power levels of −115 dBm or higher.

Field tests were performed on Prototype A to evaluate its performance in real-world conditions. The tests were conducted in typical residences where ATSC TV receivers were being received. The tests were performed in several locations within the residences to see how it performed in different conditions. Tests were performed on channels where an ATSC signal was present to make sure the prototype could identify the channel as occupied. Tests were also performed on channels where ATSC signals were not present to see if the prototype could identify these channels as unused. In addition to testing the prototype on ATSC channels, the prototype was tested on NTSC† channels to see how well it sensed for NTSC signals also. In general, Prototype A did not reliably sense for an ATSC signals in a number of cases where an ATSC signal was present. The prototype inaccurately identified the channel as unused. Details of the test results are available in the report.

The wireless microphone sensing capability of both prototypes was tested in the laboratory. No field tests were done on wireless microphone sensing. Prototype A was generally unable to sense wireless microphones in the laboratory tests. The prototype was tested with wireless microphone signals at a variety of power levels. Prototype B was able to sense wireless microphone signals as low as −120 dBm when the signal is in the middle of the TV channel; however, the prototype incorrectly identified a wireless microphone on one channel when no wireless microphone was present. When a strong −36.6 dBm wireless microphone was placed on one channel, the prototype incorrectly sensed a signal on six other channels. This may be due to intermodulation distortion in the sensing receiver front-end, causing distortion to be present in the other channels. The performance of the sensor decreased somewhat when the wireless microphone signal was moved from the middle of the TV channel toward one of the edges of the channel.

---

* ATSC: Advanced Television Systems Committee.
† NTSC: National Television System Committee.

In addition to the sensing test the FCC also tested the susceptibility of the wireless microphones to interference.

In 2008 the FCC tested a second set of prototypes. The test report was issued in October 2008 [6].

In the second set of tests, five prototypes were submitted for evaluation. The companies that submitted prototypes were: Adaptrum, the Institute for Infocomm Research (I2R), Microsoft Corporation, Motorola Inc., and Philips Electronics North America (Philips). Not all tests were performed on all prototypes, because not all were available for all tests. For example, one prototype was provided after initial tests were completed.

TV sensing was performed in both the laboratory and the field. In the laboratory the prototypes were tested on clean (i.e., no multipath) signals and also on captured real-world signals which include multipath fading. In addition, tests were run with adjacent channel blockers. Finally, tests were run in the field using over-the-air (OTA) signals.

Tests were run on the three prototypes whose providers (Microsoft, Philips, and I2R) indicated that the prototypes were capable of sensing for wireless microphones. Wireless microphone sensing testing was performed both in the laboratory and in the field.

One prototype included transmit capability, so laboratory tests were performed on the transmitter signal characteristics. In addition some limited field tests were done to evaluate the impact of the transmitter on TV receivers.

We will provide a brief summary of the test results. More detail is provided in the test report [6].

All the devices were able to sense a clean ATSC signal (i.e., no multipath) in the laboratory in a cabled-up test, in the range of –116 and –126 dBm.

Similar tests were done with captured real-world signals (i.e., with multipath), with the sensing threshold being in the range of –106 to –128 dBm. It is expected that the multipath channel would degrade the performance of the prototypes.

Additional laboratory cabled-up tests were performed with an adjacent channel blocker on one of four possible channels: $N-2$, $N-1$, $N+1$, and $N+2$, where the channel being sensed is channel N. The adjacent channel blocker was at one of three power levels: –28 dBm (strong), –53 dBm (moderate), and –68 dBm (weak).

The reliable sensing level for many of the prototypes was degraded significantly in the presence of either a strong or moderate adjacent channel blocker. It is likely that the prototypes were not designed to handle this level of adjacent channel blocker and as a result performed quite poorly.

Two of the devices were designed to sense for wireless microphones (both analog and digital) and hence were tested in the laboratory on wireless microphones. These two prototypes sensed the wireless microphones over the range of –103 to –129 dBm, depending on the type of microphone and the devices.

TV sensing field tests were run on four of the prototypes. In most cases the prototypes properly detected the TV signals when the devices were within the protected contour and the TV signal was decodable by a TV receiver. In a few cases, three of the prototypes did not detect the TV signal. In general, the devices continued to detect the TV signal when placed outside the protected contour. One of the prototypes was geo-location/database capable, which when enabled properly determined if the device was inside the protected contour. Given the accuracy of a typical GPS system, it is expected that it would be able to reliably determine the device is within the protected contour with very good accuracy.

Two of the devices were tested in the field for over-the-air wireless microphone sensing, at two different locations. The tests were run first with the wireless microphones off to see if the device could determine that no wireless microphone was present in the channel. Subsequently, the wireless microphone was turned on to see if the devices could detect the presence of the wireless microphone signals. One of the devices reported that all of the channels were occupied by a wireless microphone signal whether the microphone was on or not. The other prototype did not detect a few channels when the wireless microphone signal was on. Clearly, more work was needed on wireless microphone sensing.

Several tests were performed to see the impact of the transmitting prototype on a TV receiver. The tests showed that, for a 7 dBm transmitter, co-channel interference can cause harmful interference at up to 360 meters from the TV receiver. This is of course due to the required desired-to-undesired ratio (D/U) at the TV receiver. It is straightforward to predict this using standard path loss models.

As we can see, the prototype test results were mixed. The prototypes seemed to work reasonably well on sensing TV signals, while they tended to perform poorly on sensing wireless microphones.

1.6 FCC Report and Order

The FCC Report and Order (R&O) [1] provides the rules which an unlicensed device must follow in order to be certified to operate in the TV white space. These rules have been revised and augmented with the more recent Second Memorandum Opinion and Order which will be explained in Section 1.7. The rules are more complex and detailed than the rules for other operations in other bands, like the 2.4 GHz Industrial, Scientific, and Medical (ISM) band. The R&O includes a discussion section which explains some of the rationale for the final rules as well as comments received and how the comments impacted the final rules. The final rules are included in an appendix of the R&O. This section will provide an overview of the rules.

1.6.1 Terminology

The term used by the FCC for an unlicensed device operating the TV white space is *TV Band Device* (TVBD).

The R&O includes a number of definitions which must be understood in order to understand the remainder of the rules. Some of those terms will be described here.

An *available channel* is a TV channel that is not being used by a licensed service at or near the location of the TVBD. So an available channel is a channel that the TVBD can use. An *operating channel* is a TV channel being used by the TVBD.

Some of the definitions relate to the capabilities of these TVBDs.

Geo-location is the capability to determine the device's location to within a specified accuracy. *Spectrum sensing* is the capability to observe the RF spectrum and determine if a given TV channel is occupied by an incumbent signal.

An external entity is the *TV bands database*, which is a database of the location and channel usage of protected services. This database is used in concert with the geo-location capability to determine which are the available channels.

There are several classes of TV band devices which have different functions and capabilities.

A *fixed device* is a TVBD that is located at a fixed location. In the rules there are specific requirements about fixed devices, including that their antenna is located outdoors, so this class of devices can be thought of as a base station. A personal/portable device is a TVBD that operates while in motion or at an unspecified location. A personal/portable device can be stationary like a Wi-Fi access point, or it can be a device that can be moved around freely like a laptop computer or PDA.

Now the class of personal/portable devices is partitioned into two modes of operation: Mode I and Mode II. A personal/portable device operating in *Mode I operation* may only operate on an available channel identified to it by either a fixed or personal/portable device operating as a Mode II device. Mode I operation does not require geo-location capability or access to the TV band database. In contrast to that, a personal/portable device operating in *Mode II operation* operates on an available channel which it identified through use of its geo-location capability and access to the TV band database. The document also refers to this mode as the *master mode*. In summary, fixed and personal/portable Mode II devices have geo-location capability, and access to the TV band database and use those capabilities to identify available channels, while a personal/portable Mode I device does not have geo-location capability or access to the TV band database and hence must rely on either a fixed or personal/portable Mode II device to identify an available channel.

1.6.2 Transmission Restrictions

There are specific restrictions on the transmission for both fixed and personal/portable devices.

Which TV channels can be used depends on whether the devices are fixed or personal/portable. For example, only fixed devices may operate in the very high frequency (VHF) bands. In the VHF bands Channels 3 and 4 cannot be used

Table 1.1 TV White Space Channels

TV Channel	Frequency (MHz)	TVWS Devices Permitted
2 (VHF-Lo)	54–60	Fixed
5-6 (VHF-Lo)	76–88	Fixed
7-13 (VHF-Hi)	174–216	Fixed
14–20 (UHF)	470–512	Fixed
21–35 (UHF)	512–602	Fixed & personal/portable
36 (UHF)	602–608	Personal/portable
38 (UHF)	614–620	Personal/portable
39–51 (UHF)	620–698	Fixed & personal/portable

because they are used when connecting to a VHS player, DVD player, or a set top box. In the ultrahigh frequency (UHF) band personal/portable devices may not use channels 14–20 since they are used for private land mobile radio in some locations. Neither fixed nor personal/portable devices can used channel 37, which is used for radio astronomy. Also, fixed devices cannot use channels 36 and 38, which are the two first adjacent channels to channel 37. A summary of the TV white space channels is provided in Table 1.1.

Within the protected region around a TV broadcast station, neither fixed nor personal/portable devices can transmit on the same TV channel as the TV broadcast station. For fixed devices, there is an additional restriction that within the protected region a fixed device cannot transmit on either of the first adjacent channels (channels $N-1$ and $N+1$). Personal/portable devices can transmit on adjacent channels, with a restriction described below.

The maximum conducted transmit power for fixed devices is 1 watt (30 dBm). Fixed devices can also have antenna gain of up to 6 dBi without modification in the conducted transmit power. However, if the antenna gain exceeds 6 dBi, then the conducted transmission power must be decreased by the amount that the antenna gain exceeds 6 dBi. Therefore, the maximum effective isotropic radiated power (EIRP) is 36 dBm for fixed devices.

The maximum conducted transmit power for personal/portable devices is 100 mw (20 dBm). However, if the device transmits on one of the first adjacent channels to a TV broadcast channel, while within the protected region, then the maximum conducted transmit is reduced to 40 mw (16 dBm). If the antenna gain is larger than 0 dBi then the conducted transmit power must be reduced by the amount in dB that the antenna gain exceeds 0 dBi.

All TV white space devices must include transmit power control and must limit the transmit power to the minimum necessary for operation.

There are additional restrictions on the antenna used. For personal/portable devices the antenna must be permanently attached so that a user cannot replace it with a higher gain antenna.

For fixed devices there are height restrictions on both the transmit and receive antennas. The receive antenna must be at least 10 meters above ground, which is intended to ensure reception of the signals from incumbent systems, to provide good sensing. The transmit antenna must be no higher that 30 meters above ground, which is intended to limit the propagation distance so as to not cause harmful interference to incumbent systems. Clearly, if the transmit antenna and the receive antenna are the same antenna, which is a practical implementation, then the height restriction combines into the antenna being between 10 and 30 meters above ground.

1.6.3 Out-of-Band Emissions

There are three sets of out-of-band emission (OOBE) requirements that must be met. The first requirement is the power in an adjacent TV channel relative to the power in the band of operation. The second is the requirement to meet FCC Section 15.209 requirements beyond the two adjacent channels. And the third requirement is the emissions in channel 37, and its two adjacent channels (channels 36 and 38).

In the adjacent TV channels the out-of-band emissions shall be at least 55 dB below the highest average power in the band of operation, measured using a minimum resolution bandwidth of 100 kHz.

Beyond the two adjacent TV channels the out-of-band emissions beyond two adjacent channels shall meet FCC Section 15.209 requirements. The primary requirement within Section 15.209 is that in the band 216–960 MHz the electromagnetic field strength measured at 3 meters must not exceed 200 microvolts/meter, when measured in a 120 kHz bandwidth.

The third requirement is for the field strength in channel 37 and also the two channels adjacent to that channel, channels 36 and 38. The field strength limit is lowered from the boundary between channels 35 and 36 from 120 dBu (dB microvolts/meter) measured at 1 meter, down to 30 dBu within channel 37. Figure 1.1 illustrates the field strength limits for channels 36–38.

In order to determine the spectral mask, it is necessary to relate the transmit power to the electromagnetic field strength measured at a distance, since some of the out-of-band emission requirements are in terms of field strength.

From [7] we know that for free space propagation,

$$\frac{PG}{4\pi d^2} = \frac{E^2}{R_{FS}} \tag{1.1}$$

where P is the transmit power, G is the transmit antenna gain, d is the distance in meters, E is the electric field strength, and R_{RS} is impedance of free space and is given by $R_{FS} = 120\pi$. Solving for the field strength we get,

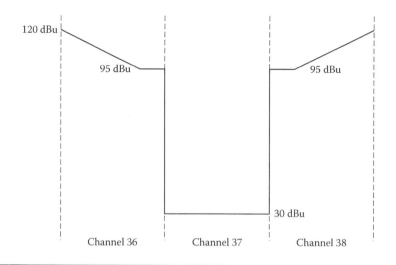

Figure 1.1 Field strength limits for Channels 36–38 measured at 1 meter.

$$E^2 = \frac{30\,PG}{d^2} \tag{1.2}$$

Taking logarithms and multiplying by 10 we have,

$$20\log(|E|) = 10\log(P) + 10\log(G) + 10\log(30) - 20\log(d) \tag{1.3}$$

We would like to represent the field strength in dBu (microvolts/meter) and the transmit power in dBm. To make these conversions we let,

$$FS_{dBu} = 20\log(|E|) + 120 \tag{1.4}$$

$$P_{dBm} = 10\log(P) + 30 \tag{1.5}$$

This gives the following formula for the electromagnetic field strength in dBu for free space propagation,

$$FS_{dBu} = P_{dBm} + 10\log(G) + 10\log(30) + 90 - 20\log(d) \tag{1.6}$$

After simplification we get,

$$FS_{dBu} = P_{dBm} + 10\log(G) + 104.77 - 20\log(d) \tag{1.7}$$

Now this formula gives us the field strength for free space propagation. We would like to convert back from field strength to transmit power. To be more specific, because the FCC specifies the in-band transmission limits in terms of EIRP, we can combine the transmit power and antenna gain into the EIRP though the following formula, $EIRP_{dBm} = P_{dBm} + 10 \log (G)$. Utilizing this in the formula for the field strength and solving for EIRP in terms of the resulting field strength we get,

$$EIRP_{dBm} = FS_{dBu} - 104.77 + 20 \log(d) \qquad (1.8)$$

We can now determine the permitted out-of-band emissions EIRP beyond the second adjacent channel, when measured in 120 kHz bandwidth. As we recall, the field strength must be below 200 microvolts/meter. When converted to dBu this is 23.01 dBu. Because this is measured at 3 meters we have $d = 3$. Therefore the maximum EIRP measured in 120 kHz beyond the second adjacent channel is,

$$EIRP_{dBm} = 46.02 - 104.77 + 9.54 = -49.21 \text{ dBm} \qquad (1.9)$$

In order to relate this to our in-band emissions we must calculate the EIRP in 120 kHz for our in-band emissions.

The in-band EIRP in 120 kHz depends on the total EIRP and the signal bandwidth. We will consider the EIRP limits for fixed and portable devices: 36 dBm and 20 dBm. The bandwidth depends on the air interface technology. In this chapter we will assume a 5 MHz bandwidth, which is a reasonable choice for a 6 MHz channel bandwidth; however, other bandwidths are possible.

Under these conditions the in-band EIRP in 120 kHz for fixed devices is,

$$EIRP_{dBm} = 36 - 10 \log \left(\frac{5}{0.12} \right) \approx 19.8 \text{ dBm} \qquad (1.10)$$

while the in-band EIRP in 120 kHz for portable devices is,

$$EIRP_{dBm} = 20 - 10 \log \left(\frac{5}{0.12} \right) \approx 3.8 \text{ dBm} \qquad (1.11)$$

The ratio of the in-band to out-of-band emission for fixed devices is thus,

$$\Delta = 19.8 - (-49.21) = 69 \text{ dB} \qquad (1.12)$$

and for portable devices is,

Table 1.2 Spectral Mask Requirements

	First Adjacent Channel	*Beyond First Adjacent*	*Channel 37*
Fixed	55 dB	69 dB	95 dB
Portable	55 dB	53 dB	79 dB

$$\Delta = 3.8 - (-49.21) = 53 \text{ dB} \tag{1.13}$$

There is a field strength limit in channel 37 of 30 dBu in a 120 kHz bandwidth, measured at 1 meter. The out-of-band EIRP in channel 37 must then be less than

$$EIRP_{dBm} = 30 - 104.77 + 20\log(1) = -74.77 \text{ dBm} \tag{1.14}$$

The ratio of the in-band to channel 37 out-of-band emission for fixed devices is thus,

$$\Delta = 19.8 - (-74.77) = 94.57 \approx 95 \text{ dB} \tag{1.15}$$

and for portable devices is,

$$\Delta = 3.8 - (-74.77) = 78.57 \approx 79 \text{ dB} \tag{1.16}$$

We can summarize these spectral mask requirements in Table 1.2.

1.6.4 Interference Avoidance Mechanisms

Protection of licensed systems shall be provided by a combination of geo-location with database access and spectrum sensing.

1.6.4.1 Geo-Location and Database Access

Fixed TV white space devices shall have a known location to within ±50 meters, through either an incorporated geo-location capability or through a professional installation. A Mode II portable device shall incorporate a geo-location capability to within an accuracy of ±50 meters. A Mode II device is required to reestablish its location after power cycling.

Fixed devices must access a TV white space database prior to initial transmission and may only operate on TV channels indicated as available at that location. Fixed devices must subsequently access the database at least once a day.

Mode II portable devices must access the TV white space database prior to initial transmission. Mode II devices must also access the database after every power cycle. They must also access the database at least once a day.

Mode I portable devices shall obtain a list of available channels from a Master device. A Master device can transmit without receiving an enabling signal. Fixed and Mode II portable devices can both be Master devices.

1.6.4.2 Spectrum Sensing

All TV white space devices, both fixed and portable, must be capable of detecting ATSC, NTSC, and wireless microphone signals down to –114 dBm. This detection level is referenced to a 0 dBi sensing antenna. If the minimum antenna gain is less than 0 dBi, then the detection level shall be reduced by the difference between a 0 dBi antenna gain and the minimum antenna gain.

A TV white device must determine that the TV channel is unoccupied by a wireless microphone above the detection threshold for 30 seconds before it can utilize the TV channel. If a TV signal is detected above the detection threshold on a channel indicated as available by the TV database, the TV white space device shall notify the user to allow the user to optionally remove the channel from the list of available channels.

A TV white space device must check the availability of the channel in use at least every 60 seconds.

If a wireless microphone is detected on the operating channel, the TV white space device shall cease transmissions within 2 seconds.

TV white space devices either communicating with one another, or linked via a base station, must share channel occupancy determined by sensing. If any device in the network determines that the current channel is occupied by a wireless microphone then the entire network must stop transmission on this channel within 2 seconds.

1.6.4.3 Additional Requirements

A TV white space device must display available channels based on sensing results and its operating channels.

A fixed TV white space device must transmit identifying information that conforms to a standard format recognized by an industry standards-setting organization.

A fixed TV white space device without database access can transmit on channels indicated from another fixed TV white space device. This slave fixed device must check every 60 seconds with the master fixed device for the list of available channels.

A fixed device may not operate as a client to another fixed device.

A Mode I portable device may only transmit upon receiving a transmission from either a fixed device or a Mode II portable device. The Mode I portable device may only transmit on channels indicated by the fixed or Mode II portable device as being available.

1.6.4.4 Protection Contours

The FCC rules provide a table of protection contours for analog and digital TV in the lower VHF, upper VHF, and UHF bands. Portable devices are only permitted in the UHF band. In the UHF band the protection contour is defined as 41 dBu for ATSC and 64 dBu for NTSC.

The devices must be outside these contours by a specified distance. For devices with antenna heights below 3 meters (e.g., portable devices), the separation distance is 6 km for co-channel operation and 0.1 km for adjacent channel operation. For devices with antenna heights of 10 meters or more (e.g., fixed devices), the separation distances are 14.4 km for co-channel and 0.74 km for adjacent channel operation.

The rules provided detailed requirements regarding operation outside the cable head-ends, fixed broadcast auxiliary services, and portable land mobile radio systems (PLMRS). See [1] for all the details.

TV white space devices are not permitted to operate within 1 km of registered wireless microphones.

1.7 FCC Second Memorandum Opinion and Order

On September 23, 2010, the FCC issued the second memorandum opinion and order [2] to address the petitions for reconsideration that they received on the original R&O. The first issue addressed by the FCC was to appoint the chief of the FCC office of engineering and technology (OET) to handle issues related to the TVWS database. The OET chief is charged with developing a method of selecting database managers and then using that specified method to select the database managers. In addition he is charged with developing procedures to be used by the database managers to ensure that they comply with the FCC regulations. The OET chief, along with the chief of the Wireless Telecommunications Bureau, is assigned to deal with requests from administrators of large events to reserve channels for wireless microphones.

In the subsequent subsections we will describe the revised or new regulations.

1.7.1 Permissible Channels

The FCC took two actions to provide additional protection for wireless microphones. The first action was to reserve two open channels for wireless microphone use in all areas of the country. They could not designate the same pair of channels in all locations due to the different channels used by TV broadcasts in different areas. So the rules now allocate the first channel below channel 37 that is not used by a TV broadcast signal to be kept free of TV white space devices so that it can be used by wireless microphones. Similarly, the first channel above channel 37 that is not used by a TV broadcast is reserved for wireless microphone use. As an example, if channel 36 is occupied by a TV signal but channel 35 is not occupied, then channel 35 is reserved for wireless microphones and is not available for use by TVWS devices. Now, in the rare case there is not a single unused channel below channel 37, then two channels are reserved above channel 37. Similarly, in the rare case that there is not a single channel available above channel 37, then two channels are reserved for wireless microphones below channel 37.

This rule provides reserved channels that can be used by wireless microphones for cases when they could not be first reserved by registering in the database. For planned events, like a sporting event, the wireless microphone operators can register in the database several days in advance and reserve the channels they need for that event. Now with the new rule there is a method for providing reserved channels for planned events and also a method of providing reserved channels for unplanned events.

1.7.2 Power Limits, Antenna Restrictions, and Out-of-Band Emissions

The FCC made several clarifications and modifications regarding power limits. The first item is just a clarification. The power limits for both fixed and portable devices are the same whether they operate on one or more channels. In other words if a device uses two channels, for example, it cannot use twice the power.

The second item is a change to address narrow-band transmissions. There was a concern that if a device was designed to transmit at the maximum power, but its bandwidth was only a fraction of the 6 MHz TV channel bandwidth, then multiple TVWS devices could use the same channel and the total emitted power from all these devices would be higher than if the bandwidth was 6 MHz. As an example, if the TVWS devices were 600 kHz in bandwidth and ten of these devices operated in the same channel, then the total emitted power would be 10 dB higher than if the TVWS devices had a 6 MHz bandwidth. To deal with this concern the FCC imposed power spectral density (PSD) limits in addition to the total power limits.

The PSD limits are such that, if the total bandwidth is 6 MHz, then the PSD limit is the same as the total power limit. Because of this, if the bandwidth is say 3 MHz instead of 6 MHz, then the total power allowed is 3 dB below the total power limit. There are PSD limits on four different types of devices: fixed, portable devices not operating adjacent to a TV signal, portable devices operating on a channel adjacent to a TV signal, and sensing only devices. The PSD limit is specified as the maximum power measured in 100 kHz. Since the difference between 100 kHz and 6 MHz is 17.8 dB, that is the difference between the total power limits and the PSD limits for the various classes of devices.

The PSD limits are given in Table 1.3.

In addition to total power limits and PSD limits, the rules require that the device incorporates transmit power control to limit its transmit power to the "minimum

Table 1.3   Maximum Transmit Power Measured in 100 kHz

Device Class	Maximum Power in 100 kHz
Fixed	12.2 dBm
Portable not operating in channel adjacent to a TV broadcast	2.2 dBm
Portable operating in channel adjacent to a TV broadcast	−1.8 dBm
Sensing only	−0.8 dBm

necessary for successful communication." The FCC does not explain exactly what it means to be the minimum necessary for successful communication, but one reasonable interpretation is that the transmit power is lowered to the value required to meet the required data rate with several dB margins. In addition to incorporating the transmit power control when submitting the device for FCC certification, the device manufacturer must also submit a document to the FCC describing the transmit power control mechanism.

The FCC introduced one additional rule for antennas of fixed devices. Recall that the antenna for a fixed device cannot be placed more than 30 meters above the ground. Now the FCC has added an additional restriction on locations where fixed device antennas can be placed. The antennas cannot be placed in locations where the ground height is more than 76 meters above the average terrain. This limits the propagation possible for fixed devices because it will eliminate placing fixed device antennas on hills that are more than 76 meters above average terrain. The method used to measure the height above average terrain (HAAT) is specified in another FCC document [8] used to predict wireless coverage.

The FCC modified the way the out-of-band emissions are measured. The out-of-band emissions are compared to a reference power. Previously the reference power was the power measured in 100 kHz, which is also the bandwidth used to measure the out-of-band emissions. So previously the rules were 55 dB lower in the adjacent channel compared to the in-band channel, both measured in 100 kHz. Now the FCC changed the reference power level to be the power in the entire 6 MHz channel of operation. Since the difference between 100 kHz and 6 MHz, as mentioned above, is 17.8 dB, the out-of-band emissions measured in 100 kHz is now 72.8 dB. So, for the case where the in-band signal has a flat PSD over the entire 6 MHz TV channel the new regulations and the previous regulations are virtually the same.

For all practical purposes, the power measured in 6 MHz and the total power are virtually the same, because the out-of-band emissions are so low. Therefore, we can restate the out-of-band emissions requirement for the adjacent channel as follows. The power measured in any 100 kHz band within the first adjacent channel must be at least 72.8 dB less than the total transmit power.

1.7.3 Interference Avoidance Methods

A number of changes were made to the interference avoidance methods. These changes affect both geo-location/database and spectrum sensing. The primary change is that originally both geo-location/database and spectrum sensing were required. With the new rules a TVWS device can use either geo-location/database or spectrum sensing. It is likely that geo-location with database access will be implemented in the initial products, with "sensing only" devices entering the market later.

The geo-location requirement stayed the same, with a requirement for 50 meter accuracy which can be accomplished by professional installation. For portable Mode II devices they are required to have an integrated geo-location capability with 50 meter accuracy. These devices now have to determine their geo-location every 60 seconds, and if they have moved more than 100 meters they must request a set of available channels in case the available channel set has changes due to moving. This does allow the Mode II device to move from one location to another. They still need to maintain Internet connectivity when moving. During sleep mode a Mode II device does not need to check its geo-location, but must check its geo-location once waking from sleep mode. One additional feature has been included to allow for more mobility. A Mode II device may request a set of available channels for a region, and then it can roam around in that region without re-accessing the database. For example, a device could request a set of available channels for a circle around the current location with a 1 km radius, and then roam around in that circle without needing to reaccess the database. One must keep in mind that the set of available channels for a region may be smaller than the set for a single location, so there may be a trade-off between the number of available channels and database access for a Mode II device that is in motion.

If the Mode II device loses access to the database, it can continue to operate until 11:59 p.m. of the next day, at which point it must cease operation until it can reestablish contact with the database.

Some additional requirements for the Mode I devices have been introduced. These devices are permitted to transit initially to a fixed or Mode II device on the channel on which the fixed or Mode II device is transmitting. This initial transmission is to be used to request from the fixed or Mode II device a list of available channels. In that request message it must include its FCC ID. The fixed or Mode II device provides that FCC ID to the database so that the ID can be validated. Once validated the database notifies the Mode II device, which may then provide the available channel list to the Mode I device. The Mode I device may now use any of the available channels. The Mode II device also provides information to the Mode I device on how to decode a message referred to as a contact verification signal (CVS). Subsequently, the Mode II device transmits the CVS on a regular basis. The Mode I device must receive the CVS from the same Mode II device that originally provided the available channel set. If it does not receive the CVS, it must attempt to contact the Mode II device. If it fails to contact the Mode II device, it must cease operation. If the Mode I device is in sleep mode, then it is not required to receive the CVS. The contact verification signal is used to ensure that the Mode I device is still within reception range of the Mode II device and hence can still use the available channel set provided by the Mode II device. If the Mode I device is in motion and moves beyond the reception range of the original Mode II device but is within range of another Mode II device, it can establish a connection with the new Mode II device and begin the process over again. So we can see that hand-off between Mode II devices now involves the message exchange for providing the Mode I device FCC ID and obtaining the available channel list and the method of decoding the CVS. The CVS must be encoded so as to ensure that the signal is sent from the original fixed or Mode II device. The Mode I device must validate that the signal came from the original fixed or Mode II device, or it cannot operate.

It is also required that the fixed or Mode II device use a secure connection to the database and be able to verify that the database is certified by the FCC.

A few changes were made which are implemented in the database. For one, there are no longer any exclusion regions with the Canadian and Mexican borders. This is addressed by allowing Canadian and Mexican TV broadcasters to register their TV stations near the U.S. borders.

The protection zones around wireless microphones are 1 km for fixed devices and 400 meters for portable devices.

The regulations now explicitly provide protection for multichannel video pro-gramming distributor (MVPD) reception sites by permitting them to be registered with the database.

Finally, the FCC changed spectrum sensing from a mandatory requirement to an optional requirement. The likely reason for this is due to the inconsistent per-formance of the spectrum sensing prototypes tested by the FCC. Many of the TV white space devices are likely to rely on geo-location and database access and not use sensing. The number of devices relying on spectrum sensing may be less than those that use geo-location and database access. Sensing may be useful in cases where geo-location is difficult (deep indoors) or do not have Internet access.

Besides making sensing optional, the FCC made sensing of wireless micro-phones easier by changing the sensing level from the original value of −114 dBm to a less stringent value of −107 dBm. The levels for ATSC and NTSC remain at −114 dBm. The following is a summary of the sensing levels:

- ATSC: −114 dBm, averaged over 6 MHz
- NTSC: −114 dBm, averaged over 100 kHz
- Wireless Mics: −107 dBm, averaged over 200 kHz

1.7.4 Final Comments on FCC Regulations

The FCC has made an unprecedented move to open up new spectrum using new spectrum access rules. The final rules rely heavily on control through the database and having the devices be location aware. In the future we may even see some devices relying only on spectrum sensing. Other regulatory agencies are studying the actions of the FCC and have been developing rules that are similar in many ways but different in some specific ways. Subsequent sections discuss regulations for TV white space that are under development outside the United States.

1.8 Outside the United States

This section surveys the TVWS regulatory development outside the United States. Due to historical and other reasons, the progress to regulate TVWS spectrum varies significantly from region to region. It also depends on when the analog-to-digital TV transition is scheduled to be completed, which is a complicated issue, and countries approach it differently. In this section, we report recent development around Europe, Canada, and some Asian countries.*

1.8.1 The European Conference of Postal and Telecommunications Administrations (CEPT)

In Europe, the switchover from analog to digital terrestrial television (DTT) is expected to finish by the end of 2012. This will free up the 470 to 862 MHz (the so-called *Digital Dividend* spectrum). Within this band, 790–862 MHz (corresponding to UHF channels 61–69) has been allocated to fixed and mobile communication networks on a primary basis, and hence is not within the scope of this chapter. For the band of 470–790 MHz, CEPT Report 24 [9] has indicated that cognitive radio techniques could be allowed in this band (called *interleaved spectrum* or *white spaces*) subject to not causing interference to incumbent radio services. CEPT ECC then decided to create a new project team—WG SE43—to define the technical and operational requirements for the possible operation of cognitive radio systems in the band of 470–790 MHz. Currently, SE43 is working on a draft report [10],† which is expected to be completed in January 2011.

In this section, we will review the technical development of the SE43 draft report from the June 2010 Mainz meeting.‡

In this report, three illustrative scenarios of cognitive devices are envisioned.

■ *Personal/portable.* These devices are of low height (e.g., 1.5 meters) and mobile. Access to Internet is likely a common feature.

---

* This section is, however, by no means a complete survey of this topic.
† In the CEPT report, a TV white space device is termed WSD instead of TVBD.
‡ [10] is the latest version at the time of writing.

- *Home/office.* These are more likely to be devices that are fixed in one position, e.g., in office or home. These devices could be used to support high quality video services or content sharing.
- *Public access points or base station.* These are also fixed devices but with high power and high antenna to provide a gateway to Internet. They might be commonly located in rural areas to extend broadband services. Market volumes are likely to be much lower than other categories.

With these application scenarios, this report discusses three topics:

- Protection of incumbent users, including Broadcasting Services (BS), Program Making and Special Event (PMSE), Radio Astronomy Services (RAS), Aeronautical Radionavigation (ARNS), and Mobile/fixed services in the bands adjacent to the 470–790 MHz
- Operational and technical requirements for white space devices (WSD) in 470–790 MHz
- Estimated amount of WS spectrum in 470–790 MHz potentially available for WSD

For the purpose of regulatory survey, we will focus on the first two topics in this chapter.

1.8.1.1 Protection of Incumbent Users

Before a device can transmit on a WS channel, it must first decide whether this channel is already occupied by any incumbent user. Generally there are three techniques that have been considered for this task.

- *Spectrum sensing.* A WSD tries to detect the presence of incumbent users on a given channel based on the conducted observations and measurements.
- *Geo-location database.* A WSD measures its location and communicates with a geo-location database to determine which channels are available at this specific location.
- *Beacon transmitter.* A local beacon transmitter will broadcast a signal indicating which frequencies are available for cognitive use in the vicinity. A WSD tunes in to this channel and then uses the information it receives to select its operation channel.

For protection of broadcast services, the current draft focuses on how to derive the sensing threshold. The general method is to start with the assumption that the electric field strength at the WSD height follows the Gaussian distribution, and then further reduce the threshold for a given detection probability and assume Gaussian distribution. Depending on the WSD being outdoor or indoor, two sensing threshold calculations are proposed:

$$P_{th,out} = (E_{m,pl} - L_{pl,H}) - \mu_{sen}\sigma_H - 20\log(f_{sen}) - 77.2 + G_{sen} - L_{pol}$$

$$P_{th,in} = (E_{m,pl} - L_{pl,H}) - BPL - \mu_{sen}\sqrt{\sigma_H^2 + \sigma_{BPL}^2 + \sigma_{in}^2} - 20\log(f_{sen}) \quad (1.17)$$

$$-77.2 + G_{sen} - L_{pol}$$

where $E_{m,pl}$ is the median electric field strength at the height at which TV reception is planned, $L_{pl,H}$ is the loss due to different height between TV reception (e.g., on the rooftop) and WSD (e.g., handset on the ground), μ_{sen} is determined by the required sensing performance, σ_H is the signal standard deviation at the WSD height, f_{sen} is the sensing frequency in MHz, G_{sen} is the antenna gain of WSD, and L_{pol} is the polarization loss. For indoor WSD, the building penetration loss is given by *BPL* and its standard deviation σ_{BPL}, and the indoor signal strength variation is characterized by a standard deviation σ_{in}. Based on these methods, [10] computed sensing thresholds ranges from −149.5 dBm to −85.0 dBm, depending on the deployment scenario. It is further commented that hidden node margin, possible caused by the height difference, propagation difference, as well as the use of directional antennas, is to be understood and will be added to the sensing threshold computation.

The PMSE systems, mainly wireless microphones, also require protection. Two approaches have been proposed to compute the sensing thresholds. The U.K. Ofcom input suggested that −128.15 dBm is a reasonable threshold, while the other input proposed −119 dBm. Meanwhile, several difficulties in wireless microphone sensing have been identified. Unlike with TV signals, there could be mobility of wireless microphones, which requires more frequent sensing. Also, if a wireless microphone operates on a channel that is adjacent to high-power DTT transmission, then sensing for this wireless microphone becomes more challenging and needs to be addressed. A similar problem happens when there is severe man-made noise on the channel. Body absorption is an issue with wireless microphones, and there could be many people in between a wireless microphone transmitter and a WSD. Another issue is when a wireless microphone uses a directional antenna so that the WSD cannot pick up the signal. On the other hand, the geo-location database approach also has some challenges. If a wireless microphone is used on a stationary site (e.g., news studio, corporate headquarters), registration in database is not a problem, and hence we can define a protection zone around the site to account for mobility within the site. However, if it is used on a temporary site at short notice, it would be difficult to register in the database on time to get protection. There is also discussion on designating a reserved channel for PMSE. However, this faces the challenge that not all PMSE devices are capable of operating on the same channel.

Table 1.4   Key Parameters of RRBS Equipments

	Co-Channel	*Adjacent Channel*
Autonomous WSD	Avoid Channel 38 unless geographical location is known and the regulation of the country where it is located allows the use of this channel.	Avoid Channels 37 and 39 unless geographical location is known and the regulation of the country where it is located allows the use of this channel.
Geo-location controlled WSD	Exclusion zones can be defined if the database management is on European level; otherwise subject to bilateral/ multilateral agreements.	Exclusion zones can be defined for each RAS site, depending on the observation mode.

The Radio Astronomy Services occupy the band 608–614 MHz (Channel 38) in a number of European countries. Naturally compatibility assessment is needed if a WSD device plans to operate on the same channel. The recommendations of [10] on this issue are summarized in Table 1.4.

There are limited discussions on protection of Aeronautical Radionavigation in 645–790 MHz and Mobile Service/IMT systems in the adjacent 790–862 MHz. Details are expected to be filled once the CEPT report is finished.

1.8.1.2 Operational and Technical Requirements for WSD

The following operational and technical issues have been considered in this report.

- Emission limits
- Spectrum sensing
- Geo-location database
- Combine spectrum sensing and geo-location database
- Beacon transmitter
- Additional considerations

For WSD emission limits, two main approaches that have been discussed are:

1. *Location-specific output power.* In this approach, the allowed transmit power at WSD is decided by the operating location as well as the device type. Intuitively this is beneficial because the same device can transmit at larger power if it is far away from incumbent users. Another benefit is that it could allow WSD to operate with lower power in some locations where it would not be allowed by the next approach. However, this may increase the complexity

of the devices. Also some data, such as the precise terrain information, may not be easily available.

2. *Fixed-output power.* This is the same as the U.S. FCC approach. The maximum allowed transmit power is determined by the device type. It may also depend on the use of adjacent channels to DTV service.

As we have mentioned, there are many challenges to implement a sensing-only WSD. This report has concluded that "sensing only is not a preferred solution to protect the broadcast service." As a side note, this conclusion is consistent with the study from the U.K. Ofcom [11]. Due to this reason, geo-location database becomes the focus of the following study.

The principle of using geo-location database is that the WSD is prohibited to transmit in the WS spectrum until it has successfully determined from the database that the channel it intends to operate on is empty at the device's location. This requires the WSD to measure its location and exchange information with the database. This report proposed all geographical areas being represented as "pixels." Each pixel is associated with a list of available channels and other relevant data. The choice of the pixel size is a trade-off. If it covers a large area, there will be less available channels; if it is too small, there will be a large amount of calculation and information transfer between WSD and database. A WSD then may cover one or more pixels, depending on location accuracy and uncertainty of WSD (e.g., speed of movement).

With the geo-location database being used, the WSD shall be capable of both communicating to the database its geographical location and receiving, from the database, information on frequencies that could be used by the WSD in its current location. The WSD shall stop transmitting on the channel immediately if it cannot re-consult the database by the end of the validity period, or if it fails to monitor its location, or if it moves outside the determined area for which the frequency information received from the database is valid.

There are applications where WSD deployment follows a master–slave mode, in which the slave device is fully controlled by the master. In this mode, the master will be responsible for communicating with database, and the slave is controlled by the master and only receives operational parameters from the master.

One essential problem with the geo-location database, as we have mentioned before, is that it might not suffice to protect PMSE systems. This problem becomes even worse for some countries where the access to the spectrum for PMSE is not driven by authorization. In this scenario, spectrum sensing probably is the only solution to protect PMSE. Thus, there is a need to consider the combination of spectrum sensing and geo-location database. In the CEPT draft report, two potential combination options are under consideration. The first one is to implement an adaptive sensing threshold based on database decision. The idea is to set a more stringent sensing threshold if the channel is indicated as occupied in the database, and to relax the threshold otherwise. In this way, the database decision is served as *a priori* information, and the final decision is still made by spectrum sensing. The other approach takes a different direction. The basic idea is to let the database decide channel occupancy, and use spectrum sensing results as side information to improve the database.

There is very limited discussion on the beacon approach. Among the three candidates, this one is of the least interest. In fact, U.K. Ofcom has decided [11] that it will drop further consideration on beacons and focus only on spectrum sensing and geo-location database.

1.8.2 The Office of Communications (Ofcom), United Kingdom

The United Kingdom is among the earliest countries that took regulatory actions to enable cognitive access in the 470–790 MHz band. On February 16, 2009, it started consultation [12] on enabling license-exempt cognitive use of interleaved spectrum without harmful interference to licensed users. The consultation ended on May 2009, and Ofcom published the final statement [11] in July 2009.

The digital switchover (DSO) started in 2007 in the United Kingdom and is expected to finish in 2012. After the DSO, there will be 256 MHz UHF spectrum (470–550 MHz and 614–790 MHz) reserved for DTT, but it will be "interleaved" for other usage, including cognitive devices. Figure 1.2 illustrates the band plan per current proposals.

During the consultation, three approaches have been proposed for study: spectrum sensing, geo-location database, and beacon transmitter. The Ofcom statement

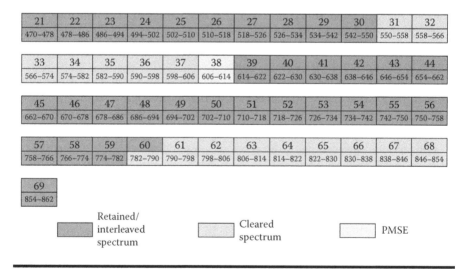

Figure 1.2   Current U.K. UHF band plan after DSO; original figure appearing in [11, Figure 1].

Table 1.5 Key Parameters for Spectrum Sensing for a Sensing-Only Device

Cognitive Parameter	Value
Sensitivity assuming a 0 dBi antenna	−120 dBm in 8 MHz channel (DTT)
	−126 dBm in 200 kHz channel (wireless microphone)
Transmit power	4 dBm (with adjacent channel) to 17 dBm
Transmit power control	Required
Bandwidth	Unlimited
Out-of-band performance	<−46 dBm
Time between sensing	<1 second

concludes that beacon transmitter is the least appropriate approach and will be dropped from further investigation. As for spectrum sensing and geo-location database, no final conclusion is made in the statement. However, it does suggest that "the most important mechanism in the short to medium term will be geo-location," and hence this seems to be the focus of the follow-up work. As for spectrum sensing, key parameters have been proposed in the consultation. There were different responses, and Ofcom updated the parameters in the final statement. Table 1.5 and Table 1.6 summarize the final key parameters [11] for spectrum sensing and geo-location database, respectively.

1.8.3 Canada

In March 2010, Industry Canada* issued a Radio Standard Specification (RSS-196, Issue 1), "Point-to-Multipoint Broadband Equipment Operating in the Bands 512–608 MHz and 614–698 MHz for Rural Remote Broadband Systems (RRBS) (TV Channels 21 to 51)" [13] and technical requirements (SRSP300-512, Issue 1), "Technical Requirements for Remote Rural Broadband Systems (RRBS) Operating in the Bands 512–608 MHz and 614–698 MHz for Rural Remote Broadband Systems (RRBS) (TV Channels 21 to 51)" [14], to set standards for the Remote Rural Broadband Systems (RRBS) based on TVWS. These specifications and requirements substituted the earlier issued (in March 2007) interim technical guidelines for TVWS RRBS.

Compared with the U.S. regulations, the key distinction of the Canadian rules is that the TVWS RRBS is a license-based system, allowing high-power

---

* Industry Canada is the department of the Government of Canada with responsibility for regional economic development, investment, and innovation/research and development.

Table 1.6   Key Parameters for Geo-Location Database

Cognitive Parameter	Value
Location accuracy	100 meters
Frequency of database access	To be determined
Transmit power	As specified by the database
Transmit power control	Required
Bandwidth	Unlimited
Out-of-band performance	<–46 dBm

transmission over long distance in rural areas. The licensing is on a first-come, first-served (FCFS) basis. There are two types of equipment: base station, and subscriber, the latter of which is termed Customer Premises Equipment (CPE). A base station communicates with multiple CPEs. Equipment is classified as Category I equipment, and requires a Technical Acceptance Certificate (TAC) issued by the Certification and Engineering Bureau (CEB) of Industry Canada or a certificate issued by a recognized Certification Body (CB). RRBS is for fixed wireless access only, and currently no nomadic or portable system is under consideration. The application of RRBS is only for subscriber-based broadband Internet access, while in-band backhaul and other subscriber-based services are permitted at the discretion of Industry Canada's regional office, on a case-by-case basis, provided that the main application of the network is broadband access.

A few key parameters of equipment in the TVWS RRBS are summarized in Table 1.7.

The operating frequency is from 512 MHz to 698 MHz, i.e., TV Channels 21 to 51, except Channel 37 (608–614 MHz), which is allocated to the Radio Astronomy Service, similar to that in the U.S. regulations. The channel arrangements are identical to those of the TV broadcast channels, as displayed in Table 1.8.

In normal situations, for a frequency division duplex (FDD) system, a pair of channels is assigned, and for a time division duplex (TDD) system, one channel is assigned. When contiguous TVWS channels are available, and when extra capacity is required, a pair of up to two contiguous channels may be assigned for an FDD system, and up to two contiguous channels may be assigned for a TDD system. So a FDD system may have bandwidth of 12 or 24 MHz, and a TDD system may have bandwidth of 6 or 12 MHz.

RSS-196 gives a detailed description of the spectral mask to control the transmitter unwanted emissions; see [13, Table 1, Figure 1]. The graphical illustration of the spectral mask is displayed in Figure 1.3, in which it is shown that at the channel edge the required attenuation is approximately 45 dB.

In SRSP300-512, the regulations further specify the NTSC/DTV broadcast protected contours and the RRBS base station guard distance, in order to protect

Table 1.7 Key Parameters of RRBS Equipment

Operating frequency	512–608 MHz and 614–698 MHz
Nominal channel bandwidth	6 MHz and 12 MHz (two contiguous 6 MHz blocks)
Occupied bandwidth	≥500 kHz, and ≤ nominal channel bandwidth
Modulation	Digital
Frequency stability	± 10 ppm
TX power limit	For CPE:
	≤1 W/6 MHz, PSD ≤ –7 dBW/100 kHz, e.i.r.p. ≤ 500 W
	For base station equipment:
	≤125 W/6 MHz, PSD ≤ 14 dBW/100 kHz, e.i.r.p. ≤ 4 W
	See [14, 5.2.1] for details
TX antenna height limit	For CPE: 30 meters, see [14, 5.2.2] for details
	For base station equipment: see [14, 7.2.1] for details
Antenna polarization	For CPE: Only vertical polarization for TX, no restriction for RX
	For base station equipment: No restriction
Antenna directivity	For CPE: Cross-polarization isolation ≥ 14 dB, front-to-back ratio ≥ 14 dB
	For base station equipment: Both omnidirectional and sectoral, no minimum front-to-back discrimination requirement
Receiver spurious emission	Comply with the limits specified in RSS-General

Table 1.8 **Radio Frequency Channel Arrangement**[a]

Channel	Frequency (MHz)	Channel	Frequency (MHz)
21	512–518	37	—
22	518–524	38	614–620
23	524–530	39	620–626
24	530–536	40	626–632
25	536–542	41	632–638
26	542–548	42	638–644
27	548–554	43	644–650
28	554–560	44	650–656
29	560–566	45	656–662
30	566–572	46	662–668
31	572–578	47	668–674
32	578–584	48	674–680
33	584–590	49	680–686
34	590–596	50	686–692
35	596–602	51	692–698
36	602–608		

[a] Original table appeared in [14, Table 1].

TV service from interference of RRBS. There is no explicit protection of RRBS from broadcasting TV stations, because RRBS is a secondary service. It should be noted that, in RRBS, there is no autonomous broadcasting protection means like spectrum sensing or geo-location database access. Instead, the TVWS channel allocation and coordination are directly conducted and monitored by Industry Canada, on a case-by-case basis. This is in sharp contrast with the U.S. regulations. Finally, SRSP300-512 also specifies rules for operation near the Canada–United States border.

1.8.4 Asia

1.8.4.1 Japan

The digital terrestrial TV standard in Japan is Integrated Services Digital Broadcasting (ISDB-T). Analog terrestrial TV broadcasts will end by July 24, 2011 [15].

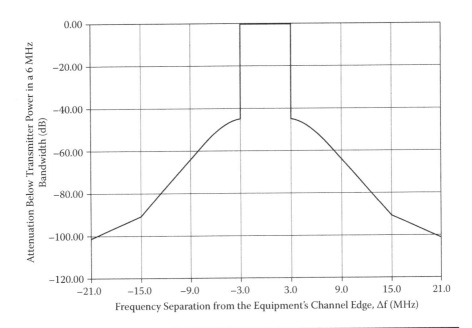

Figure 1.3 Unwanted emissions limits with Δ < 18 MHz; original figure appearing in [13, Figure 1].

Since December 2009, the Ministry of Internal Affairs and Communications (MIC) has established a round-table conference titled "Examination Team for New Radio Usage Vision," to carry out examination for the promotion of newly effective frequency usages, such as white space usage [16]. The team is made up of the State secretary for Internal Affairs and Communications and various experts in the field. The examined topics include usage models, system and technology issues, and other issues.

National Institute of Information and Communications Technology (NICT) has been researching and developing enabling technologies for cognitive radio and dynamic spectrum access. One of its currently ongoing projects, ASTRA (Advanced Spectrum-management Technology for Radio Access innovation), covers the PHY/MAC design and coexistence of secondary networks in the TVWS spectrum [17].

1.8.4.2 Korea

With the forthcoming completion of DTV transition in Korea in 2012, the Korea Communications Commission (KCC) is very interested in the use of the TVWS spectrum. The Korean government has been developing the core technologies of cognitive radio and seeking the applications of these technologies, especially personal/portable devices [18]. Home-life type of use cases, such as video streaming in home, have been envisioned, and a license-exempt approach appears to be under consideration.

1.8.4.3 Singapore

In order to evaluate the potential utility of white space spectrum for future opportunistic channel access like cognitive radio, the Institute for Infocomm Research (I²R) and Infocomm Development Authority of Singapore (IDA) in 2007 jointly conducted a spectrum survey to measure and analyze the 24-hour spectrum usage pattern in Singapore in the frequency bands ranging from 80 MHz to 5850 MHz, including the VHF and UHF TV spectrum [19]. It was shown that most of the allocated spectrum, except that for cellular and broadcasting systems, was heavily underutilized, with an overall occupancy rate as low as 4.54%.

As a consequence, recently IDA initiated a call for trials of white space technology for accessing VHF and UHF bands in Singapore [20]. Any interested party is invited to conduct white space technology trials in Singapore, in order to "explore the various spectrum environments and regimes that white space technology could operate in." The trials will be conducted over a period of six months, strictly within the premises of five specified test sites called the CRAVE (Cognitive Radio Venue) zones. The trials may be conducted jointly with IDA, or on one's own without the involvement of IDA.

The TVWS devices used for the trials are required to transmit only in UHF TV Channels 41–49, 51, and 53–54 (i.e., 622–694 MHz, 702–710 MHz, and 718–734 MHz). Note that, in Singapore, UHF TV channel bandwidth is 8 MHz. Spectrum sensing is a mandatory function, required for detecting analog/digital TV broadcast services as well as analog/digital wireless microphones in both VHF band (TV Channels 2–12) and UHF band (TV Channels 21–62).

A few key parameters of the TVWS devices for the trials are summarized in Table 1.9.

Table 1.9   Key Parameters of TVWS Devices for the Trials[a]

Parameter for Detection	Value
Sensitivity	–120 dBm
Transmit power	4 dBm adjacent channel
	17 dBm (N + 2, …)
Transmit-power per channel	100 mW
Bandwidth	8 × N MHz, N= max # of contiguous TVWS channels
Out-of-band performance	<–48 dBm
Time between sensing	<1 second
Maximum continuous transmission	400 milliseconds
Minimum pause after transmission	100 milliseconds
Parameter for Geo-Location	
Location accuracy	< 50 m
Transmit power	As specified in database (up to 100 mW)
Bandwidth	8 × N MHz, N = max # of contiguous TVWS channels
Out-of-band performance	<–48 dBm
TVBD refresh rate	1 second
Database refresh rate	24 hours
Maximum continuous transmission	400 milliseconds
Minimum pause after transmission	100 milliseconds

[a] Original tables appeared in [20].

References

1. Federal Communications Commission, "Second Report and Order and Memorandum Opinion and Order in the Matter of Unlicensed Operation in the TV Broadcast Bands, Additional Spectrum for Unlicensed Devices Below 900 MHz and in the 3 GHz Band," Docket Number 08-260, November 14, 2008.
2. Federal Communications Commission, "Second Memorandum Opinion and Order in the Matter of Unlicensed Operation in the TV Broadcast Bands, Additional Spectrum for Unlicensed Devices Below 900 MHz and in the 3 GHz Band," Docket Number 10-174, September 23, 2010.
3. Federal Communications Commission, "Spectrum Policy Task Force Report," Docket Number 02-135, November 2002.
4. Federal Communications Commission, "In the Matter of Additional Spectrum for Unlicensed Devices Below 900 MHz and in the 3 GHz Band," Docket Number 02-380, December 11, 2002.
5. Technical Research Branch Laboratory Division Office of Engineering and Technology, Federal Communications Commission, "Initial Evaluation of the Performance of Prototype TV-Band White Space Devices," FCC/OET 07-TR-1006, July 31, 2007.
6. Technical Research Branch Laboratory Division Office of Engineering and Technology, Federal Communications Commission, "Evaluation of the Performance of Prototype TV-Band White Space Devices Phase II," FCC/OET 08-TR-1005, October 15, 2008.
7. T. Rappaport, *Wireless Communications: Principals and Practice,* Upper Saddle River, NJ: Prentice Hall, 1996.
8. Federal Communications Commission, "Prediction of Coverage," Code of Federal Regulations—Title 47: Telecommunication, Part 73.684.
9. CEPT Report 24, "A Preliminary Assessment of the Feasibility of Fitting New/Future Applications/Services into Non-Harmonised Spectrum of the Digital Dividend (Namely the So-Called 'White Spaces' Between Allotments)," Report C from CEPT to the European Commission in response to the Mandate on: "Technical considerations regarding harmonisation options for the Digital Dividend," Final Report on June 27, 2008.
10. CEPT ECC SE43 Draft Report, "Technical and Operational Requirements for the Possible Operation of Cognitive Radio Systems in the White Spaces of the Frequency Band 470–790 MHz," Annex 3 to Doc. SE43(10)103, 6th meeting, Mainz, June 15–17, 2010.
11. Ofcom, "Statement on Licence-Exempting Cognitive Devices Using Interleaved Spectrum," Digital Dividend: Cognitive Access, Publication date: July 1, 2009.
12. Ofcom, "Consultation on Licence-Exempting Cognitive Devices Using Interleaved Spectrum," Digital Dividend: cognitive Access, Closing date for responses: May 1, 2009, Publication date: February 16, 2009.
13. Industry Canada, Spectrum Management and Telecommunications, Radio Standard Specification, RSS-196, Issue 1: "Point-to-Multipoint Broadband Equipment Operating in the Bands 512-608 MHz and 614–698 MHz for Rural Remote Broadband Systems (RRBS) (TV Channels 21 to 51)," March 2010.

14. Industry Canada, Spectrum Management and Telecommunications, SRSP300-512, Issue 1: "Technical Requirements for Remote Rural Broadband Systems (RRBS) Operating in the Bands 512–608 MHz and 614–698 MHz for Rural Remote Broadband Systems (RRBS) (TV Channels 21 to 51)," March 2010.
15. Broadcasting Digitization Schedule: http://www.dpa.or.jp/english/schedule/index.html.
16. Communications News, "'Examination Team for New Radio Usage Vision' Starts Work," Ministry of Internal Affairs and Communications (MIC), Japan, Vol. 20, No. 20, January 2010.
17. Hiroshi Harada et al., "Research, Development, and Standards Related Activities on Dynamic Spectrum Access and Cognitive Radio," in *Proc. IEEE DySPAN,* Singapore, April 2010.
18. C.-J. Kim, J. Kim, and C. Pyo, "Dynamic Spectrum Access/Cognitive Radio Activities in Korea," in *Proc. IEEE DySPAN,* Singapore, April 2010.
19. M. H. Islam et al., "Spectrum Survey in Singapore: Occupancy Measurements and Analysis," in *Proc. CROWNCOM,* Singapore, May 2008.
20. Infocomm Development Authority of Singapore (IDA), "Trial of White Space Technology Accessing VHF and UHF Bands in Singapore," Information Pack, online: http://www.ida.gov.sg/doc/Policies%20and%20Regulation/Policies_and_Regulation_Level2/WST/WhiteSpaceRegFW.pdf.

Chapter 2

Projecting Regulatory Requirements for TV White Space Devices

Benoît Pierre Freyens and Mark Loney

Contents

2.1 Overview

The desirable propagation characteristics of spectrum in the UHF band tradition-ally used for broadcasting services in conjunction with the continuing growth of the Internet and increasing demand for broadband services are fuelling demand for secondary usage of the UHF band by an ever larger array of devices. In particular, there has been sustained interest in, and increasing regulatory support for the devel-opment and use of "white space" devices that are intended to provide wireless broad-band services. Despite opposition from broadcasters, regulators in several countries have tended to authorize white space devices on a secondary or "unlicensed" basis. However, as regulators reallocate UHF spectrum released by the digital switchover to new services requiring a high degree of license certainty (e.g., cellular networks) the nature of the relationship between the new licensees and unlicensed white space users is likely to become much more challenging. What becomes of entrenched secondary usage rights if additional broadcast spectrum is eventually reallocated to telecommunications and relicensed on far more exclusive conditions than those emerging for white space devices operating on a secondary basis to broadcasting services? As shown by the program making and special events (PMSE) controversy in the United Kingdom, well-established secondary users may have de facto rights that have to be taken into account when valuable spectrum is reallocated to a new primary use (in that case, from broadcasting to telecommunications). If white space devices achieve the goals of their proponents and become an important component of the national broadband infrastructure, their widespread deployment could have a paradoxical impact on the ability of regulators to reallocate the primary use of spectrum in the UHF band from broadcasting to higher-value uses. While white space devices have not been authorized to provide wireless broadband services in the UHF band in Australia, this article considers Australian regulatory arrange-ments in light of this issue and suggests licensing reforms required to manage com-peting white space usage rights in the future. We do so because regulators in many regions and countries will be confronted by the challenge of establishing regulatory arrangements for white space devices that are technically sustainable and economi-cally efficient over the short, medium, and longer terms.

2.2 Introduction

There has been increasing interest in telecommunications use of "white spaces" in the UHF bands that have traditionally been used for television broadcasting. In more and more countries this interest is fuelled by the now inexorable move toward digital terrestrial television broadcasting, the consequential and looming end of ana-logue terrestrial television broadcasting, continuing growth in Internet usage, and increasing demand for wireless broadband access. In the United States, the Federal Communications Commission (FCC) has established regulatory arrangements for white space devices, and the United Kingdom's spectrum regulator and communications competition authority (Ofcom) is in the final stages of developing white space arrangements for the United Kingdom. However, emerging scenarios in digital switchover policy are challenging these developments by implicitly earmarking additional portions of the UHF band for purposes other than broadcasting in the longer term: namely mobile phone networks and wireless broadband services in the 10- to 15-year timeframe. In such a context, there is a risk that enabling the widespread deployment of white space devices in the UHF bands on a secondary basis will compromise the ability of regulators to achieve higher value use of the UHF band in the medium to long term.

In 2009, one important and emerging scenario for the UHF band was the complete cessation of terrestrial television broadcasting in the UHF band. This development was foreshadowed in Finland [1], studied by the European Union, and the subject of a legislative proposal in the United States. If white space devices are allowed in the UHF band, their widespread use could seriously constrain reallocation of more of the UHF band from broadcasting to higher value uses such as telecommunications should there be a "digital dividend 2.0" in the 10-to-15-year time frame.

This would particularly be the case if the regulatory requirements for and the technical characteristics of white space devices are based on the assumption that the UHF band will continue to be used for digital terrestrial television. Because of the significant differences between broadcasting and telecommunications use of the spectrum, white space devices capable of successfully operating on a secondary basis in a broadcasting environment may have little or no capacity to operate in a telecommunications environment without causing harmful interference or suffering from degraded performance and capacity. Alternatively, and more optimistically, the regulatory regime for white space devices could incentivize the development of white space devices with technical characteristics that enable cooperative sharing with telecommunications services. In a best-case scenario, white space devices operating on a secondary and ad hoc basis would collaboratively operate in conjunction with telecommunications networks deployed on a planned basis by network operators using advanced technologies to provide services such as mobile broadband access. In this scenario, the deployment of white space devices in the UHF band in the short term could allow early realization of the telecommunications potential of the UHF band and lead to increased producer and consumer benefit in the medium to long term.

In this chapter we discuss the regulatory requirements and the licensing structures better suited to the realization of this objective—the establishment of forward-looking regulatory arrangements that will help enable a commercially viable and economically efficient transition of the UHF band from analog terrestrial television services to digital terrestrial television and secondary telecommunications usage and, finally, to high value telecommunications use. Section 2.3 describes the most commonly encountered types of white spaces and white space-using devices. Section 2.4

describes radiocommunications licensing arrangements for spectrum-using services in Australia. Section 2.5 describes proposals for a licensing regime allowing secondary unlicensed usage in "private" spectrum spaces (these spectrum spaces are typically used for mobile phones in Australia but the regulatory set-up in 2010 does not allow secondary usage by *any* device). Finally, Section 2.6 evaluates this proposition with regard to the entrenched rights it might confer on secondary unlicensed users, particularly if more of the UHF band is ultimately to be reallocated from broadcasting to higher value services. Some final comments conclude the chapter.

2.3 TV White Space and White Space Devices

2.3.1 White Spaces

TV white spaces (TVWS) are those channels that have been *allocated* for terrestrial television broadcasting but which have not been *assigned* to the provision of television services in a particular license area (a more expansive definition would assert that TVWS also exist in channels that have been allocated to the provision of broadcasting services in license areas, but for the purposes of this discussion, we limit our consideration of TVWS to channels that have not been so assigned).

TVWS arise for three reasons. The first was the need for guard spaces between analog TV (ATV) services in the same license area. If ATV services in the same license area operated on immediately adjacent channels, the result would be mutual interference between the two services. Consequentially, planning for ATV services typically provided for a vacant channel (a "guard space") between ATV services in the same license area. Because of the superior performance characteristics of digital terrestrial television technologies, the need for guard spaces between DTV services in the same license area can be significantly reduced or eliminated.

TVWS also arise from the need for geographic separation between TV services that are in different license areas but are broadcasting on the same channel. Regardless of whether planning for TV services is noise-limited or interference-limited, there will be areas where the channel is unable to be used for TV services (if a noise-limited planning model is used, the TVWS will typically be larger). Geographic separation occurs in both ATV and DTV planning and deployments.

Finally, TVWS arise in areas where channels are not allocated to broadcasters for TV services, either because of the limited supply of broadcasting services (because only a small number have been authorized or deployed) or because there is limited demand for broadcasting services (typically because of low population density, but more commonly because of the increasing range of technologies that can be used to deliver broadcasting services). In this context, TVWS occur regardless of whether ATV or DTV technology is in use.

A device that opportunistically uses these available channels is commonly referred to as a "white-space device" (WSD). WSDs have so far come in two types:

(i) "symbiotic" devices, which broadcasters have long tolerated, and (ii) "invasive species," which have acquired secondary (unlicensed or class licensed) rights against the wishes of broadcasters through controversial regulatory decisions.

2.3.2 "Symbiotic WSD Devices"

Despite the controversy surrounding recent proposals for WSD, there is a long history of radiocommunications devices that provide nonbroadcasting services coexisting with broadcasters in TVWS. We have chosen to describe these precursor devices as "symbiotic WSD," because their ability to share successfully with broadcasting services was without doubt and based on compliance with a small set of relatively simple rules that were not controversial and which did not require advanced technologies to be implemented.

From a regulatory perspective, even though symbiotic WSD were uncontroversial, their use was often, but not always, based on tolerance rather than legitimacy. Wireless microphones used for program making and special events (PMSE) are an example of symbiotic WSD that have attracted a high profile as a result of the loss of their ecological niche—unused guard channels between ATV services in heavily populated areas where PMSE activities often occur. Wireless microphones are mostly used by stage performers (in concerts, in sports stadium, in megachurches, etc.), but also by TV performers (e.g., reality shows), using idle television channels in given locations. In a common scenario, wireless microphones operated in the UHF band but were able to be programmed to operate on a range of channels. The channels chosen depended on the particular area where the microphones were to be used. This approach relied on the avoidance of channels that had been assigned to TV services in a license area (for some time, an automated version of this approach was proposed for what we have characterized as "invasive WSD"). Although the wireless microphones were not cognitive in the sense proposed by Mitola, they were operated in a "cognitive" fashion as their operators determined the idle channels that they could safely operate in (thereby able to be characterized as *overlay* signal devices). At the same time, wireless microphones are quite low power compared to terrestrial television transmissions and operate below the noise floor of ATV services in adjacent channels (and are thereby also able to be characterized as *underlay* signal devices). Depending on the service, the band, and the country's regulatory framework, secondary PMSE usages of TVWS were managed under different types of licensed or unlicensed regimes. For example, in Australia as in the United Kingdom and elsewhere in Europe, the use of wireless microphones for PMSE usage in white space is licensed.

In the United States however, wireless microphones operating in TV white space have been floating in something of a legal limbo. Officially part of the Broadcast Auxiliary Service (BAS), they are theoretically subject to licensing. However, due to the overlay/underlay nature of the equipment, interferences have been rare, mild, and have proved impossible to detect in real time anyway. Consequently, licensing requirements never saw much enforcement by the FCC, and a well-funded corporate media constituency, politically difficult to disenfranchise, has emerged. To a large extent, this constituency was not perceived as harmful and quietly tolerated by broadcasters (perhaps because one important use of wireless microphones is to provide content for television broadcasters).

For many years, the two services peacefully coexisted on licensed broadcast spectrum. The status quo was challenged in 2008, when the FCC signaled that BAS devices would no longer be allowed to operate in the 700 MHz bands (which of course they were never authorized to do in the first place) due to the reallocation of that band to telecommunications services by way of auctions in February and March 2008.

2.3.3 *"Invasive WSD Species"*

The successful occupation of TVWS by "symbiotic" PMSE devices was one factor* that led to increased interest in the exploitation of TVWS by other services and technologies, again on an underlay/overlay basis. We describe this second generation of WSD as "invasive WSD," because their ability to share with broadcasting services was strongly contested, and their operation depended on the use of advanced technologies to implement a significantly more sophisticated set of rules than symbiotic WSD.

The first waves of invasive WSD were proposed to work on an underlay rather than overlay basis. In 2002, the FCC amended its Part 15 rules to permit the unlicensed operations of some services using Ultra-Wide Band (UWB) technology [2]. UWB was a relatively new communications technique that relied on the development of integrated circuits with high computing capacity, low power consumption, and small size. UWB devices transmit at extremely low power across very large bandwidths (hence the name) and thus transmit across a number of bands used by a wide range of services. The power density (power per MHz) of UWB devices is so low that the proponents of UWB technologies argued that they could not cause harmful interference to other services using more traditional technologies that operated at higher powers in narrower bandwidths. Broadcasters, along with many other established users of the spectrum, did not accept these arguments and had significant concerns about the potential for UWB devices to cause harmful interference to their services. Ultimately, the use of UWB has been allowed on a limited basis in a number of bands in various spectrum-reforming countries including the United States, the United Kingdom, and some Continental European nations. The 2002 FCC decision, followed in 2007 by similar decisions in the United Kingdom [3], in the European Union [4], and in the Asia-Pacific region explicitly authorized the usage of UWB devices in

---

* Other factors include increasing sophistication and reduced cost of radiocommunications technologies and increasing demand for mobile communications services.

TV broadcasting bands and thus in TVWS (Australia has authorized the use of ground-penetrating radars that use UWB technology but which operate at frequencies above 2 GHz).*

In opposition to their relatively benign approach to PMSE services, it is fair to say that traditional broadcasters were very wary of UWB technology and have successfully lobbied their respective regulatory authorities to limit the range of frequencies it can use. Consequently, and considering that the UWB decision was not specific to TVWS in the first place, broadcasters and other incumbents eventually won the day, and simmering tensions over the matter of secondary usage of TVWS by UWB WSDs somewhat lapsed over time.

Since then however, another debate has emerged with the FCC's decision to open up TVWS channels to cognitive radio technologies on an unlicensed basis (i.e., overlay rather than underlay technologies) to provide wireless broadband access. For instance, one early FCC suggestion included the use of GPS receivers in all invasive WSDs in conjunction with a database that contained details of all TV stations broadcasting in any given geographical area (essentially an attempt to automate the "sense and avoid" approach that was so successfully used on a manual basis with symbiotic WSD such as wireless microphones).

Pressure to allow secondary use of TVWS intensified with increasing concern about the "digital divide," including the failure of well-publicized efforts by municipalities to use WiFi technologies in the 2.4 and 5.8 GHz bands to provide ubiquitous access to the Internet and increasing recognition that the propagation characteristics of the UHF band were far better suited to provide wireless broadband access. Unsurprisingly, the next wave of invasive WSD were intended to use TVWS to provide wireless broadband access. The National Broadband Plan unveiled by the FCC in March 2010 identified TVWS as an ideal input to provide broadband services to rural and remote communities. In particular, TVWS was identified as ideally suited to meet the shortfall in middle mile infrastructure in rural areas.

Earlier in 2008, the FCC famously agreed to open up TVWS for use by invasive WSD on an unlicensed basis. This decision was strongly opposed by the TV broadcasting and wireless microphone industries, which teamed up to defend their previous regulatory and quasi-regulatory arrangements. Broadcasters again argued that using this portion of "their" licensed spectrum would interfere with the reception of their television broadcasts by consumers. In February 2009, the National Association of Broadcasters asked a federal court to wind down the 2008 FCC authorization for TVWS usage by invasive WSDs. The case built upon the argument that the testing of white space sensing devices had not been thorough enough. Initial tests performed by the FCC in early 2008 showed that portable, unlicensed personal devices operating in the same band as TV broadcasts could indeed cause interference. The

---

* Note that in the 2002 FCC decision, only ground-breaking radar UWB applications were authorized to operate below 960MHz. All other UWB devices were instructed to operate above 1.99, GHz, 3.1 GHz and higher frequencies.

FCC subsequently agreed on some device restrictions, including using geo-location technology to allow invasive WSDs to compare their location against an established database of broadcast services. The FCC stressed that test devices using a combination of spectrum sensing and geo-location managed to successfully identify all occupied TVWS channels. Invasive WSDs that did not incorporate geo-location capabilities would be considered for authorization on a case-by-case basis and have to submit to a rigorous application process for FCC certification.

By the third quarter of 2010, little had happened since the FCC's 2008 decision. The deployment of invasive WSD in TVWS appeared bogged down by uncertainties on future spectrum allocation in TV bands, uncertainties about the certification of spectrum-sensing technologies intended to identify TVWS (idle or vacant channels), and little progress on the establishment of a system of geo-location-facilitated database consultations.

2.3.4 Digital Switchover and TVWS

So far so good. If the FCC and other regulatory agencies see fit to enable the coexistence of broadcast services and underlay–overlay wireless devices, the regulatory framework needs little more than a standard secondary usage easement for unlicensed usage such as commonly found for many bands in allocations made by the International Telecommunication Union (ITU), including specification of the rules that precisely establish the TVWS "niche" that is to be inhabited by invasive WSD. Alternatively, regulators could specifically reallocate TVWS, either in whole or in part, from broadcasting to unlicensed services as was done on ISM (Industrial Scientific and Medical) and UNII (Unlicensed National Information Infrastructure) microwave bands. If broadcast spectrum is indeed large enough to accommodate both types of services or tolerate these kinds of subdivisions, then the regulatory choices are relatively simple.

These choices are, however, more complicated if one factors in uncertainty about the future use of that part of the UHF band that is being used for DTV after digital switchover. In the United States, digital switchover enabled the reallocation of the 700 MHz band. Australia has announced that it will also have a digital dividend in the 700 MHz band, while countries in the European Union are aligning around a smaller digital dividend from 790 to 862 MHz. In these countries the digital dividend has already been, or is expected to be allocated predominantly to advanced mobile cellular and mobile broadband networks using technologies such as HSPA, WiMAX, or LTE.

Until recently, it appeared that the situation for DTV (and thus WSD, both symbiotic and invasive) would stabilize after digital switchover, with a habitat that was reduced in size but still ecologically viable for both the primary service and secondary users. But increasing interest in the potential for further reallocation of UHF spectrum from broadcasting to other services—which we describe as a "digital dividend 2.0"—has emerged since 2008 when the Finnish national plan of action identified that the achievement of high speed broadband by 2015 would allow consideration of the future of terrestrial DTV by 2017. In August 2009, the European Commission received a report on exploiting the digital dividend that identified two options for making more spectrum available for wireless broadband or other non-broadcasting purposes in the European Union. One option was for reallocation of 694–790 MHz—which would be more consistent with the American and Australian digital dividends—but the second option was to completely clear DTV from the UHF band to realize a digital dividend from 470–790 MHz. While the report indicates that neither of these options is realistic in the short term, it identifies that they may well be achievable in the medium (beyond 2015) to long (beyond 2020) term. Finally, in December 2009 the FCC issued a public notice that sought data on future use of spectrum used for terrestrial television broadcasting, including the impact of reallocating more spectrum—a digital dividend 2.0—away from DTV [5].

Further reallocation of UHF spectrum from broadcasters to other uses in these timeframes clearly has major implications for regulators (as well as industry and consumers). One of those implications is that regulatory frameworks that allow the use of invasive WSD must be considered very carefully. This is because the widespread use of invasive WSDs on a secondary basis is likely to limit the ability of telecommunications operators to successfully deploy high-capacity networks in spectrum that has been vacated by broadcasters. This is, in turn, a consequence of the different technical characteristics of broadcasting stations and telecommunications networks.

Broadcast services such as DTV are characterized by the use of a small number of carefully planned sites that transmit at extremely high power (kilowatts or even megawatts) and use a limited number of channels to provide services to a relatively large area. Broadcast infrastructure is typically stable over time because an increase in the number of receivers (TV sets) in the license area does not normally require an increase in the number of transmission sites.

In contrast, telecommunications networks involve the deployment of a much larger number of base stations, typically on a cellular basis to maximize frequency reuse, which operate at power levels well below one kilowatt. Base stations both transmit to and receive signals from user devices. Importantly, an increase in the number of user devices in the license area often results in an increase in the number of base stations as network operators increase the intensity of use of the spectrum allocation (which is typically the limiting resource in overall network capacity and performance).

In summary, broadcasters typically operate from a small number of sites that transmit at very high power with a low rate of change in transmitter locations and transmission frequencies. Telecommunications networks typically operate from a much larger number of sites that transmit and receive at low power. Network deployments change quickly with increasing numbers of base stations, with frequent changes in frequencies of operation for individual base stations as a result of the ripple effect that new base stations can have on the frequencies used by base stations in their vicinity. These contrasting technical characteristics mean that invasive WSD that operate successfully in a broadcasting environment on a secondary basis may cause harmful interference or suffer degraded performance if the primary use of spectrum is reallocated from broadcasting to telecommunications.

As a result, there is a significant risk that the widespread deployment of invasive WSD in a broadcasting environment may deter or frustrate the further reallocation of the UHF band from broadcasting to other services. This is because prospective new users of the spectrum may be reluctant to bid for licenses (or pay the prices sought by governments) as a result of constraints (actual or perceived) in their ability to deploy new networks because of interference from invasive WSDs that turn out to be harder to dislodge from their ecological niche than broadcasters.

While broadcasters have an impressive record of achieving favorable regulatory outcomes from government, they are ultimately businesses that comply with legal requirements and regulatory obligations. The majority of citizens are also law-abiding, but as consumers, may not recognize, or may be reluctant to accept, that the highly functional wireless device in their house that has operated flawlessly for several years is an invasive WSD that may need to be modified or replaced because of the reallocation of additional UHF band spectrum in a "digital dividend 2.0" that is intended to further improve the availability of spectrum for wireless broadband access. To further complicate matters, network operators have increasingly enjoyed more flexible usage rights than broadcasters (in their ability to deploy new technologies and offer new services) and, as wireless telecommunications services have become increasingly ubiquitous and more important, sought more stringent license conditions to protect their large investments in network infrastructure.

If these emerging scenarios for a digital dividend 2.0 are combined with the fact that suitable arrangements for PMSE have yet to be finalized in countries that are in the process of implementing "digital dividend 1.0," we can only wonder what would happen if regulatory authorities did not have to deal with a relatively small PMSE community using symbiotic WSD but a host of citizens enamored of the invasive WSD that provide them with crucial connectivity to, and improve the functionality of nationally deployed high-speed broadband networks. How would a regulatory agency keep the necessary flexibility to prioritize the rights of new licensees and enable invasive WSD to successfully adapt and contribute an environment that is making a step change from high-power broadcasting services to low-power telecommunications services?

To attempt an answer requires a formulation of the current regulatory arrangements for broadcasting, network operators, and unlicensed devices. A full review of these licensing structures by country of interest (United States, United Kingdom, Canada, Australia, New Zealand, Japan, Singapore, etc.) is well beyond the scope of this chapter. Instead, we briefly review the regulatory instruments in Australia, where broadcasting services are authorized by apparatus licenses,* network services are usually licensed through a property rights framework known as spectrum licenses, and "unlicensed" usage is regulated through class licenses, with all of these different types of radiocommunications license issued under the Radiocommunications Act 1992.

2.4 Radiocommunications Licensing in Australia

As an intangible resource, spectrum cannot be purchased and owned in physical units as with other factors of production, nor can it legally be used without regulatory consent.† Spectrum users therefore "acquire" spectrum by purchasing licenses and/or adhering to license conditions. In Australia, all broadcasting and radiocommunication licensing regimes are regulated through two acts of Parliament: the Radiocommunications Act 1992 (the Act) and the Broadcasting Services Act 1992 (for our purposes, the Act is of principal relevance, and the majority of the following discussion refers to that Act).

Licenses are renewable access rights issued by the Australian Communications and Media Authority (ACMA), the Australian regulator. The licenses specify a variety of regulatory arrangements, mainly to govern interference management.

The legislative framework flowing from the Act (which draws heavily from ITU regulations) fully authorizes and defines the specifications and operations of three categories of licensing regimes for radiocommunications; apparatus, spectrum, and class licensing (somewhat imperfectly reflecting the so-called "trichotomic" approach of "command and control," "property rights," and "open access"). These regimes set out the rights and obligations of users. They differ in many respects, but they also have common denominators. For instance, they all impose a minimal set of "core" specifications about frequency, bandwidth, area, power, and emission types, although the way these conditions are integrated in practice vary among these licensing regimes.

In general, apparatus and spectrum licensing have more in common than class licensing. Both licenses are typically used to provide exclusive access rights, can be auctioned, leased, or traded, and are subject to regulatory approval for license renewal—though on very different terms. The common and divergent features of these regimes are outlined below, as are the underlying legal provisions and practical aspects of the implementation of both apparatus and spectrum licenses.

---

* Apparatus licenses are issued under the Radiocommunications Act. Where broadcasting services are concerned, the apparatus licences are issued after the services are planned under the Broadcasting Services Act.

† Sections s.46 and 47 of the Act prohibit both unlicensed use and possession of radiocommunications devices, unless there are reasonable expectations that the device will be needed for emergency and safety of life operations in the future (s. 49).

2.4.1 Apparatus Licenses

Section s.96 of the Act authorizes the operation of radiocommunications devices by virtue of issuing an apparatus license. Apparatus licensing is the natural heir to the administrative licensing regimes predating the Act. This regime focuses on the operations of radiocommunications devices of a specified kind through transmitter licenses, receiver licenses, or both. Most licenses are issued by the ACMA after a frequency assignment process in which either the ACMA or an industry professional accredited under s.263 of the Act certifies that the operation of the device will satisfy technical conditions.* There are currently 21 types of apparatus licenses (mostly transmitter licenses), and they are capable of authorizing the operation of devices across most of the radiofrequency spectrum.

Most of the time, this "device-centric" regime is highly prescriptive, specifying the types of authorized primary services, authorized technology, the device to be used, and the equipment standards. As is the norm among all licensing regimes, apparatus licensing also enforces tight technical conditions about site, band, power, and emission types. Despite the highly prescriptive nature of the apparatus licensing regime, there is flexibility in the sense that different "molds" are accommodated. Section s.98 of the Act authorizes the ACMA to determine by written instrument (a determination disallowable by Parliament) the type of transmitter and receiver licenses it issues. This provides the ACMA with significant discretion in setting the regime terms of an apparatus license. For instance, and as opposed to many other countries' approaches to command and control licensing regimes, apparatus licenses can be traded and leased to other users (section s.114 of the Act allows apparatus licensees to authorize third-party users). They can also be auctioned by the ACMA under s.106 of the Act (price-based allocation), and have been on many occasions (e.g., ongoing auctions for broadcasting services such as LPON—low-power open narrowcasting—and HPON—high power open narrowcasting—services).

The issue and use of apparatus licenses is also closely constrained by the Spectrum Plan that the ACMA makes under s.30 of the Act. For particularly crowded bands, their use is further detailed in frequency band plans. Issued for a maximum of five years, apparatus licenses are renewable on a basis varying between 1 day and 5 years. Apparatus licensing maximizes the technical benefits of pooling similar services and technologies over a specific frequency range, particularly in view of minimizing interferences. The apparatus licensing regime is also very effective at controlling interferences. It does so through the frequency assignment process: spectrum engineers develop processes, which if followed correctly (transmissions over specified frequency, bandwidth, site, area, with given technology and devices) will result in users not suffering harmful interference. Because interference management is so tightly controlled, apparatus licenses can authorize the use of much

---

* Note that nonassigned apparatus licenses (such as amateur radio licenses) are not subject to a frequency assignment process.

narrower bandwidths than spectrum licenses. The main drawback of the approach is its prescriptive nature; by limiting the freedom of spectrum users to change the types of services and technologies they would like to develop or deploy on the bands, the question of whether spectrum supply indeed meets demand ambles in a blurred landscape. In high-value bands for which demand has increased at an accelerating pace over the years, this weakness of the apparatus licensing approach has become a severe problem.

2.4.2 Spectrum Licenses

A key reform of the Act in 1992 was to establish a legal framework for a regulatory alternative to apparatus licensing that was an economic response to the limitations identified above. Spectrum licensing is a market-oriented, private ownership, ser-vice- and technology-neutral response to this concern. Typically, spectrum licenses are allocated by auction* and issued for fixed terms of up to 15 years. Unlike appa-ratus licenses, which are typically issued following a frequency assignment process for a particular device, spectrum licenses are normally issued after price-based allo-cation and authorize the use of a spectrum-space rather than the use of a specific device at a specific location on a specific frequency (although it is the case that many devices operated under spectrum license are subject to a registration process).

The Act provides two legislative pathways to the issue of spectrum licenses in particular parts of the spectrum. Under s.36 of the Act, which has been in place since 1992, the Minister may *designate* parts of the spectrum for spectrum licens-ing (in which case only the nature and tenure of the band changes, and ACMA prepares a conversion plan to convert existing apparatus licenses into spectrum licenses†). Under s.153B, which was included in the Act in 1997, the Minister may declare a *reallocation* of encumbered spectrum (in which case the band is cleared of apparatus licenses, and new users are expected to bid for the spectrum released). The latter is the mechanism expected to be used for the digital dividend from 698 to 820 MHz that the Australian Government identified in 2010 or for any further reallocation of UHF spectrum from broadcasting to telecommunications services.

There are currently about 10 spectrum-licensed bands (known as "spectrum spaces") in Australia, each defined by their own specific technical framework. Overall, these spectrum licenses cover a relatively small portion of the entire spec-trum, albeit a highly valuable one, because the most commercially valuable bands have progressively been spectrum licensed since the first spectrum licenses were issued in 1997.

---

* Note that section s.60 of the Act also authorizes allocation by tender and negotiated prices which are less commonly used.
† Note that the conversion plan is used only if there are existing apparatus licences using the relevant spectrum; otherwise ACMA will instead use a marketing plan.

For various reasons, spectrum licenses offer a more appropriate approach to the development of spectrum markets than apparatus licenses. The Act sets up a legislative regime where spectrum licenses are regulated as if they were property (albeit leasehold rather than freehold) rather than treated as if they were permissions granted on an administrative basis. Consequently, the duty of managing interferences also passes to the licensee. The private property-like attributes of spectrum licenses confers some degree of certainty to their holders, although there are issues with license expiry that are beyond the scope of this discussion. Second, spectrum licenses are "service-neutral" and "space-centric," i.e., they permit users to operate *any* radiocommunications devices in a given "spectrum space" (the equivalent of a large private real estate), subject to respecting the requirements of a technical framework. The degree of service neutrality of the license varies along a continuum depending on the exact formulation of the technical framework; too much service neutrality weakens the technical framework and may lead to inefficient uses—i.e., there is a fundamental trade-off between flexibility and efficiency.*

The space-centric approach defines big blocks of electromagnetic spectrum based on a specific array of standard trading units (STUs)—"atomic" blocks of spectrum specified by geographic and bandwidth limits. As the smallest spectrum unit in use, an STU rests on a *minimal bandwidth unit* (since bandwidths vary along the spectrum, there is no absolute general value for this minimal requirement) and a *minimal geographical unit* (a cell in Australia's Spectrum Map Grid—SMG†). An STU can then be aggregated geographically or physically by combination with other STUs, defining nonoverlapping "spectrum spaces."

A spectrum space comes with an underlying population density, incumbent or prospective services, and is specified through the development of a technical framework. Only fully specified spectrum spaces are marketable to private or public operators. Because this approach defines spectrum space over large areas and bandwidths, it is much more compatible with a higher degree of freedom in usage and is generally preferred by large operators seeking the deployment of large-scale networks within the geographic boundaries and technical framework specified in spectrum licenses.

---

* In practice, regulators make some assumptions about the most likely use for the licence (often this is mobile networks or wireless broadband services), and these assumptions are integrated in the technical framework through system models, reference technology, and deployment constraints. It would be unexpected and certainly suboptimal for users to deploy a radically different type of technology or service on the band. Although there would be no legal impediments to it, in practice, Australia is a nonmanufacturing country. So holders of spectrum licences (usually companies) will always find it in their interest to use the licences for purposes that the technical rules make them most efficient for. It would be more efficient for companies to use their licences in a harmonized way with other countries, because not doing so would make operations much harder and more expensive.

† Here too there is no absolute minimal standard because cell sizes on the SMG vary with location (higher population density means lower cell size).

Device registration and certification under sections s.145 and s.262 of the Act has proved a bit controversial, with some users regretting an unnecessary intrusion of red tape in an otherwise very secluded private spectrum space [6].* This debate also relates to the type of interference management model undertaken for spectrum licenses, notably the transmitter-by-transmitter approach and the use of block-edge masks to determine out-of-band emissions. To an extent, this approach emphasizes legal certainty over flexibility.

The United Kingdom has very recently started experimenting with its own approach to defining property rights in spectrum (receiver licensing, also known as SURs) in which in- and out-of-band emissions are controlled by power flux density, and interferences are considered on an aggregate rather than transmitter-specific way [7, 8]. Because this approach has only recently been experimented with, and because interference conditions and allocative issues vary widely between a high-density manufacturing country like the United Kingdom and a low-density, technology-taker country such as Australia, there is no particular consensus that Australia should reconsider its own property rights model [9].

More than ten years down the track, there is no doubt that spectrum licenses have been a success story in Australia. They provide a workable compromise between maximizing flexibility and certainty in usage, channeling spectrum supply toward market demand and minimizing the need for ex post regulatory intervention. A good recent example of the success of spectrum licensing and the durability and utility of the technical frameworks was the fact that Telstra was able to replace its 2G CDMA2000 network in the 850 MHz band with its 3G "NextG" network without any need for the ACMA to change the spectrum licensing framework for the band.†

---

* See, for example, presentations and discussions in Chapter 2 of the proceedings of the 2003 ACA Workshop on Spectrum Licensing Approaches that took place in March 2003 [6].

† Telstra was required by a Carrier Licence Condition imposed by the Minister under the Telecommunications Act 1997 to keep its 2G CDMA network operating until the Minister was satisfied that equivalent or better coverage had been achieved with the NextG network. This condition did, theoretically, restrict the way in which Telstra could deploy its new network, because the Minister would most likely not have approved 2G switch off if ACMA, which provided a report to the Minister about coverage of the two networks, had found that Telstra had built the NextG network in such a way that equivalent coverage had not been achieved— for example, that the 2G and 3G networks were not substantially coextensive. In addition Telstra is still subject to an ongoing license condition requiring same coverage, restricting what Telstra can do with the spectrum. However, this coverage issue was and remains mainly political. The radiocommunications licensing arrangements did not need to change with the move to 3G. ACMA made an assessment for the government about coverage equivalence, and some performance issues with a few of the new handset models had to be fixed. When that was done, coverage was assessed as equivalent, and the Minister allowed Telstra to shut down its old CDMA network. The whole process was only indirectly related to the ability of Telstra to deploy a new network under its spectrum licences.

Achieving such a high degree of flexibility in network self-redeployment was truly an outstanding achievement for Australian spectrum policy, particularly when compared to the regulatory restrictions in other countries that have delayed similar outcomes for lengthy periods. However, there are several issues currently mitigating this otherwise resounding success:

1. *No presumption of license renewal:* Expiring licenses are reissued by auction or other price-based means unless the spectrum license is renewed by the ACMA "in the public interest" under section s.82(1) of the Act. Because no spectrum license has expired so far (the first will expire in 2012 in the 500 MHz band, and the first high-value spectrum licenses will expire in 2013), there is much uncertainty as to which legal rules will govern the reissuing process.

2. *Spectrum space idleness:* Some bands that were allocated by spectrum licensing have not seen much activity (e.g., 500 MHz, 2.3 GHz, 27 GHz), and although the most valuable spectrum spaces (800 MHz, 1800 MHz, 2.3 GHz, 3.4 GHz) have seen a reasonable amount of trading, there is still concern that trading on the secondary market is at a level that is suboptimal from an economic efficiency perspective. There is some contention as to whether this is due to the complexities of the technical framework (which may put off some average users and potential secondary users), to "option value" strategic behavior,* or whether this is simply due to a lag in the technology and business models.

3. *Unused regime options:* A spectrum license offers its owner considerable discretion to tailor the spectrum space in order to maximize its economic profitability. If a portion of the spectrum space is unused for some period of time, various options exist to put this spectrum to productive use by letting "tenants" (such as white space devices) lease the unused spectrum under interference-controlled conditions for a fee. These potentially profitable options have, however, seen little or no take-up.

2.4.3 Class Licenses

Class licensing is a "technology-centric" approach to spectrum management with the distinct feature that it operates on an open and shared access basis. There is no exclusivity in usage, and for that reason it is regularly referred to as the "public park" or "open access" regime. It is probably the simplest approach to assigning and allocating spectrum, both for regulators and users. A class license sets out some minimal operational conditions. These conditions consist of:

---

* Under this argument, the licence has a value for its holder even if it is not being used. Spectrum licensees may keep idle spectrum holdings with a view of using them at a later stage, when there may be an expectation of better technology for use in the band (opportunistic holdings), or higher demand for the band (speculative holdings).

1. Technical parameters somewhat akin to the basic technical conditions of apparatus licenses (operation band, maximum transmitting power, types of device and emissions)
2. Requirements of equipment compliance with standards, with the onus generally on manufacturers
3. Geographic deployment constraints (where relevant)

Operation of a device under a class license does not require that an application be made to the regulator, and a class license is not "issued" to an individual organization (although the ACMA is able to impose conditions about the class of user as well as the class of the device). As such, Australian class licenses are the equivalent of "unlicensed" approaches used in other countries. Class licenses work under the understanding of free, shared, noncoordinated user access to specifically designated common bands. Since class licenses are not issued to individuals, the license conditions are not applied nor tailored to individual users (as is the case, e.g., through frequency assignment certification under apparatus licensing, and through device registration under spectrum licenses). Unlike apparatus and spectrum licenses, the term of a class licenses is not limited by the Act, although it is open to the ACMA to revoke or amend class licenses after satisfying its statutory obligations to consult about changes. Class licensing is a highly efficient regime in cases where the conditions are appropriate for the use to which the licensee wants to put his/her license [10].

On the other hand, class licensees effectively operate on a secondary basis and are not authorized to interfere with the transmissions of other (apparatus and spectrum) licensees or provided protection from interferences generated by other services. For this reason, this licensing regime is generally not popular with larger radiocommunications operators, such as telecommunications carriers, whose business model and reputation may largely depend on maintaining a minimum guaranteed level of service quality.

Hence as the services and devices authorized under class licenses must present a low potential for interference, they are most often only used to authorize for short-range, low-power applications (however, class licenses have been used to authorize earth stations in several frequency bands which do operate using high powers, but which, importantly, have a low potential for interference because of the coordination required for satellite networks). As briefly mentioned earlier, some bands, such as the 2.4 GHz (ISM) and 5.2 GHz (UNII) bands are specifically set aside for class licensing, but most class licenses operate as noninterfering secondary usage easements in apparatus licensed bands (e.g., symbiotic WSD such as biomedical telemetry devices that have long operated in TVWS in Australia).

Notwithstanding what has just been said, some class licenses are very prescriptive and will restrict a frequency to the sole use of a specified device or service (e.g., wireless microphones, which unlike their usage in the United States are more strictly regulated in Australia). After a regulatory program in the last decade to simplify and consolidate the class license regime, there are now 12 class licenses in force across Australia. It is worth noting that, although the Act provides quite detailed constraints on the regulator in relation to apparatus and spectrum licenses, this is not the case (except in one important respect discussed further below) for class licenses.

The class license of most interest to the white space debate is the *Low Interference Potential Devices Class Licence 2000* (the LIPD Class Licence—pronounced "lipid") which governs the operation of tens of thousands of low-power, short-range radiocommunications devices using cutting edge technologies such as the IEEE 802.11 standard (Wi-Fi) and Ultra-Wide Band (UWB) services. LIPD-licensed devices are ubiquitous, operating on a disparate array of frequency bands such as broadcasting bands or radio navigation bands (but never in spectrum licensed mobile phone bands). They operate mostly with secondary usage rights over apparatus-licensed portions of the spectrum, but are not authorized in spectrum-licensed spaces where decisions regarding secondary usage rights are entirely at the discretion of the owner.* They are also found in specifically allocated spectrum bands such as the ISM and UNII bands.

The legal rules for the LIPD Class Licence are relatively basic. Class licenses are subordinate legal instruments directly enabled under the Act. As opposed to apparatus licenses, which can be administratively issued by ACMA staff, the issue of class licenses is a statutory decision that cannot be delegated to staff and must be made by the Authority.† The ACMA can include any relevant conditions on the license but must consult with stakeholders before doing so. The LIPD Class Licence stipulates a schedule of frequencies or frequency ranges for different classes of transmitters and sets associated radiated power conditions (in terms of EIRP limits) with additional conditions (such as frequency modulation, bandwidth limits, or field strength) where necessary. There are no coordination requirements and no interference protection for the devices operating in the frequency schedule.

2.5 Regulating Unlicensed White Space Services after Spectrum Reallocation

Unlicensed secondary usage is commonly encountered on licensed bands characterized by a relatively mild degree of service exclusivity. For instance, there are various illustrations of unlicensed WSDs operating on a secondary usage basis in

---

* The only exception to this rule relates to the situation where a class license is in place before a frequency band designated or reallocated for spectrum licensing. While the class license would continue in force, the ACMA would be unable to vary it after a designation notice or reallocation declaration was made that included the frequency ranges in the class license.

† The Authority is made up of statutory officers appointed by the Governor-General of Australia and is akin to the board of a company.

the United States, ranging from PCS* cellular services in the 1.9 GHz band, to a variety of short-range, indoor UWB applications such as imaging systems (surveillance systems, ground-penetrating radars, etc.), precision vehicular radar, and other personal networking and communication devices mostly using bands between 3.1 and 10.6 GHz. In 2007 EU countries also authorized the use of unlicensed UWB technology on bands already licensed for other services and technologies.

In Australia, a secondary usage regime for WSDs corresponds to a situation in which the regulator—the ACMA—would be allowed to issue class licenses (aka unlicensed spectrum) in the same spectrum space that is occupied by apparatus licenses (e.g., as currently held by broadcasters) or spectrum licenses (e.g., as currently held by 3G and WiMAX network operators). While the former is commonly encountered, such as symbiotic WSD operating in TVWS and sharing the UHF band with apparatus-licensed broadcast services, the Act does not permit the latter. The current rule only allows that, if an existing class license is used in a band over which a spectrum license is auctioned, the class license arrangement remains unchanged, but this "theoretical" situation has never been seen in practice. This means that, if TVWS spectrum was ever to be sold as private spectrum space and acquired by a network operator, the regulatory authority would be faced by an unusual dilemma regarding the respective usage rights of each user. If unresolved, the uncertainty would seem likely to reduce the value of a spectrum space to prospective acquirers.

This standoff is part of a decade-long property right vs. shared access dilemma gripping the policy debate in the United States and elsewhere [11]. Proposed solutions have converged toward an easement regime, in which primary licensees ("owners") have clear broadcast rights, but secondary users (unlicensed or differently licensed users) may use the band so long as they never purposely interfere with the owner's transmission and reception rights. Compliance would be enforced through identification signals and compensation schemes for owners-incurred monitoring costs [12,13].

The idea of allowing noninterfering easements or at the very least some forms of coexistence formula for other users within spectrum-licensed (i.e., private property) spaces was proposed by the Australian Communications Authority (the ACA), which preceded the ACMA. In a submission to a 2002 government inquiry, the ACA suggested allowing noninterfering class-licensed services into spectrum spaces, which is the matter at stake for continued use of symbiotic WSD should TVWS be reallocated to telecommunications network operators. Given that class licenses operate rather successfully on noninterfering conditions in apparatus licenses bands, why would that option not be available in spectrum-licensed spaces as well? This was the

---

* When new PCS systems were allocated in the 1.9 GHz band, the spectrum was still occupied by private microwave systems. Yet, PCS licensees were quickly able to operate on the band, rolling out their networks by coordinating with incumbents and managing interference prior to incumbent relocation. In practice, PCS licensees effectively operated on the band on an informal secondary basis for some period of time.

question posed by the ACMA when it released a discussion paper in 2006 about a proposal to allow the authorization of devices under class licenses in spectrum designated for spectrum licensing. The initiative was concerned with the likely expansion of spectrum licenses (and exclusion zones that would result for class licensing) at a time when new technologies needed larger and larger bandwidth to be experimented and/or deployed (e.g., UWB applications such as ground-penetration radar and dynamic channel selection technologies).

The prevention of easement opportunities for class licenses (either directly or through ambiguity) is a particularly severe shortcoming of the current system of legal rules in Australia. The easement regime offers an important promise to alleviate spectrum congestion in some bands and satisfy unmet demand in others. More importantly still, and as surmised by numerous legal scholars, "private property" exclusion zones impose a large welfare cost to society by preventing or slowing the development, testing, and deployment of cutting-edge new wireless technologies, which require fast and easy access to specific spectrum bands [14–16].

A case of the "tragedy of the anticommons"* was also in regulators' minds. In the case of class licenses, the tragedy of the anticommons means that device operators would, under the current legal provisions of the Act, face "Swiss cheese" arrangements (to excise frequencies protected by s.36) in order to gain access to various areas of the spectrum while having to negotiate access with the owners of each spectrum-licensed space. Because many low-power devices trade power for bandwidth, this situation is counterproductive and onerous both for class-licensed device operators† and band owners (who would face countless applications for third party authorizations). Consequently, the ACMA suggested repealing section s.138 of the Act, explicitly allowing class licenses into spectrum-licensed spaces in s.36 and including ancillary provisions into the Act to make statutory consultations about each easement proposal. Spectrum licensees strongly opposed such a move because they were very concerned that their networks would suffer interference— which they argued would infringe upon their property rights. One area of concern was the cumulative impact on noise levels by a large number of low-powered devices operating in the same spectrum band, even if only one device would not produce detectable interference.

Last year, the matter of secondary easement rights for class licenses emerged once again in a government discussion paper about the definition of public interest criteria for spectrum license renewal [18]. In the discussion paper, the Department of Broadband, Communications and the Digital Economy (DBCDE) alluded to

---

* The "anticommons" or "hold up" problem arises when there exist multiple rights to exclude access to a resource along with incentives for strategic behavior [17]. If spectrum property rights are held by more than one user, new technologies and applications may be "held up" (i.e., barred from spectrum access) by rent-seeking behavior [11].

† Note that this is also relevant if class-licensed operators want to exceed power or other conditions and wish to proceed by getting a third party authorization from a spectrum license holder.

possible easement regimes to the second generation of (reissued) spectrum licenses. The Department clearly foresaw the need to cater to coexistence plans for future technologies such as Ultra-Wideband and cognitive radio and to correctly antici-pate future technological trends among unlicensed wireless devices.

Why indeed keep class license out of spectrum-licensed spaces when they have proved so effective at generating highly valued uses without interfering with the receivers of primary band users? In the meantime, and wherever possible, the ACMA has shifted the load of allowing some class-licensed devices to the apparatus license regime (e.g., UWB services such as imaging or ground-breaking radar under scientific licenses). To an extent this interim approach has worked well because the size of the total spectrum occupied by spectrum licenses is still rather small, so the scope for inserting class licenses as easements to apparatus licenses is rather large, and the ACMA has been able to tailor apparatus licensing solutions for devices that have challenged the overall licensing framework established by the Act.

Yet, as recognized in the 2009 discussion paper, the ACMA will not be able to indefinitely continue with ad hoc and interim measures. If the portion of the total spectrum occupied by spectrum licenses is called to grow, as is frequently proposed in Australia, the regulatory squeeze on new devices, such as symbiotic WSD oper-ating in TVWS, will become unbearable.

Responding to this challenge, the government introduced legislation to amend the Act into Parliament in June 2010 [19]. The Radiocommunications Amendment Bill 2010 (the Bill) proposes several amendments of the Act in relation to spectrum licensing. Of those amendments, the most significant would be the amendment of section 138 to give the ACMA the power to issue class licenses in spectrum spaces (whether established by conversion or reallocation) subject to a requirement spectrum licensees be protected from unacceptable interference. A new subsection 136(1A) is also proposed that would allow the ACMA to vary a class license that existed before a spectrum space was established. The Bill passed through the House of Representatives in October 2010, but only after the Opposition expressed con-cern that the amendments to allow coexistence between class and spectrum licenses did not provide enough clarity about what is or is not unacceptable interference and potential for adverse impacts on the value of the networks established by spectrum licensees. The Bill was referred to the Environment and Legislation Committee of the Senate for inquiry but was passed without amendment and became law in December 2010.

2.6 Regulatory Requirements

Amending the Act to allow the ACMA to issue and vary class licenses in spectrum spaces is a significant reform to the regulatory framework and one that holds out the promise of licensing arrangements that support the full range of spectrum manage-ment regimes available to the regulator [13,20]. However, as demonstrated by the debate in the House of Representatives, there is some concern about how that ACMA will implement that regulatory framework when it comes to consider the potential for class licenses to authorize devices in highly valued spectrum spaces. While this is an important issue, our interest is in a different and more general scenario—the establishment of licensing arrangements for invasive WSD after digital dividend 1.0 that do not constrain the later realization of an additional digital dividend 2.0.

One approach to this challenge would be for the regulator to determine that invasive WSD will not be allowed to continue to operate once broadcasting ceases to be the primary service and it is reallocated for other purposes. While this approach may have some appeal, it suffers from some significant limitations. Consumer devices that are not tightly bound to particular networks and technologies will tend to continue in use even if the regulatory permission to operate them is withdrawn. If invasive WSD are able to flourish and thrive in TVWS, then they may prove to be a persistent feature of the spectrum landscape. Secondly, this risk may lead regulators to decide against authorizing invasive WSD or to place stringent limitations on their use with consequential reductions on total welfare. Finally, and most importantly, this approach would forego the opportunity to create regulatory arrangements that maximized total welfare by developing technical frameworks that would allow invasive WSD to flourish and thrive in both broadcasting and telecommunications environments.

We suggest that an alternative approach would be for regulators and industry to identify the technical and operating characteristics that would allow invasive WSD to successfully transition from broadcasting environments where there are a small number of high-powered transmitters and where technology changes little over time, to telecommunications environments where there are large numbers of lower-powered transmitters and rapid change in both technologies and network topologies as operators seek to intensify their spectrum use. Ideally, invasive WSD would be able to adaptively collaborate with telecommunications networks using advanced technologies.

Importantly, there is time for regulators and industry to develop an understanding of the technical and operating characteristics that are required for optimal invasive WSD. This is because it seems likely that it will be at least 5 years and more likely 10 to 15 years before the realization of a digital dividend 2.0. That period of time is sufficient for the development of increasingly sophisticated invasive WSD and the consequential retirement of earlier generations of invasive WSD that will be suboptimal in a telecommunications environment.

The operation of WSD (either symbiotic or invasive) in spectrum that is privately owned by large communications carriers and other network operators raises a number of questions. What are the limitations in the Act if ACMA wishes to proceed with a bespoke approach, and what sort of legislation would be needed? What type of current licensing models could be amended and how? Although apparatus licenses are already conducive to a tailored licensing approach, spectrum licenses by contrast are not. Spectrum licenses currently constitute a very monolithic regime under the Act. They occupy large portions of spectrum in very exclusive ways. Although spectrum licenses were designed as a tool to remove market barriers, the large spectrum-licensed spaces they control are too often dictated by the "spectrum-guzzling" technologies (such as 3G, and soon LTE) deployed on them. How could licensees be expected to subdivide and trade quantities of spectrum with smaller mobile broadband wireless services when they need the spectrum in a large amount to guarantee the workability of their technical frameworks? Under these conditions, the main alternative available to increase the yield of spectrum-licensed spaces is to allow noninterfering secondary usage such as by low-power, spread-spectrum, and cognitive spectrum-sensing devices. Concretely, in an Australian context, the Parliament agreed to the proposal to amend s.138 of the Act to allow secondary usage rights for class licenses within spectrum-licensed spaces—in order to improve the allocative efficiency of spectrum licenses and enable secondary usage of TVWS by WSDs.

2.7 Conclusions

Digital switchover, and the consequential realization of a digital dividend, has been an important focus of governments and regulators over the last decade. Digital switchover has either been achieved or is underway in a majority of advanced economies with significant economic, commercial, and social benefits expected to result from the reallocation of valuable spectrum in the UHF band from broadcasting to telecommunications. The last decade has also seen increasing advocacy for, and interest in the use of TVWS in the UHF band—spectrum that has been allocated to the broadcasting service but not assigned to a particular broadcaster—by new generation white space devices—invasive WSD—that are intended to be deployed on a large scale but operate without affecting the reception of broadcast services because of their intrinsic ability to automatically adapt to their specific local spectrum environment.

However, now that there is an emerging scenario of a second digital dividend in the medium to long term—a digital dividend 2.0—we identify the potential for the widespread deployment of invasive WSD to impact adversely on the realization of any additional digital dividend. This is because the broadcasting environment is characterized by high-power transmitters operating from a small number of sites with little change over time, while the telecommunications environment is characterized by a much larger number of transmitter sites that operate at lower powers and which change relatively quickly over time. White space devices capable of successfully operating on a secondary basis in a broadcasting environment may have little or no capacity to operate in a telecommunications environment without causing harmful interference or suffering from degraded performance and capacity.

In this context, we set out the Australian regulatory framework governing the licensing operations of the three main players (or future potential players) in the

TV white space debate: broadcasters (governed by apparatus licenses issued after detailed planning for broadcasting services), unlicensed wireless service devices (typically governed by the LIPD Class Licence), and mobile phone/mobile broadband network services (governed by a property rights approach known as spectrum licenses). The latter group consists of future claimants to broadcasting UHF spectrum released by digital switchover and other anticipated deregulatory policies in Australia.

At this time, there is no conflict between these players in Australia, due to (i) a focus by the government and the regulator on digital switchover issues, including the realization of digital dividend 1.0 ahead of consideration of the potential for invasive WSD and (ii) regulation prohibiting coexistence of unlicensed underlay/overlay wireless devices in private property spectrum (spectrum-licensed spaces).

We recognize that amendments to the Radiocommunications Act proposed in 2010 by the Australian Government will remove this regulatory impediment and provide the Australian regulator with a flexible licensing framework that is better suited to the complex task of modern spectrum management. In particular, if the proposed amendments are successful, the Australian legislative framework will allow the development of regulatory arrangements that enable the development of invasive WSD with technical characteristics that enable cooperative sharing with telecommunications services.

In order to avoid complex ambiguities between invasive WSD that become established in the UHF band in a broadcasting environment in the short to medium term and the realization of an additional digital dividend 2.0 in the medium to longer term, we establish that regulators must have the requisite legislative flexibility—as the Australian Government now proposes for its regulator—and also recognize the need to project regulatory requirements for invasive WSD into the longer term. There are several ways to provide clarity and certainty, but all ultimately reside in either outlawing unlicensed devices in UHF bands (an economically inefficient option) or relaxing the laws ruling property right approaches. This article suggested the latter. Class licenses governing the use of unlicensed devices provide considerable flexibility to regulators to curtail or expand secondary unlicensed usage, but this flexibility is currently prevented by law in private spectrum spaces.

If regulators fail to successfully clarify and mend these uncertainties, it will prove very difficult and costly to expand or claw back the rights of white space devices whenever governments eventually reallocate broadcasting bands to mobile network services. Although the analysis was conducted here in the context of Australian regulations, the issues easily translate to the regulatory frameworks of other countries, such as the United States and the United Kingdom. Unless regulators recognize the potential for invasive WSD to impact on the realization of a digital dividend 2.0, confrontation between the rights of unlicensed White Space Devices and future telecommunication owners of the UHF band is likely to become a near-universal matter.

References

1. MTC Finland, "Making Broadband Available to Everyone—The National Plan of Action to Improve the Infrastructure of the Information Society," Rapporteur's proposal, 15 September 2008, Ministry of Transport and Communications Finland 2008.

2. FCC, "Revision of Part 15 of the Commission's Rules Regarding Ultra-Wideband Transmission Systems," FCC 02-48, Federal Communications Commission, Washington, DC, ET Docket 98-153, 2002.

3. OFCOM, "Decision to Make the Wireless Telegraphy (Ultra-Wideband Equipment) (Exemption) Regulations 2007," 2007.

4. EC, "Harmonisation of the Radio Spectrum for Equipment Using Ultra-Wideband Technology in a Harmonised Manner in the Community," EU Commission Decision of 21 February 2007, 2007/131/EC, 2007.

5. FCC, "Data Sought on Uses of Spectrum," Federal Communications Commission, Washington, DC Docket DA 09-2518, 2009.

6. ACA, *ACA Workshop on Spectrum Licensing Approaches.* Melbourne: Australian Communications Authority, 2003.

7. W. Webb, "Licensing Spectrum: A Discussion of the Different Approaches to Setting Spectrum Licensing Terms," OFCOM, London, 2008.

8. W. Webb, "An Optimal Way to Licence the Radio Spectrum," *Telecommunications Policy,* 33, pp. 230–237, 2009.

9. A. Kerans, G. Philips, and W. Webb, "Flexible Spectrum Licensing: Comparing the UK and Australia," *Policy Tracker,* 10 February 2009.

10. Productivity Commission, "Radiocommunications: Inquiry Report," Productivity Commission, Canberra Report Nr 12/2002, *AusInfo,* 1 July 2002.

11. G. R. Faulhaber and D. Farber, "Spectrum Management: Property Rights, Markets, and the Commons," in *Rethinking Rights and Regulations: Institutional Responses to New Communication Technologies,* L. F. Cranor and S. S. Wildman, Eds. Cambridge, MA: MIT Press, 2003.

12. W. A. Leighton, "Models for Spectrum Allocation: Which Is Most Efficient, and How Do We Achieve It?" presented at 15th Biennial Conference of the International Telecommunications Society, Berlin, 5–7 September, 2004.

13. B. P. Freyens, "A Policy Spectrum for Spectrum Economics," *Information Economics and Policy,* 21, pp. 128–144, 2009.

14. W. H. Melody, "Radio Spectrum Allocation: Role of the Market," *American Economic Review,* 70, pp. 393-397, 1980.

15. Y. Benkler, "Some Economics of Wireless Communications," in *Rethinking Rights and Regulations: Institutional Responses to New Communication Technologies,* L. F. Cranor and S. S. Wildman, Eds. Cambridge, MA: MIT Press, 2003.

16. J. H. Snider, "Spectrum Policy Wonderland: A Critique of Conventional Property Rights and Commons Theory in a World of Low Power Wireless Devices," 30 September 2006, George Mason University, Arlington, VA. Paper prepared for the Telecommunications Policy Research Conference, 2006.

17. M. A. Heller, "The Tragedy of the Anticommons: Property in the Transition from Marx to Markets," *Harvard Law Review,* 111, pp. 621–688, 1998.

18. DBCDE, "Public Interest Criteria for Re-issue of Spectrum Licenses," Discussion paper for public consultation, Department of Broadband, Communications and the Digital Economy, Commonwealth of Australia, Canberra 2009.

19. Parliament of Australia, "Radiocommunications Amendment Bill 2010," House of Representatives, The Parliament of the Commonwealth of Australia 2010.

20. B. P. Freyens, M. Loney, and M. Poole, "Wireless Regulations and Dynamic Spectrum Access in Australia," presented at IEEE DySPAN Conference, Singapore, 6–9 April 2010.

Chapter 3

TVWS System Requirements

Randy L. Ekl, David P. Gurney, and Bruce D. Oberlies

Contents

3.1 Introduction

The U.S. Federal Communications Commission (FCC) authorized TV white space (TVWS) devices to use the television (TV) broadcast spectrum as unlicensed secondary users of the spectrum in their Second Report and Order in November 2008 [1]. Further modifications were made to the TVWS rules in the Second Memorandum Opinion and Order released in September 2010 [2]. Allowing unlicensed secondary operations in the vacant spectrum enables innovators to develop new communication opportunities for broadband wireless users, including wireless Internet service providers (WISPs) and wireless LAN users, among others. However,

69

the TV spectrum incumbents operate wireless services that could be negatively impacted by TVWS devices transmitting on a secondary basis, unless appropriate steps are taken to avoid such conflicts. Accordingly, this ruling set forth technical requirements that help ensure TVWS solutions operate where the spectrum is not in use by licensed services, to prevent harmful interference to those incumbent services using the spectrum.

The incumbents using this spectrum include several primary licensed types of users and some secondary licensed types of users. Primary users include full-power digital TV broadcasting services, analog and digital low-power TV stations, boosters, and translators, and two-way land-mobile radio systems using channels 14–20 in 11 major metropolitan areas (described below). The secondary licensed users include low-power broadcast auxiliary stations, including wireless microphone systems, which in the United States are regulated under Part 74 of the FCC rules. In addition to these licensed services, there are numerous businesses and religious institutions that utilize generally unlicensed wireless microphone systems that may be entitled to protection in certain limited cases. Households and businesses using cable set-top boxes should also be considered as users that must be protected from harmful interference by TVWS devices. The TV white space devices transmit signals which can be picked up at a cable headend receiver or potentially by poorly fitted coax cable running through the home or business. A TVWS device can interfere with more than one cable channel because the cable channels are typically offset from the TV broadcast channels.

The regulations, along with other practical technical and business considerations, form a set of requirements for a TV white space system to be developed and deployed, which can provide significant benefits to end users. To be meaningful, requirements must be testable, which leads to examining relevant technologies, tradeoffs, and selections. This chapter will also address the technology used to implement the requirements and show both feasibility and testability.

3.2 TVWS System Requirements and Implications

This section will discuss system requirements and implications of the FCC TVWS rules. It is comprised of four subsections: Regulatory Development of Requirements, Geo-Location Database Protection Requirements, Additional TVWS Device Requirements, and finally Latest TVWS Regulatory Developments.

3.2.1 Regulatory Development of Requirements

The U.S. Federal Communications Commission (FCC) began formally supporting unlicensed use of the television (TV) broadcast bands in its May 2004 Notice of Proposed Rule Making (NPRM) [3]. The proposed rule making considered secondary unlicensed use in the TV bands (i.e., in TV white space, or TVWS, from channels 2–51) in areas where the spectrum is underutilized by incumbent services. The proposal noted the advantageous propagation characteristics of the TV bands, as well as the opportunity for new and innovative unlicensed devices, including the benefits to wireless Internet service providers (WISPs). A major goal of the proceeding was to improve broadband wireless access for underserved users, for example, those in rural communities.

In the NPRM, the FCC allowed for the potential use of spectrum sensing techniques to determine available (i.e., unoccupied) TV channels, or the use of geo-location databases to protect incumbent services. Geo-location databases use the knowledge of unlicensed device location, as well as the knowledge of licensed incumbent transmitters and services, to determine which channels are available for secondary unlicensed usage. Spectrum sensing relies on local monitoring of the radio frequency (RF) environment to determine the presence of incumbent operations. Single-node spectrum sensing techniques must deal with several real-world RF effects, such as fading and shadowing (i.e., hidden node effects), as well as potentially low antenna gains, due to antenna polarization mismatch, low antenna heights, and body losses for portable devices. These factors combine to drive the required incumbent detection levels to very low values, which make the use of sensing techniques challenging from a manufacturing and cost perspective and can make TVWS devices prone to falsing [4]. For example, the FCC set incumbent sensing level requirements at –114 dBm in the initial TVWS operating rules to detect wireless microphones.

The low incumbent signal detection levels required also result in potentially large protected service areas, which could result in inefficient spectrum usage. Figure 3.1 shows an example where there can be a large amount of lost secondary usage opportunities due to the very low required sensing levels (e.g., –114 dBm), compared to geo-location database-driven protection. In this case, nearly three times the normal operating area is unnecessarily protected using outdoor-based sensing, compared to geo-location techniques, which only protect a TV station's protected service area (or contour) plus a small buffer area. In other words, roughly twice the potential usable geographical area can be wasted (i.e., go unused) using sensing-only techniques compared to geo-location database techniques, due to the very low required

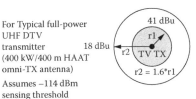

For Typical full-power UHF DTV transmitter (400 kW/400 m HAAT omni-TX antenna)

18 dBu   41 dBu   r1   r2   TV TX   r2 = 1.6*r1

Assumes –114 dBm sensing threshold and use of F(50,90) propagation curves

– r1 = 104 km radius TV station protected service area via geo database (98 km to 41 dBu contour +6 km required separation distance)
– r2 = additional wasted protected area due to sensing (@-114 dBm)
– Typ. protected contour area ~34,000 km² (104 km radius)
– protected area using sensing: 91,000 km² (~2.7x coverage area)
– Up to 57,000 km² of usable area wasted using sensing.

Figure 3.1   Area potentially wasted using sensing-only techniques compared to geo-location techniques for an outdoor scenario.

incumbent detection levels. Implementing sensitive (e.g., −114 dBm) sensing detectors also requires very linear receiver lineups (to detect very weak signals in the presence of strong adjacent channel TV signals), as well as significant signal processing power to accurately resolve the weak signals. These requirements tend to significantly drive up the implementation costs for TV band devices.

Geo-location databases offer the potential for highly reliable and repeatable incumbent protection, because available channels are determined by the unlicensed and incumbent transmitter locations, and other key operating parameters (e.g., incumbent and unlicensed transmitter antenna height, incumbent antenna pattern, effective radiated power level). Geo-location-based protection is well suited for the TV bands, because most licensed incumbent transmitters are fixed in their location, and their service areas are well defined.

A complicating factor in the use of TVWS for the United States was the ongoing transition of full-power analog TV broadcasting services to digital operation, using the Advanced Television Standards Committee (ATSC) 8-VSB modulation format. The ATSC 8-VSB modulation format provides for more efficient use of the spectrum compared to the analog National Television Standards Committee (NTSC) format, because several program streams can be multiplexed onto a single TV channel due to the use of MPEG-2 video encoding. The ATSC modulation format also has improved co- and adjacent channel interference tolerance compared to NTSC modulation, allowing theoretically a tighter (i.e., closer in separation distance and frequency) packing of broadcast TV transmitters. The spectrum effectively gained from the improved digital ATSC broadcasting techniques is often termed the "digital dividend." During the transition, most full-power TV stations were broadcasting their programming in both digital ATSC and analog NTSC formats. This resulted in the TV bands being very crowded. The digital transition was scheduled to be completed in February of 2009, but was delayed, and was finally completed in June 2009. At that time, full-power analog NTSC stations ceased operation, though low-power analog NTSC television transmitters were not required to shift to ATSC digital modulation. As part of the digital television (DTV) transition and digital dividend process, TV spectrum ranging from channels 52–69 (698 MHz–806 MHz) was cleared of full-powered TV stations, which relocated into the lower portion (channels 2–51) of the TV bands. The 700 MHz band (698–806 MHz) portion of the digital dividend spectrum was auctioned for licensed commercial services, such as mobile broadband, and a portion was made available specifically for public safety operations.

The FCC TVWS proceeding further evolved over a period of several years [5]. Numerous parties commented throughout the proceeding. In fact, over 35,000 comments were officially filed in the proceeding, which put the entrenched incumbents (especially the TV industry) and the technology producers and device manufacturers (e.g., Silicon Valley companies) on opposite sides of the issue. In October of 2006, a further notice of proposed rulemaking (FNPRM) was released [6] on the subject, and in December of 2006, the FCC announced the first round of prototype unlicensed device testing. The first round of device testing involved sensing-only units. In July of 2007, the FCC released the first round device test report [7]. In general, the sensing-only units that were tested performed below expectations and were unable to consistently detect incumbents. In March of 2007, the FCC OET released an extensive report [8] on DTV interference rejection levels that was an important step in determining actual interference tolerance levels for consumer DTV receiver equipment. In November of 2007, the FCC announced the second round of prototype unlicensed device testing. Several companies participated in this round, which involved both indoor lab testing and outdoor field testing at several locations, which was completed during the January to August time period of 2008. Four of the companies involved in the testing provided sensing-only equipment, while the fifth company provided a prototype geo-location database-driven unit. The FCC released the second round test report in October of 2008. During testing, it was found that the geo-location database-driven unit was the best overall performer and provided highly reliable protection to incumbent TV transmitters operating in the band [9].

In November of 2008, the FCC released the initial rules for unlicensed device operation in the TV bands [1]. The initial rules provide the beginning framework for TVWS devices, and required the use of geo-location databases for secondary use of TV band spectrum. Sensing-only TVWS devices were subject to a further (undefined) testing and verification process. Geo-location capability of unlicensed TV band devices is required to be accurate to within 50 meters, though fixed devices can rely on professional installers to enter the fixed device's location upon installation (a one-time process). Devices must access a geo-location database to determine available operating channels on a periodic basis (i.e., daily, under the current rules) and must also access the database upon any change in location or at device power-up. If an unlicensed device loses access to the geo-location databases, it must cease operation by 11:59 p.m. the following day. TVWS devices that move must requery the database upon any movement of more than 100 m, or upon device power-up. Industry recommended the FCC authorize multiple TVWS database administrators to provide high levels of reliability, innovation, and competitive pricing for consumers. In November of 2009, the FCC released a Public Notice soliciting applications for TVWS Database administrators. Nine potential database vendors responded.

The FCC allowed the use of fixed unlicensed devices on TV channels 2–51 operating at up to 4 W effective isotropic radiated power (EIRP) levels (excluding TV channels 3, 4, 36–38), and personal portable unlicensed devices operating on channels 21–51 (excluding channel 37) up to 100 mW EIRP levels (with some additional restrictions—see below). Antenna gains are generally limited to 6 dBi for fixed devices, and 0 dBi for personal/portable devices. Personal portable devices are allowed to operate inside a TV station's adjacent channel contour at a reduced power level of 40 mW EIRP. Fixed devices are not currently allowed to operate inside adjacent channel contours. Both fixed and personal/portable device operation is allowed on a co-channel basis with licensed incumbents well outside the incumbents' protected service areas (see below).

Personal/portable devices were further subdivided into *Mode II* master devices (with geo-locating capability), and *Mode I* client devices (that lack geo-locating capability and must rely on a master device to determine available channel lists). A master device may be either a (Mode II) personal/portable or a fixed device. Under the November 2008 rule set, all unlicensed devices were required to sense for wireless microphone incumbent users and vacate the channel if one was detected. Sensing was also required for NTSC and ATSC TV signals, though channels did not have to be vacated if a TV signal was detected, because the geo-location database approach provided the ultimate determination of protected TV services.

3.2.2 Geo-Location Database Protection Requirements

Geo-location databases are responsible for adequately protecting incumbent services and have many of their own critical operating requirements. Although many people do not realize it, there are a wide range of licensed or authorized incumbent services that need protection in the TV bands, including roughly ten different types of TV transmitters (e.g., low-power analog transmitters, translators, and boosters, as well as low-power and full-power digital transmitters), commercial mobile radio services (CMRS), private land mobile radio services (PLMRS), special TV receive sites (including TV translators, Broadcast Auxiliary Services, and cable headend receive sites), Part 74 licensed wireless microphones, radio-telephone use, medical telemetry (on channel 37), and radio astronomy. CMRS and PLMRS services are contained in the T-band (channels 14–20) in 11 major metro areas in the United States (see FCC Section 90.303(a)) and include mission-critical two-way mobile radio use by public safety agencies. All of the above services must be protected from interference from unlicensed TV band device usage.

In the case of TV transmitters, their protected service areas have been long defined by the FCC to be the Grade-B service contour for analog TV stations, or the Noise Limited Contour (NLC) for digital TV stations. These areas define regions where TV signals are generally receivable with an outdoor TV antenna, and the area where incumbent TV receivers must be protected from harmful interference. Table 3.1 shows the typical protected contour levels for various TV services.

The protected contour level varies based on whether the signal is analog or digital, and in what frequency band the TV transmitter operates (both of which relate to the minimum sensitivity level of the TV receiver). The service contour level is defined in terms of dBμV per meter or dBμ. Note that low-power TV services are currently treated with the same protection as if they were full-power TV services (i.e., the protected service contour level is lowered by 15, 12, and 10 dB at low-VHF, high-VHF, and UHF channels, respectively) for the purposes of computing protected service contours.

Table 3.1   Typical Protected Contour Levels and Propagation Models Used for TVWS

	Protected Contour		
Type of Station	*Channel*	*Contour (dBu)*	*Propagation Curve*
Analog: Class A TV, LPTV, translator and booster	Low VHF (2–6)	47	F(50,50)
	High VHF (7–13)	56	F(50,50)
	UHF (14–69)	64	F(50,50)
Digital: Full service TV, Class A TV, LPTV, translator and booster	Low VHF (2–6)	28	F(50,90)
	High VHF (7–13)	36	F(50,90)
	UHF (14–51)	41	F(50,90)

FCC-defined statistical propagation models are utilized for predicting TV broadcast service areas. These R-6602 F-curve propagation models are based on empirical measurements taken by the FCC. The F-curve models use both a location and time-reliability variable (e.g., 50% of locations, 90% of time), which represents the statistical percentage of locations and time that service is available at least at the specified signal (i.e., field strength) level. The models rely on each TV station's transmitter parameters (e.g., operating frequency, effective radiated power level, service type, antenna height above average terrain, and antenna pattern), which are gathered from the FCC's Consolidated DataBase System (CDBS). A sample F-curve plot for UHF DTV transmitters is shown in Figure 3.2.

The FCC also utilizes standard co-channel and adjacent channel interference tolerance levels for TV receivers in their TV band planning process. These ratios are derived from both the ATSC TV receiver performance recommendations [10] and standard FCC TV-band planning practices [11]. These ratios help ensure generally interference-free reception of TV services within their protected contours. Finally, note that estimates of over-the-air (OTA) TV reception generally range from 10% to 15% of U.S. households. In most markets, 85% to 90% of U.S. TV households obtain their television service through cable or satellite distribution systems [12].

Based on the above predicted protected service contours, and the interference tolerance levels of incumbent TV receivers, the complete co-channel and adjacent channel protected service areas are determined for licensed incumbent services. In the case of TV services, a buffer zone is provided around the TV station's Grade B or NLC protected contour, based on antenna height of the unlicensed device transmitter, as shown in Table 3.2. This approach ensures an acceptable interference level to TV receivers at the edge of a TV station's protected service contour.

Estimated Field Strength Exceeded at 50 Percent of the Potential Receiver
Locations 90 Percent of the Time, at a Receiving Antenna Height of 9 Meters

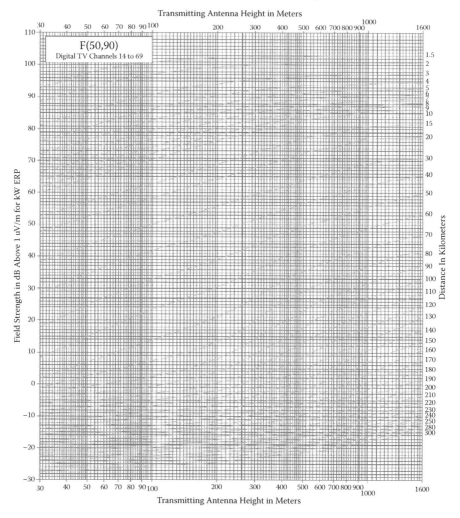

Figure 3.2   Sample R-6602 F-curve plot for DTV service in the UHF band.

As can be seen, the required separation distance is based on the unlicensed TV band device's transmitting antenna height for fixed devices. Personal portable devices are always assumed to be operating in the *less than 3 meter* category. As is expected, higher antenna heights necessitate larger required separation distances from licensed TV services. Since TV receivers are much more tolerant of adjacent channel interference than co-channel interference (e.g., −33 dB desired-to-undesired adjacent channel interference tolerance vs. a 16 dB desired to undesired co-channel interference tolerance for DTV receivers), the required separation distance

**Table 3.2   Required Separation Distances for
Unlicensed Devices from TV Contours**

Antenna Height of Unlicensed Device	Required Separation (km) from Digital or Analog TV (Full Service or Low Power) Protected Contour	
	Co-Channel	Adjacent Channel
Less than 3 meters	6.0 km	0.1 km
3 to less than 10 meters	8.0 km	0.1 km
10–30 meters	14.4 km	0.74 km

for unlicensed device operation adjacent to a TV channel is much smaller than for unlicensed device operation on a TV co-channel frequency.

Licensed PLMRS and CMRS systems are authorized to operate on certain channels in the T-band (UHF channels 14–20) in 13 major metro areas (although only 11 of the 13 market areas have active operations at this time). These mobile radio systems are protected by enforcing a 134 km co-channel keep-out region and a 131 km adjacent channel keep-out region around the center of those metro areas. Life-saving mission-critical mobile two-way radio communications may take place on these channels, necessitating the large keep-out zones. Other smaller mobile radio operations (listed in the FCC's Universal Licensing System) also need to be protected, with 54 km co-channel and 51 km adjacent channel keep-out regions.

Special TV-receive sites also need to be protected from interference from unlicensed white space devices. These include TV translator receive sites, cable head-end OTA receive sites, and Broadcast Auxiliary Service (BAS) sites. These sites are currently protected by a keyhole region that covers a ±30 degree arc and reaches a maximum distance of 80 km. Figure 3.3 illustrates an example of the keyhole protection region.

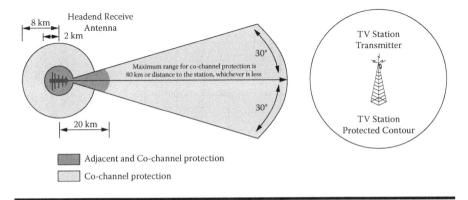

Figure 3.3   Keyhole projection region for special receive sites.

Authorized wireless microphones (e.g., Part 74 wireless microphones) are protected from fixed TVWS device with a 1-km-radius circular keep-out region in the database. Part 15 wireless microphones are entitled to no special interference protections, except in very limited cases (e.g., at large concert halls and sports stadiums). It is estimated that well over 200,000 unlicensed wireless microphones are in operation today [13]. Secondarily licensed Part 74 wireless microphones are allowed to transmit up to 250 mW in the UHF band, anywhere within an adjacent channel contour, with a channel bandwidth of roughly 200 KHz. This translates into roughly a 23 dB higher wireless microphone power spectral density level than a 40 mW personal/portable unlicensed device, giving the microphones a performance advantage in such cases. In the 13 major CMRS/PLMRS market areas discussed above, at least two channels on each side of channel 37 were required to be reserved for exclusive wireless microphone use under the initial TVWS rules. This was later extended to include all markets.

3.2.3 Additional TVWS Device Requirements

The required transmit spectral mask for unlicensed TV band devices is currently specified to be a very tight –55 dBr/100 KHz (essentially a –55 dBr brick wall mask, or a –73 dBr mask relative to the 6 MHz on-channel power level). Power levels in the passband and stopband must be taken as the highest value measured in a 100 KHz resolution bandwidth for the –55 dBr mask. Note that the specified mask is much tighter than other commercial/consumer communications systems (e.g., as shown in Figure 3.4, the specified mask at the near edge of the adjacent channel is about 35 dB tighter than the IEEE 802.11a/g mask, and about 30 dB tighter than the WiMax mask). Also note that fixed unlicensed devices must also meet

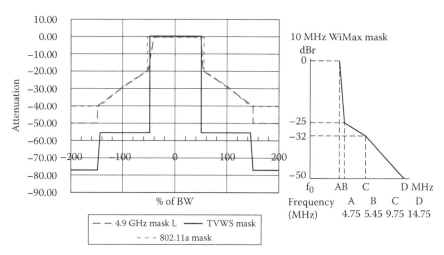

Figure 3.4   Unlicensed device transmit spectral mask comparison.

Table 3.3 Unlicensed Device Emissions Limits for Channels 36 and 38

Frequency (MHz)	Field Strength dBµV/meter/120 kHz
602–607	120 – 5[F(MHz) – 602]
607–608	95
608–614	30
614–615	95
615–620	120 – 5[620 – F(MHz)]

this mask in the adjacent channel, even though they are not currently allowed to transmit within adjacent (TV) channel contours.

Unlicensed device emissions in alternate channels and beyond must meet Part 15.209(a) requirements, which require absolute emissions (measured within a 120 KHz bandwidth) to be below 200 µV/m at a distance of three meters. Emissions must also be further restricted when operating on channels 36 and 38, as described in Table 3.3. For example, at F = 603 MHz, the field strength would be 120 – 5[603–602] = 115 dBµV/meter/120 kHz. Note that the emissions measurements for channels 36–38 are taken at a one meter distance (as opposed to three meters in Part 15.209).

Given all of the above operating restrictions and incumbent protection requirements, spectral availability in major metro areas can be relatively scarce. It turns out that, in some major metro areas (Chicago, Dallas, etc.), only adjacent channels are unoccupied by incumbents, meaning that only low power (40 mW) personal/portable device operation is allowed in the UHF band. Note however, that fixed devices also can operate on channels 2–20, which provides additional spectral availability, because many full power TV stations transitioned out of the low-VHF band. In rural areas, there may be several tens of megahertz of spectrum typically available for unlicensed device use. Note that the FCC's National Broadband Plan also calls for the repurposing of up to 120 MHz of TV band spectrum, although there is the possibility that some of that spectrum could be dedicated for unlicensed uses.

3.2.4 Latest TVWS Regulatory Developments

Numerous parties petitioned the FCC to reconsider the initial 08-260 TVWS rules released in November 2008. In March 2009, 18 parties submitted formal petitions for reconsideration for the FCC's 08-260 Report and Order. Numerous parties requested the removal of incumbent (e.g., wireless microphone) sensing requirements, because the geo-location database could be utilized to protect those incumbent users. Several parties requested relaxing the antenna height restrictions for fixed devices, which are currently limited to 30 m above ground level. Some parties requested higher powered (e.g., 20 W EIRP) uses for open alternate channels, because such situations frequently occur in wide-open rural areas anyway. Numerous manufacturers also requested that multiple independent database administrators be named by the FCC to encourage reliability and competition in pricing and services. Some parties filed to relax the stringent unlicensed device transmit spectral mask (especially for fixed devices). Some parties also filed to reduce the power limits for personal/portable devices. FCC resolution on these reconsideration issues will ultimately determine the business viability for TVWS.

In September of 2010, the FCC released a Second Memorandum Opinion and Order addressing some of the issues in the pending petitions for reconsideration [2]. The Order removed all sensing requirements for unlicensed TV band devices that utilize geo-location databases, increasing the viability of the TVWS device market. It also supported naming multiple database providers to encourage competition in database services, though the actual database providers will be named at a later date. Localized caching of database results for multiple potential operating locations is allowed in TV band devices, which improves reliability in mobility scenarios. Two TV channels were reserved for exclusive wireless microphone usage in every market (as opposed to only in the 13 major metro markets where PLMRS/CMRS services operate), and fixed venues (e.g., sporting stadiums, Broadway, concert halls) that require more wireless microphone channels than are available may apply to receive additional database protection, subject to a public review process. Sensing-only devices were also allowed to operate at 50 mW transmit power levels, subject to further testing and public review, and are required to detect wireless microphone operations down to −107 dBm levels (as opposed to the −114 dBm levels for ATSC and NTSC TV signals). The Order did not, however, modify the fixed device antenna height constraints or relax the transmit spectral mask requirements. It also added new uniform power spectral density limits (i.e., 12.2 dBm/100 KHz conducted power for fixed devices, and 2.2 dBm/100 KHz for personal/portable devices), as well as fixed device transmitter location constraints (i.e., transmitter sites must be below 76 m height above average terrain, as defined in FCC Part 73.684). It is expected that further petitions for reconsideration will be filed to modify the rules.

3.3 TVWS System Requirements to Yield Useful Systems

There are other considerations beyond what the FCC has laid out which would be needed to make a TVWS system useful. Typical wireless system requirements, such as manageability, reliability, and security, must be revisited for TVWS systems. New requirements, such as coexistence, must also be considered. This section will discuss requirements in several of these areas, including cost, geo-location capabilities, security, and efficiency. The next section will discuss other future requirements, such as coexistence, beacon technologies, spectrum management, and reliability.

Many TV band devices require geo-location capability. This needs to be cost-effective, so that the end-customer cost is reasonable. This can readily be accomplished (in outdoor settings) with low-cost GPS receivers. The location accuracy requirement of ±50 m can be met in most cases, and device movement of more than 100 m (which can readily be detected) can trigger the required database access.

Furthermore, the latest FCC rule set allows TV band devices to locally cache database information, which can reduce the required database access rates (and required bandwidth) in mobility scenarios or in cases where a master device is serving multiple client devices. The less a channel is used for database updates, the more area of the channel is available for user information (i.e., the most efficient use of the channel). Note that the localized database information still needs to be updated at the required rates (i.e., daily). Client (Mode I) devices do not need to have geo-location capability, as long as they can communicate with a master device that has geo-location database access capability and sends an enabling (contact verification) signal. Though the exact format of the contact verification signal is not defined, it must meet certain security and verification requirements. It is likely that the definition of these signals will be left to the standards bodies (e.g., IEEE) and vendors themselves. If these additional items are carefully defined and described, they could be considered as additional requirements. In particular, customers could itemize these additional items in RFPs (request for proposals) and similar documents with system providers or product vendors.

Efficiency in the utilization of the spectrum is not currently a requirement for TVWS systems, but the overall TVWS ecosystem is likely to evolve to make this a market requirement. Several near-term steps can be taken to then satisfy this requirement. Most simply, allocation of only a subsection of a TVWS channel would also lead to a higher percentage of the overall spectrum being used, because this would allow an otherwise unusable 6 MHz channel to be partially used. For example, if an incumbent, such as a licensed wireless microphone, is detected in the top half of a channel and is not utilizing the bottom half of the channel, the current FCC rules state that the entire channel is unusable. Beacon technology, database technology, or detection (sensing) technology could all currently be able to determine that only half the channel is being used, and then direct the TVWS system to use the other half, driving up the utilization of the overall spectrum.

In addition, the geo-location databases in the system could be utilized to coordinate TV band spectral usage among unlicensed devices (similar to the role played by traditional frequency coordinators). Coordination among systems could take place geographically and via time-sharing of channels. This would have the benefit of reducing interference between unlicensed systems and increasing overall spectral efficiency and utilization. The IEEE 802.19 coexistence working group is currently studying some of these issues.

In addition to the technical requirements detailed in this chapter, there are several nontechnical requirements which would foster a viable TVWS system. The two primary nontechnical requirements of TVWS Systems are the low cost of development and the requirements which could be placed around a reasonably-sized customer base.

To keep development costs down, reuse of other system components, such as radio designs and transmission protocols, would be beneficial. In particular, the WiMax (IEEE 802.16), TD-LTE (3GPP) and WiFi (IEEE 802.11) systems at 2.4 GHz, 2.5 GHz, and 3.5 GHz are "obvious" choices for extension into the TV bands, due to their physical layer technology, prevalence in the industry, and channelization scheme. However, the required spectral mask may be difficult to meet. These protocols are based on orthogonal frequency division multiplexing (OFDM) that allows these protocols to operate in a variety of channel bandwidths. LTE, WiMax, and IEEE 802.11a have defined 5 MHz channels which align well with the 6 MHz TV channels, though they would need additional filtering to meet the tight spectral mask. While the majority of these protocols can be reused, other minor modifications may be required to operate in the significantly lower frequency TV band. Note that the removal of sensing requirements for TV band devices significantly eases these modifications. Other proprietary broadband systems in use today at 900 MHz or other unlicensed frequencies could be rebanded. Also, an open standard, IEEE 802.22, is being developed for WISP operation in TVWS. Standardized protocols are a plus for customers, because it is more likely that multiple manufacturers would develop interoperable equipment for the system based on these protocols. This would also drive equipment costs down.

A reasonably sized customer base provides economies of scale for equipment development, which benefits both the manufacturer and the end customer. A reasonably sized customer base is, unfortunately, not a testable requirement, as "reasonably sized" is not a concrete value and depends on many business and development factors. A better approach for this "requirement" would be to ensure that enough customers (users and/or devices) utilize the spectrum to some quantifiable value over a given period of time.

The regulatory requirements, detailed in Section 3.2, above, are concise and testable, and make the most sense when considered as part of an overall system scenario or real-world deployment. For example, a manufacturing plant, a transportation hub for courier operations, or a refinery could use available TVWS spectrum to help implement wireless video surveillance on its campus. Some parties have proposed using TVWS spectrum for the non-mission-critical automated meter reading communications in Smart Grid systems. As a specific example, consider a power plant which could use TVWS for plant maintenance. During routine maintenance and inspection of the plant, an issue may be found. The inspector could take a digital video of the issue and stream it over long ranges, via a wireless device (laptop) on the TVWS system, to workers back in the main office, who could look up schematics, "as built" documents, or other information required for the repair.

Instructions could then be sent back to the inspector or maintenance staff to affect the repair. As long as the maintenance staff stays within the coverage area of the fixed TVWS access points and maintains communications (i.e., remains tethered) with the *access point* (AP), the specific channel can be utilized for this operation, at typically longer ranges than today's WiFi WLAN communications systems [14].

The above described systems capitalize on the excellent propagation characteristics of TVWS spectrum. It is also a good fit for outdoor systems. In particular, for outdoor TVWS systems, benefits to WISPs are long communications ranges, ample throughputs, reduced interference levels, and lower total cost of ownership than competing systems. With the 4 W power limit, several miles of coverage are possible, as has been demonstrated in Spectrum Bridge trial systems in Virginia and California [15].

Indoor systems could also capitalize on the excellent propagation capabilities of TVWS. For example, instead of needing multiple WLAN access points at 2.4 GHz or 5.1 GHz to cover a facility, with its myriad of conference rooms, offices, labs, and work areas, significantly fewer TVWS access points could accomplish the same coverage. This same concept could apply to coverage for big-box retailers, malls, offices, and industrial parks, and so forth. These examples show benefit to various customer operations.

A primary concern of customers, not specifically mentioned in the FCC rulings, is that of transport security. Security is a broad area, typically thought of in either layers (e.g. layer 2 (MAC), layer 3 (IP), layer 7(Application)), physical security vs. cyber-security, or security functions, such as confidentiality, integrity, authentication, authorization, access control, availability and nonrepudiation. Since the regulatory rules do not detail anything above layer 2, and do not fully describe a layer 2, any of a number of choices could be utilized: IEEE 802.11, IEEE 802.16, IEEE 802.22, and so forth. Then the requirements for cyber-security of the TVWS system could simply follow the protocol chosen at the lower layers and the best practices in the industry at the upper layers. Similarly, requirements and solutions for security functions (authentication, etc.) could also follow the lower layer protocol choice as well as the policies by the owning and/or operating agencies.

It is appropriate that the FCC or other regulatory bodies do not dictate the exact mechanism for security on a TVWS system, but security is needed on these systems. Even though the regulatory rules do not specify TVWS requirements regarding security, it is imperative, to have a robust security solution on the TVWS system, depending on the operational scenario and sensitivity of the information being transported.

3.4 Potential Future TVWS System Requirements

The FCC commented in their Second Report and Order that the actions were "a conservative first step that includes many safeguards to prevent harmful interference to incumbent communications services [1]." There are a number of cognitive radio (CR) technologies that are being developed within universities and industry that could become future requirements for TVWS solutions. These future CR technologies could improve the protection of incumbent services and improve the efficient use of the spectrum.

To overcome some of the difficulties of system coexistence and protection of wireless microphones and other incumbents (which have a variety of technologies and waveforms), a potential additional method of protecting incumbents and coordinating spectral usage is through the use of a beacon signal. This is the proposed method for protecting licensed wireless microphone systems and has been an area of development in the IEEE 802.22.1 group. The disabling beacon broadcast in the area of the licensed wireless microphone or other incumbent system would indicate to any TVWS device which is able to detect the beacon signal to not operate on that channel. If the beacon is able to be decoded instead of simply detected, then it would further augment the relatively static geo-location database with an over-the-air database, providing near-real-time supplemental information about that incumbent—location of the system, transmit power, coverage range to be protected, channel, and so forth. The TVWS system would receive the beacon and choose operational parameters to not interfere with this incumbent. The beacon signal is designed to be much easier to detect than a general wireless microphone. It can be at a known offset within the channel, will be a narrowband signal, and will have a known waveform. The beacon defined in IEEE 802.22.1 is placed at the same frequency location as the DTV pilot tone and is comparable in bandwidth to a wireless microphone but generally higher power to enhance detection while minimizing the bandwidth occupied in the TV channel.

Information about the current operation of the local and surrounding systems can be used by multiple users to coexist in a given geographic area or provide the means for adjacent systems (TVWS, other unlicensed networks, Land Mobile Radio, etc.) to operate. Enabling beacons and cognitive pilot channels [16] can provide this information in a timely manner to allow for multiple users to coexist in a given geographic area. A cognitive pilot channel is a broadcast, with a known protocol at a known frequency or frequencies of the operational wireless systems within a certain geographic area, containing some or all of the following information: system type, protocol, power, location, and so forth. While some of this information can be provided via extension of the TVWS database, the enabling beacon and cognitive pilot channel enable nearly real-time information about the operational status of TVWS systems in the local and regional area. Including these future requirements will lead to even greater utilization of the spectrum.

More specifically, having and sharing information with neighboring systems allows for higher levels of wireless system decisions—anywhere from transmit/do not transmit, choice of channel, choice of protocol, and up through the possibilities of interoperability. For example, if a system is using a variant of IEEE 802.11 in a TVWS channel, a second co-located system could also choose to use that protocol so that both systems could have access to the channel. This places a requirement on at least one of the systems to have a mechanism to determine the protocol the other system is using, and to switch over to also use that protocol. This is a feasible future requirement.

Manageability is important in all wireless communication systems. Given the secondary use nature of TVWS, it is especially important to manage the use of the channel resources. Additional collected spectrum use data, over time and location, should make the management of the spectrum better. These data-tuples (usage, time, location, other signal information) can be gathered and analyzed to yield predictive conclusions to be used to drive up the efficient use of the spectrum. As with other technical items discussed, this mechanism is feasible, and may eventually yield additional requirements on TVWS systems.

The level of reliability in a wireless communications system depends on the anticipated end use of the system. The ability to switch to an alternate channel, and the timing of that switch, will likely be near-term requirements placed on TVWS systems. Building upon those mechanisms would be data buffering and retransmission, at a lower layer in the stack (L2), so that timing constraints for switching can be met, as opposed to waiting for the application layer (L7) to detect and retransmit missing data, or worse, dropping information.

Depending on the application, there may be a requirement for multiple transceivers in a TVWS device. This second transceiver could perform a number of functions. First, it could be monitoring the other available channels to see if there is a better operational channel than the one being used. Second, it could be gathering location/time/use information for further analysis. Third, it could be used for parallel transmission to increase the overall throughput of the system, or to provide additional reliability to the communications. It is unlikely that this would be a firm requirement going forward for TVWS devices and systems, due to the cost and secondary use nature of the spectrum, but it may be a requirement for a specific customer situation.

3.5 Conclusion

The FCC's release of the TVWS rules in November of 2008 defined the basic regulatory requirements for TVWS systems. The TVWS rules are expected to further evolve over time. The rules were a conservative first step in order to ensure protection of the multiple classes of incumbent users, which include full-power TV broadcasters, low-power TV translators, Land Mobile Radio users, and wireless microphone users. These regulatory rules drove requirements including maximum antenna heights used in the system, operation of a database, and sensing levels to protect broadcast TV and wireless microphones. Many of these items will continue to be under further reconsideration at the FCC. This chapter examined these requirements in depth. Other requirements were explored to meet the practical technical and business considerations for a viable TVWS system, and future CR technologies that can improve the spectrum efficiency and define future system requirements were discussed.

References

1. FCC, *"Unlicensed Operation in the TV Broadcast Bands,"* Second Report and Order and Memorandum Opinion and Order, ET Docket 04-186, November 14, 2008.
2. FCC, *"Unlicensed Operation in the TV Broadcast Bands,"* Second Memorandum Opinion and Order, FCC 10-174, released September 23, 2010.
3. FCC, *"Unlicensed Operation in the TV Broadcast Bands,"* Notice of Proposed Rule Making, ET Docket 04-186, released May 25, 2004.
4. Motorola Inc. filing, *"Petition for Reconsideration and Clarification,"* ET Docket 04-186, March 19, 2009.
5. Shellhammer, Shen, Sadak, *"TV White Space for Wireless Broadband: Concepts, Techniques and Application*—Chapter 2: White Space Regulations," October 20, 2010.
6. FCC, *"Unlicensed Operation in the TV Broadcast Bands,"* First Report and Order and Further Notice of Proposed Rule Making, ET Docket 04-186, released October 18, 2006.
7. FCC/OET, "Initial Evaluation of the Performance of Prototype TV-band White Space Devices," FCC/OET 07-TR-1006 Report, released July 31, 2007.
8. FCC/OET, *"Interference Rejection Thresholds of Consumer Digital Television Receivers Available in 2005 and 2006,"* FCC/OET 07-TR-1003 Report, released March 30, 2007.
9. FCC/OET, *"Evaluation of Prototype TV-band White Space Devices Phase II,"* FCC/OET 08-TR-1005 Report, released October 15, 2008.
10. Advanced Television Systems Committee, "ATSC A/74 Recommended Practice: Receiver Performance Guidelines," June 18, 2004.
11. FCC/OET, "Longley-Rice Methodology for Evaluating TV Coverage and Interference," OET Bulletin 69, February 6, 2004.
12. *"Cable and ADS Penetration by DMA,"* http://www.tvb.org/rcentral/markettrack/Cable_and_ADS_Penetration_by_DMA.asp.
13. "FCC Committee Discusses White Spaces at WISPA Regional Meeting," http://www.wispa.org/?p=2716.
14. Borth, Ekl, Oberlies, and Overby, *"Considerations for Successful Cognitive Radio Systems in US TV White Space,"* IEEE DySpan Conference, October 2008.
15. *"The Smart Grid Meets White Spaces,"* http://electronicdesign.com/article/techview-communications/the_smart_grid_meets_white_spaces.aspx, July 23, 2010.
16. E²R, *"Cognitive Pilot Channel Concept,"* IST Summit, 2006, http://tns.ted.unipi.gr/ktsagkaris/img/papers/conferences/c19.pdf.

Chapter 4

White Space Availability in the United States

Ahmed K. Sadek, Stephen J. Shellhammer, Cong Shen, and Wenyi Zhang

Contents

4.1 Overview

This chapter provides an analysis of the number of available TV white space channels in the United States. The chapter begins by describing the information available on TV broadcast stations and how that information is used to calculate the number of available channels. Both standard and low-power TV stations are considered in this analysis. White space availability is characterized based on four different criteria:

- Channel availability per location.
- Channel availability per user, taking population density into account.
- Channel availability for different duplexing schemes: time division duplexing versus frequency division duplexing.
- Channel availability depending on application and quality of service requirements taking interference into account.

The first analysis is for devices that have database access and hence can use that information to determine channel availability. The first analysis only takes into account channel availability per zip code. The second analysis considers population density. In particular, rural areas with less population tend to have more white space available, which skews the results. The third category considered in this chapter is white space availability based on the duplexing method. Time division duplexing (TDD) is simple to analyze and results in the same white space availability as the first two categories (location based and per user based). Frequency division duplexing, however, results in different white space availability depending on the duplexer method used. The reason is that the frequency map of the white space channels available varies from one location to another, and hence the number of channels that could be paired together differ. The fourth criteria for evaluating white space availability is based on quality of service requirements and the fact that interference between devices increases as the number of white space channels drop.

4.2 Introduction

White space is the term used by the Federal Communication Commission (FCC) for unused TV spectrum, and one of the issues that comes up when considering the use of this spectrum is, "How much spectrum is actually available?" The TV channels include the very high frequency (VHF) channels 2–13 and the ultrahigh frequency (UHF) channels 14–51. However, there are restrictions on which channels are permissible for use by TV band devices (TVBDs). Fixed devices are permitted in the VHF channels except channels 3 and 4 and on the UHF channels except channels 36–38. Portable devices are not permitted in the VHF band. Portable devices are permitted on the UHF channels except 14–20 and channel 37. The exclusion for channels 3 and 4 is to prevent interference with external devices (e.g., DVD players), which are often connected to a TV utilizing either channel 3 or 4. Portable devices are not permitted on channels 14–20, because in 13 metropolitan areas some of those channels are used for public safety applications. Finally, channel 37 is a protected channel, used for radio astronomy measurements.

Television broadcast signals are protected with a *protection contour*. The FCC rules provide distances that a TVBD must be outside the protected contour for it to transmit. Within the protected contour there are special rules for operation on a TV channel adjacent to the TV broadcast channel. Fixed TVBDs are not permitted to operate on channels adjacent to the TV broadcast channel. Portable devices are permitted to operate on an adjacent TV channel; however, when operating on an adjacent TV channel, the maximum allowed transmission power is 16 dBm (4 dB lower than on nonadjacent channels).

Both fixed and Mode II TVBDs are required to access a TV band database in order to determine the permissible set of operating channels. According to the FCC rules [1,2], the TV database should contain the following information on full-power television stations, digital and analog Class A stations, low-power television stations (LPTV), television translator stations, and television booster stations:

- Transmitter coordinates (latitude and longitude)
- Effective radiated power (ERP)
- Height above average terrain of the transmitter (HAAT)
- Horizontal transmit antenna pattern (if the antenna is directional)
- Channel number
- Station call sign

An analysis for white space availability that takes into account population is [3]. This study also considers white space availability for sensing-only devices and shows that for such devices the channel availability drops significantly. In [3], only fixed white space devices are considered in the analysis. In this chapter, we consider both fixed and portable devices, as well as mixed networks that consist of both fixed and portable devices.

4.3 TV Station Databases

Prior to deployment of the TV white space incumbent database we need to rely on several databases in order to estimate the number of TV white space channels that are available in any given location. The FCC maintains a database of standard power

TV stations (Media Bureau's Consolidated DataBase System (CDBS)* for post-DTV transition full-power television stations) [4]. The database includes entries for all the standard-power TV stations in the United States. For each standard-power TV station, the database includes the location of the TV station (latitude and longitude), the antenna height, the transmit power, the protected contour, and the TV channel number. From this information we can determine for any given location whether at that location that TV channel number can be used by a TV white space device. This is done by adding the necessary distance beyond the protected contour as specified in the FCC report and order (R&O) and then determining if the specified location is far enough from the TV station location so that the channel is available. By checking all the TV stations in the database, we can determine if a given location is far enough from all the TV stations. So the FCC database provides all the necessary information for calculating which TV channels are available in any given location.

However, in addition to standard-power TV stations there are also low-power TV stations, which can transmit at up to 50 kW. These TV stations will not cover an entire metropolitan area, but will cover a region within the metropolitan area. The database of low-power TV stations does not provide nearly as much information as is available in the standard-power TV station database. The lower-power TV station database only provides a list of zip codes covered by each lower-power TV station [5]. This gives us a much coarser resolution and also an increased ambiguity. The database does not specify if all or only a portion of the zip code is covered by the low-power TV station. In rural areas the size of a zip code can be quite large, and hence whether the coverage area is all or only a portion of the zip code can be important. In our analysis we will be conservative and assume that, if a zip code is listed for a given low-power TV station, then the entire zip code is covered by the lower-power TV station. This means that the entire zip code is lost to the TV white space device, which may not be the case. However, we will not know whether all or part of that zip code is lost to the TV white space device until the actual TV white space incumbent database comes online. At that point the information on low-power TV stations will be comparable to the information currently available regarding standard-power TV stations.

Given that the resolution we have for low-power TV stations is at the zip code level, we will perform our analysis at that level.

The FCC R&O also protects other entities like cable headends whose locations are not currently available, but will be included in the TV white space incumbent database. Therefore, these entities are not included in the analysis.

---

* http://www.fcc.gov/mb/cdbs.html.

4.4 White Space Availability

To identify the used channels in any city the following procedure is followed:

■ The longitude and latitude of the location is specified. This is usually a location in the middle of the zip code.
■ Within the coverage area of the transmitter plus a margin, all of the channels are identified.

The analysis in this section is based on a combination of the standard-power and low-power TV station databases. In order to understand the analysis of white space availability, we need to introduce some notation. The number of available TV channels is a random variable which can change based on location. Let us indicate this random variable as N.

The number of available white space channels in any given zip code Z is denoted by N_z. In this analysis N_z is a number and not a random variable because we know how many channels are available in zip code Z. In real life, that number could change a small amount as a function of time due to wireless microphones coming on and going off. However, in the initial analysis we are not including wireless microphones, except for including the impact of two reserved channels for wireless microphones.

4.4.1 White Space Availability Based on Location

In this section we consider the white space availability for different locations, N_z. The most interesting locations to consider are the largest metropolitan areas, because that is where the largest number of people reside. It is clear that in rural areas there is a very large number of white space channels. In this analysis we consider zip codes in all of the United States, and population density is considered in subsequent sections.

Now we need to consider fixed and portable TV white space devices separately, because the FCC rules are different for these types of devices. Given that there are two general classes of devices, there are three types of networks we can construct from these devices:

■ Fixed—Networks consisting of only fixed devices
■ Portable—Networks consisting of only portable devices
■ Mixed—Networks consisting of a combination of fixed and portable devices

Fixed devices can operate in both the VHF and UHF frequency bands, while portable devices can operate only in the UHF band. Fixed devices cannot operate in a channel adjacent to a TV broadcast channel, while a portable device can operate in

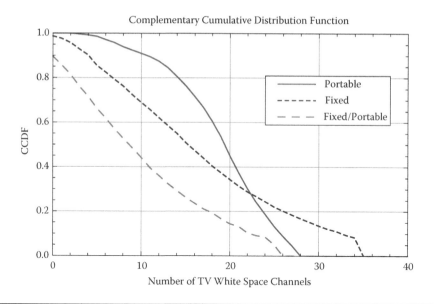

Figure 4.1 Number of available white space channels per location

an adjacent channel. A mixed network of both fixed and portable devices can only operate on channels that meet the conditions set for both the fixed and portable devices; hence a mixed network always has the least number of channels available.

Figure 4.1 shows the complementary cumulative distribution function for the number of white space channels available in any zip code in the United States.

The y-axis is the complementary cumulative distribution function (CCDF), and the x-axis is the number of available channels. In other words, the y-abscissa is the percentage of zip codes that have at least the white space channels of the x-abscissa. It is clear from the figure that there are more channels available to portable devices than fixed devices. Although the fixed devices can use both VHF and UHF channels while portable devices can only use channels above 21, portable devices can use channels adjacent to TV channels while fixed devices cannot. Channels adjacent to TV channels constitute most of the white space in most urban and dense urban locations.

Table 4.1 gives the number of channels available for more than 50% of the locations (zip codes). There are at least 16 channels available for portable networks deployments in 50% of the zip codes. This number drops to 6 channels available for fixed network deployments and 5 for mixed network deployments. Table 4.2 shows the number of channels available for more than 90% of the locations. Looking at the tail of the distribution, there are 6 or 7 channels available in more than 90% of the locations for portable networks, while there are only 1 or 2 channels available for fixed network deployments, and no channels available for mixed network deployments.

Table 4.1   Number of Channels Available in More Than 50% of the Locations

Network Deployment	Number of Channels
Fixed	6
Portable	16
Mixed	5

Table 4.2   Number of Channels Available in More Than 90% of the Locations

Network Deployment	Number of Channels
Fixed	1.5
Portable	7
Mixed	0

4.4.2 White Space Availability for a Typical User

Knowing the number of available channels at different locations is useful, but it is more useful to know the probability distribution of the number of channels for a typical user.

Here we need to do a little more probability analysis. First we define the number of channels for a typical user as N_u. This is a discrete random variable with a probability mass function (PMF) of,

$$p(n) = \sum_{k} P(N_u = n | \phi_k) P(\phi_k) \tag{4.1}$$

where f_k is the event that the user is in zip code k. From the previous section we have that,

$$P(N_u = n | \phi_k) = \delta(n - N_k) \tag{4.2}$$

where N_k is the number of available white space channels in zip code k, and $d(x)$ equals 1 if $x = 0$ and equals 0 otherwise.

The probability that a typical user is in zip code k is given by the number of people in that zip code divided by the number of people in the United States,

$$P(\phi_k) = \frac{M_k}{M} \tag{4.3}$$

where M_k is the number of people living in zip code k, and M is the total number of people living in the United States.

Therefore, we have the following formula for the probability mass function for N_u,

$$p(n) = \sum_k \delta(n - N_k) \frac{M_k}{M} \tag{4.4}$$

which is simply the total number of population in all zip codes with n free channels divided by the total number of population in the United States.

Given the probability mass function for the random variable N_u, we can calculate the cumulative distribution function $F(n)$, which is the probability that there are less than or equal to n channels for a typical user. Now what one is often more interested in is the complementary cumulative distribution function, $F_c(n) = 1 - F(n)$, which is the probability that there are more than n channels available to the typical user.

As was done in the previous section, we provide three sets of plots for the three types of networks: fixed, portable, and mixed.

Figure 4.2 depicts the probability mass function for the number of white space channels for a typical user. It is clear that, for portable networks, the PMF is almost uniform between 5 and 18 channels, while PMF peaks at a lower number of channels for both fixed and mixed networks.

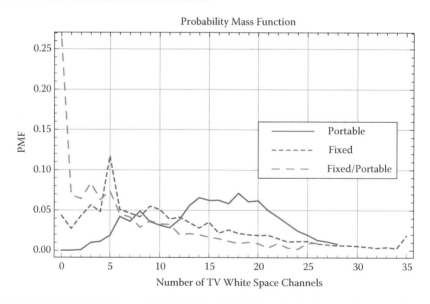

Figure 4.2 Probability distribution function of the number of channels available for a typical user.

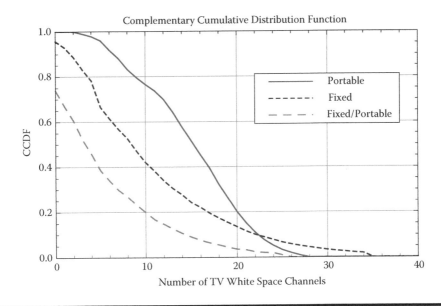

Figure 4.3 **Complementary cumulative distribution function of the number of channels available for a typical user.**

Figure 4.3 is a plot of the complementary cumulative distribution function for the number of channels available for a typical user. The y-axis is the complementary cumulative distribution function (CCDF), and the x-axis is the number of available channels. Table 4.3 gives the number of channels available for more than 50% of the population. There are at least 11 channels available for portable network deployments for 50% of the population. This number drops to 3 channels available for fixed network deployments and 1 channel for mixed network deployments. Table 4.4 shows the number of channels available for more than 90% of the population. Looking at the tail of the distribution, there are 5 channels available in more than 90% of the locations for portable networks, while there are no channels available for fixed network deployments, and no channels available for mixed network deployments.

Table 4.3 **Number of Channels Available for More Than 50% of the Population**

Network Deployment	Number of Channels
Fixed	3
Portable	11
Mixed	1

Table 4.4   Number of Channels Available for More Than 90% of the Population

Network Deployment	Number of Channels
Fixed	0
Portable	5
Mixed	0

Comparing the white space availability per location and per user, it is clear and intuitive that the number of available channels per user is much less. This is mainly because users are not uniformly distributed in the available areas, and most users are concentrated in urban and metropolitan areas with less channel availability. On the other hand, in rural areas there are only a few TV stations and hence ample white space, but only few people living there. For example, the available white space drops from 6 channels to 3 channels on average for fixed network deployment, and from 16 channels to 11 channels for portable network deployments, which is like 50% loss in the former case, and 33% loss in the latter case. The main conclusion is that there are more channels available per population for portable network deployments. There are far fewer, and almost no channels, available for fixed network deployments or mixed deployments in dense urban areas. However, there is a lot of white space available in rural areas.

4.5 White Space Availability Based on Duplexing Method

A time division duplexing (TDD) system requires a single channel to operate, while a frequency division duplexing (FDD) system requires two channels. Depending on the duplexing method, the number of available white space channels and its impact on the number of independent sets of channels (where an FDD channel is two TV channels) will differ. Moreover, depending on how the FDD duplexer is designed, the available white space channels will also change.

There are three performance metrics that are characterized and studied in this chapter:

■ Maximum number of independent networks that can utilize the white space without requiring co-channel sharing. We denote this metric by MI.
■ The minimum number of TV channels occupied by other networks before co-channel sharing is required. We denote this metric by WSA_{min} which stands for minimum white space availability.
■ The average number of TV channels occupied by other networks before co-channel sharing is required. We denote this metric by WSA where $(\cdot)$ denotes

average. The randomness here will be generated by assuming that any channel of the white space can be lost with equal probability.

In the analysis, we assume that there are N white spaces available with arbitrary distribution. The N channels take into account channels lost to satisfy guard band requirements in the FDD scenario.

4.6 Time Division Duplexing

In a TDD system, both the transmitter and the receiver use the same channel. The computations for TDD are straightforward because each white space represents an opportunity. The problem becomes more tricky with FDD because an opportunity requires a pair of white spaces that satisfies certain conditions. The FDD scenario is described in the following sections.

The maximum number of independent networks that could be established for a TDD network is thus given by

$$MI = N \tag{4.5}$$

The number of channels that have to be occupied before sharing is required are given by $WSA_{min} = N - 1$ and $WSA = N - 1$, respectively.

So, as expected, there is no difference between white space availability based on location or population for a TDD system.

4.7 Frequency Division Duplexing

In an FDD system, the transmitter and receiver channels are different. Because most FDD systems are full-duplex, simultaneous transmission and reception, an isolation between the Tx RF chain and Rx RF chain are required to prevent the Tx strong signal jamming the Rx weak signal. A guard band is thus required between the Tx and Rx channels.

The FDD duplexer could have different designs. We consider two different RF architectures for the FDD network:

■ The single duplexer approach, in which the whole band is divided into two subbands
■ The dual duplexer approach, in which the whole band is divided into four subbands

In the sections that follow, we present our analysis for the channel availability results for both RF architectures.

4.7.1 Single Duplexer Approach

A typical FDD system commonly uses a duplexer to isolate Tx band and Rx band. Denote the Tx band by A and the Rx band by B. Considering the UHF band for this analysis, which consists of channels 14 to 51, let band A contain channels from 14 to 31 and let band B contain channels from 34 to 51. Channels 32 and 33 are unavailable due to duplexer band gap. This is a conventional design where the duplexer divides the whole band into two equal parts. The number of networks that can be operated in the Bay area according to this design is only 1. This is due to the nonuniform distribution of the white space across the UHF band in the San Francisco Bay Area. Later, we consider different single duplexer designs and analyze their performance.

First, we analyze the performance of a general single duplexer. The analytical results are then used to optimize the single duplexer design. Assume that the available N white spaces are distributed in the two subbands such that the first subband has N_1 channels and the second subband has N_2 channels. Of course $N = N_1 + N_2$. The analysis is given in Appendix 4.1, and here we only report the final results. The maximum number of independent networks that could be established for the single duplexer design is given by

$$\mathrm{MI}_{\mathrm{Single}} = \min(N_1, N_2) \tag{4.6}$$

The worst case, or minimum, number of channels that has to be lost before sharing is required are given by

$$\mathrm{WSA}_{\mathrm{Single,\,min}} = \min(N_1, N_2) - 1 \tag{4.7}$$

The average number of channels that need to be lost before sharing is required is given by

$$N_1 N_2 \left[\frac{1}{N_1 + 1} + \frac{1}{N_2 + 1} \right] - 1 \tag{4.8}$$

The white space availability results for the single duplexer approach are summarized in Table 4.5.

4.7.1.1 Optimizing the Single Duplexer Design

A typical single duplexer will divide the UHF spectrum equally into two bands. More specifically, band A contains channels from 14 to 31 and band B contains channels from 34 to 51. Channels 32 and 33 are unavailable due to duplexer band gap. If the white space distribution in the UHF band was uniform, this design would be optimal in terms of white space availability.

**Table 4.5 FDD White Space
Analysis (Single Duplexer Case)**

Metric	Analytical Results
MI	$\min(N_1, N_2)$
$\text{WSA}_{\min}$	$\min(N_1, N_2) - 1$
$\overline{\text{WSA}}$	$N_1 N_2 \left[\dfrac{1}{N_1 + 1} + \dfrac{1}{N_2 + 1} \right] - 1$

However, the white space distribution is nonuniform. For example, in the Bay area most of the white space is found in the lower part of the spectrum. This is the reason that a uniform single duplexer leads to only 1 FDD network in the Bay area.

To solve this problem, one can reduce the size of Band A and increase the size of Band B. However, because the distribution of white space differs from one location to another, it is difficult to find one optimal design that leads to the best results in all locations. To overcome this problem, we study the tradeoff that different duplexer designs present. In general, one would like to increase the white space availability in areas with very few white spaces, like the San Francisco Bay Area, without hurting too much the performance in other cities. To accomplish that, we consider as our metric the generalized average of the number of independent networks in the 50 largest cities in the United States. The generalized average can be defined as follows:

$$GA_p = \left(\frac{1}{n} \sum_{i=1}^{n} x^p \right)^{1/p} \tag{4.9}$$

From the above equation, different values of p lead to different tradeoffs. For example:

- $p = 1$: Arithmetic average
- $p = -1$: Harmonic average
- $p \rightarrow \pm\infty$: Maximum/minimum
- $p \rightarrow 0$: Geometric average

The more negative p is, the more the average reflects the tendency of the smaller numbers. In our analysis we will consider the arithmetic average, $p = -3$, and the minimum.

The only variable in the single duplexer design is the size of the Tx and Rx bands. We try to optimize this parameter according to some cost function of the white space availability in the largest fifty cities in the United States, largest in terms of population. The following single duplexer designs summarized in Table 4.6 will be considered.

Table 4.6   Different Single Duplexer Designs

SD_0	Band A (14 To 31) and Band B (34 to 51)
SD_1	Band A (14 to 30) and Band B (33 to 51)
SD_2	Band A (14 to 29) and Band B (32 to 51)
SD_3	Band A (14 to 28) and Band B (31 to 51)
SD_4	Band A (14 to 27) and Band B (30 to 51)
SD_5	Band A (14 to 26) and Band B (29 to 51)
SD_6	Band A (14 to 25) and Band B (28 to 51)

Table 4.7   Analytical Results for Different Single Duplexer Designs

Design	Mean: $p = 1$	$p = -3$	Minimum: $p \to -\infty$
SD_0	10.2	2.53	1
SD_1	10.3	2.54	1
SD_2	10.3	4.86	2
SD_3	**10.1**	**6.66**	**3**
SD_4	9.56	6.64	3
SD_5	9.12	6.54	3
SD_6	8.52	7.28	4

The analytical results for the seven single duplexer designs described above averaged over the largest 50 cities in the United States are shown in the Table 4.7.

From the above, SD_3 (Band A (14 to 28) and Band B (31 to 51)) captures the best tradeoff between the minimum number of available channels and the average white space availability. The conventional design, which is SD_0, only results in a generalized average of 2.5 with $p = -3$, while the optimal design results in a generalized average of 6.7. We have chosen $p = -3$ to emphasize that higher weight is given to locations with very few white spaces (which happens to be locations with higher population density).

4.7.2 Dual Duplexer Approach

A dual duplexer can be used to enhance white space availability for FDD networks. A dual duplexer divides the band into four subbands. For example, band A contains channels 14 to 21. Band B contains channels 24 to 31. Band C contains channels

34 to 41, and band D contains channels 44 to 51. Channels 22, 23, 32, 33, 42, and 43 are used as band gaps to isolate Tx and Rx RF chains.

The dual duplexer approach divides the whole band into four subbands. Assume that the available N white spaces are distributed in the four subbands such that the i-th subband has N_i channels, $1 \leq i \leq 4$, such that

$$\sum_{i=1}^{4} N_i = N \tag{4.10}$$

Without loss of generality, assume that the available white spaces in the four subbands are related as follows: $N_1 \leq N_2 \leq N_3 \leq N_4$. Note that the results still hold for any case, because we can always relabel the subbands without any effect on the results.

The proof is kept to Appendix 4.2, and here we present the main results. The maximum number of independent networks is given by

$$\mathrm{MI}_{Dual} = N - x \tag{4.11}$$

where x is given by

$$x = \begin{cases} \left| N_4 - N_1 - N_2 - N_3 \right| & \text{if } (N_4 \geq N_1 + N_2 + N_3) \,\&\, (0 \leq N_4 - N_2 - N_3 \leq N_1) \\ \left| N_4 + N_1 - N_2 - N_3 \right| & \text{otherwise} \end{cases} \tag{4.12}$$

The worst case scenario, or minimum number of networks that have to be established before sharing between networks is required, is given by

$$\mathrm{WSA}_{Dual,min} = N_1 + N_2 + N_3 - 1 \tag{4.13}$$

The average white space that has to be lost before sharing is required is given by

$$\overline{\mathrm{WSA}}_{Dual} = \frac{N_4(N_1 + N_2 + N_3)}{N_1 + N_2 + N_3 + 1} + \frac{N_3(N_1 + N_2 + N_4)}{N_1 + N_2 + N_4 + 1} \\ + \frac{N_2(N_1 + N_3 + N_4)}{N_1 + N_3 + N_4 + 1} + \frac{N_1(N_2 + N_3 + N_4)}{N_2 + N_3 + N_4 + 1} - 1 \tag{4.14}$$

Given this information, the white space availability results for the dual duplexer approach is summarized in Table 4.8.

Table 4.8   FDD White Space Analysis: (Dual Duplexer Case)

Metric	Analytical Results
MI	$N - x$
WSA$_{min}$	$N_1 + N_2 + N_3 - 1$
$\overline{\text{WSA}}$	$\dfrac{N_4(N_1 + N_2 + N_3)}{N_1 + N_2 + N_3 + 1} + \dfrac{N_3(N_1 + N_2 + N_4)}{N_1 + N_2 + N_4 + 1}$ $+ \dfrac{N_2(N_1 + N_3 + N_4)}{N_1 + N_3 + N_4 + 1} + \dfrac{N_1(N_2 + N_3 + N_4)}{N_2 + N_3 + N_4 + 1} - 1$

4.7.2.1 Optimizing the Dual Duplexer Design: Two Single Duplexers

A drawback in the dual duplexer design is that six channels are lost because three band gaps are required. We consider using two single duplexers such that for any city the best duplexer can be used. Therefore at most two channels are lost because there is always one band gap.

From optimizing the single duplexer design, we found out that the best single duplexer design is SD_3. For our two single duplexer design, one of the single duplexers is fixed to be SD_3, while the second single duplexer design is changed. Table 4.9 gives the set of single duplexers used with SD_3.

Next we tabulate the results, and as a benchmark, the performance of the dual duplexer is also considered in the comparison. Note that for all the two single

Table 4.9   Different Dual Duplexer Designs

SD_2	Band A (14 to 29) and Band B (32 to 51)
SD_1	Band A (14 to 30) and Band B (33 to 51)
SD_0	Band A (14 to 31) and Band B (34 to 51)
SD_-	Band A (14 to 32) and Band B (35 to 51)
SD_{-2}	Band A (14 to 33) and Band B (36 to 51)
SD_{-3}	Band A (14 to 34) and Band B (37 to 51)
SD_{-4}	Band A (14 to 35) and Band B (38 to 51)
SD_{-5}	Band A (14 to 36) and Band B (39 to 51)
SD_{-6}	Band A (14 to 37) and Band B (40 to 51)

Table 4.10 Analytical Results for Different Dual Duplexer Designs

Design	Mean: $p = 1$	$p = -3$	Minimum: $p \rightarrow -\infty$
Dual duplexer	9.88	6.65	3
SD_2	10.5	6.7	3
SD_1	10.8	6.75	3
SD_0	**10.84**	**6.73**	**3**
SD_-	10.7	6.72	3
SD_{-2}	10.5	6.7	3
SD_{-3}	10.26	6.7	3
SD_{-4}	10.26	6.7	3
SD_{-5}	10.14	6.77	3
SD_{-6}	10.1	6.66	3

duplexer designs, one of them is always fixed to SD_3. These results are averaged over the largest fifty cities in the United States.

From the results in Table 4.10, the best two single duplexer design is the combination of the two single duplexer filters SD_3 and SD_0.

4.8 White Space Availability Based on Application Requirements

Interference is one of the major hurdles for operation in the white space because, if not properly managed, it can limit the achievable capacity of the TVBDs. Managing interference between nodes in the same network is generally a difficult problem, and the problem becomes more challenging when these TVBDs belong to heterogeneous networks using different air interfaces. In the previous sections, we have evaluated white space availability in terms of the number of channels available for a given user. A different approach for studying white space availability is to consider what is the number of white space channels sufficient to achieve certain quality-of-service (QoS) requirements for a given application. The impact of interference depends on the deployment model. In this chapter we only consider indoor deployment models.

We simulate an apartment complex model, and in particular we simulate a single block which has two buildings facing each other. Each building has three to six floors; the number of floors is selected randomly for each drop. Figure 4.4 illustrates the floor plan of the building, which has ten apartments per floor. Each apartment is 10

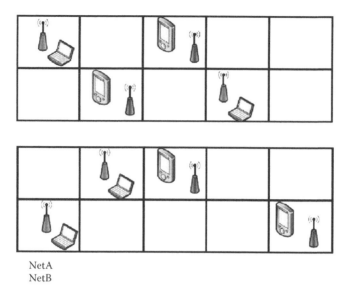

NetA
NetB

Figure 4.4 **Floor plan of the building with access points and clients from Network A and Network B dropped with uniform distribution.**

m ×10 m. There are two rows of apartments with five apartment per row. Two different indoor networks are dropped inside the apartments, Network A and Network B. Each Network consists of a multiple of links, for example a wireless local area network (WLAN) network with an access point (AP) and client are dropped in an apartment. The nodes are dropped with uniform distribution inside the apartments.

There is no coordination or cooperation between the links, so it could be considered as an ad hoc network. Devices are turned on randomly, and each device when turned on performs channel selection and selects a channel to operate on according to some channel selection algorithm.

4.8.1 Outdoor Propagation

An outdoor propagation model is needed for the scenarios in which both the transmitter and the receiver are in different buildings. We use the Hata pathloss model to model this scenario, and for urban areas it is given by [6]

$$L_H(d) = 69.55 + 26.16 \log f_c - 13.82 \log l_{tx} - a(l_{rx})$$
$$+ (44.9 - 6.55 \log l_{tx}) \log d \tag{4.15}$$

where $L_H(d)$ is the path loss in dB, f_c is the carrier frequency in MHz, l_{tx} and l_{rx} are the heights of transmit and receive antennas in meters, respectively, and d is the distance in km. In (4.15), $a(l_{rx})$ denotes a mobile antenna correction factor, which for large cities is given by

$$a(l_{rx}) = 3.2(\log(11.75l_{rx}))^2 - 4.97 f_c \geq 300 MHz \qquad (4.16)$$

4.8.2 Indoor Propagation

When both the transmitter and the receiver are indoors and in the same building, the multiwall model is used to calculate the indoor pathloss propagation [7]. The multiwall model is given by

$$L_{in}(d) = 20\log\left(\frac{4\pi f_c}{C}\right) + 20\log(d) + q_{in}W_{in} + q_{ex}W_{ex} + F_n^{\frac{n+2}{n+1}-0.46} \qquad (4.17)$$

where d is the distance in meters, f_c is the carrier frequency in MHz, C is the speed of light, q_{in}/q_{ex} is the number of internal/external walls, W_{in}/W_{ex} is the internal/external wall attenuation, F is the floor attenuation, and n is the number of floors. The total number of walls $q = q_{in} + q_{ex}$ is a uniformly distributed random variable given by

$$q \sim U\left[0, 1, \ldots, \text{floor}\left(\frac{d}{d_w}\right)\right] \qquad (4.18)$$

where d_w is the minimum separation between walls and is considered to be 2 m in the simulations.

It is worth mentioning that the multiwall model results in a pathloss model that is exponential in distance compared to the outdoor models that are usually polynomial functions of distance. This is because of the attenuation factor in the multiwall model that depends on the number of internal walls.

For indoor-to-indoor (the two nodes are inside different buildings) scenarios, both the Hata and the multiwall models are used to calculate the overall pathloss.

The parameters in Table 4.11 are kept fixed in all the simulations. In the simulations, wall penetration is 5 dB, floor penetration is 18 dB, and building penetration is 15 dB.

A simple greedy algorithm is used for channel selection, in which each AP selects the channel with minimum interference. This algorithm, though simple, has been shown in recent papers to have nice convergence properties and significant performance gains compared to randomly selecting channels at each AP.

In the simulations, the penetration of NetA and NetB is assumed to be 20%. In other words, dropping an AP in an apartment is a Bernoulli random variable

Table 4.11   Simulation Parameters

Parameter	NetA/NetB
BS/AP antenna height (m)	2
UE/Station antenna height (m)	2
BS/AP transmit power (dBm)	20 or 16
BS/AP antenna gain (dB)	–5
UE/Station transmit power (dBm)	20 or 16
UE/Station antenna gain (dB)	–5
AP noise figure (dB)	5
UE/Station noise figure (dB)	8
Bandwidth (MHz)	5

with probability 0.2. Two APs from networks A and B can be dropped in the same apartments. APs are turned on sequentially, and when each AP is turned on it runs the channel selection algorithm.

Figure 4.5 depicts the cumulative distribution function (CDF) of the signal-to-interference ratio (SINR) as a function of the number of available white space channels. Another metric that is used for performance evaluation is outage probability. Outage is defined as the event that the received SINR is lower than the minimum decodable SINR. We use 15 dB in the simulations, which is the requirement for decoding a standard definition video stream. Hence the outage probability can be defined as

$$p_o = \Pr(\text{SINR} < \text{SINR}_{\text{thr}}) \tag{4.19}$$

Figure 4.6 depicts the outage probability. It is clear that, for different SINR requirements, different numbers of channels available are required to guarantee performance.

4.9 Conclusions

This chapter presented an analysis for the white space availability in the United States. The FCC rules for determining white space availability for fixed and portable devices have been used. The analysis was performed to cover four different scenarios. First scenario is channel availability per location, which results in an overestimate of the actual available white space because it does not distinguish between rural and urban areas. The second scenario is more important to network

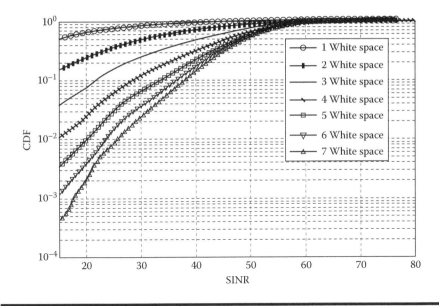

Figure 4.5 Cumulative distribution function of the SINR as a function of the number of white space channels for a portable network deployment.

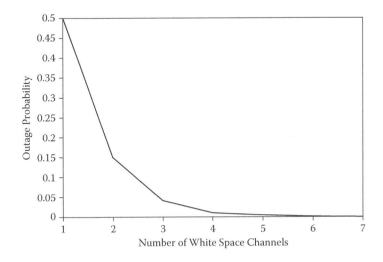

Figure 4.6 Outage probability as a function of the number of channels available for a typical user for a portable network deployment.

operators because it takes into account population density to find the available white space per person. The third and fourth scenarios consider the dependence of white space availability on the technology used, and in particular the third scenario studies the impact of the duplexing method used on white space availability, while the fourth scenario considers impact of QoS requirements on channel availability.

4.10 Appendix 4.1: Single Duplexer Analysis

For simplicity of notations we will characterize the number of channels that need to be occupied at which sharing is required.

4.10.1 Worst Case Scenario

Since there are two bands with N_1 and N_2 white spaces, respectively, it is obvious that the worst case scenario happens when the minimum of both bands is occupied first.

4.10.2 Average Case Scenario

The problem can be formulated as follows. There are two bands; the first band has N_1 white spaces and the second has N_2 white spaces. We start occupying white spaces randomly such that there is equal probability to select any white space from the two bands. The experiment ends when one of the two bands has zero white spaces left.

The problem then can be stated as follows: Find the probability that the experiment ends after n white spaces are occupied. Define this event by E_n.

Define the event that the first band is emptied first at n by A_n and the event that the second band ends first at n by B_n. Clearly we have

$$\Pr[E_n] = \Pr[A_n] + \Pr[B_n] \tag{4.20}$$

Let us first calculate the probability of A_n. Event A_n happens if at trial $(n-1)$ there is exactly 1 white space left in the first band and there are a positive number of white spaces in the second band, and at trial n the white space from the first band is selected. This probability can be written as follows:

$$\Pr[A_n] = \frac{\binom{N_1}{N_1-1}\binom{N_2}{n-N_1}}{\binom{N_1+N_2}{n-1}} \frac{1}{N_1+N_2-n+1} \tag{4.21}$$

The above expression can be simplified after some arithmetic as follows:

$$\Pr[A_n] = \frac{\binom{n-1}{N_1-1}}{\binom{N_1+N_2}{N_1}} \tag{4.22}$$

Similarly the probability of event B_n can be written as follows:

$$\Pr[B_n] = \frac{\binom{n-1}{N_2-1}}{\binom{N_1+N_2}{N_1}} \tag{4.23}$$

The average can be calculated as follows:

$$E(E_n) = \sum_{n=\min(N_1,N_2)}^{N_1+N_2-1} n\Pr[E_n]$$

$$= \frac{1}{\binom{N_1+N_2}{N_1}} \left[\sum_{n=\min(N_1,N_2)}^{N_1+N_2-1} n\binom{n-1}{N_1-1} + n\binom{n-1}{N_2-1} \right]$$

$$= \frac{1}{\binom{N_1+N_2}{N_1}} \left[N_1 \sum_{n=\min(N_1,N_2)}^{N_1+N_2-1} \binom{n}{N_1} + N_2 \sum_{n=\min(N_1,N_2)}^{N_1+N_2-1} \binom{n}{N_2} \right], \; (*) \tag{4.24}$$

$$= \frac{1}{\binom{N_1+N_2}{N_1}} \left[N_1 \binom{N_1+N_2}{N_1+1} + N_2 \binom{N_1+N_2}{N_2+1} \right], \; (**)$$

$$= N_1 N_2 \left[\frac{1}{N_1+1} + \frac{1}{N_2+1} \right], \; (***)$$

Where, to prove (*), the following identity is used:

$$n\binom{n-1}{N_1-1} = N_1\binom{n}{N_1} \tag{4.25}$$

To prove (**), the following identity is used:

$$\sum_{j=n}^{N}\binom{j}{n}=\binom{N+1}{n+1}$$

(4.26)

Step (***) requires some straightforward but tedious arithmetic.

4.11 Appendix 4.2: Dual Duplexer Analysis

The analysis is more involved with the dual duplexer case. Let E_n denote the event that at trial n exactly 3 out of the 4 bands are empty. Define A_n to be the joint event that E_n happens AND the third band to empty was the first band. Similarly define B_n to be the joint event of E_n and the third band to empty is the second band. Events C_n and D_n are defined similarly to correspond to the third band and the fourth band, respectively.

Therefore the desired probability of the event E_n can be calculated as

$$\Pr[E_n]=\Pr[A_n]+\Pr[B_n]+\Pr[C_n]++\Pr[D_n]$$

(4.27)

The event A_n happens if any of the following three events happened. At time instant $(n-1)$ the first band has only one white space and

■ the second and third band are only empty
■ the second and fourth are only empty
■ the third and fourth are only empty

and in the n-th trial a white space is picked from the first band. It is clear then that event A_n consists of three exclusive events, and since the main event E_n consists of four events, we have in total 12 events to calculate the total probability.

Here, for simplicity, we give the probability expression for event A_n only, and the rest of the probabilities can be calculated in similar ways.

$$\Pr[A_n] = \frac{N_1 \binom{n-1}{N_1+N_2+N_3-1}}{(N_1+N_2+N_3)\binom{N_1+N_2+N_3+N_4}{N_4}}$$

$$+ \frac{N_1 \binom{n-1}{N_1+N_2+N_4-1}}{(N_1+N_2+N_4)\binom{N_1+N_2+N_3+N_4}{N_3}} \qquad (4.28)$$

$$+ \frac{N_1 \binom{n-1}{N_1+N_3+N_4-1}}{(N_1+N_3+N_4)\binom{N_1+N_2+N_3+N_4}{N_2}}$$

The average can be calculated after several steps as follows:

$$\mathrm{E}(\mathrm{E}_n) = \frac{N_4(N_1+N_2+N_3)}{N_1+N_2+N_3+1} + \frac{N_3(N_1+N_2+N_4)}{N_1+N_2+N_4+1} + \frac{N_2(N_1+N_3+N_4)}{N_1+N_3+N_4+1}$$

$$+ \frac{N_1(N_2+N_3+N_4)}{N_2+N_3+N_4+1} \qquad (4.29)$$

References

1. Federal Communications Commission, "Second Report and Order and Memorandum Opinion and Order in the Matter of Unlicensed Operation in the TV Broadcast Bands, Additional Spectrum for Unlicensed Devices Below 900 MHz and in the 3 GHz Band," Federal Communications Commission, Document 08-260, November 14, 2008.
2. Federal Communication Commision, ``Second Memorandum Opinion and Order in the Matter of Unlicensed Operation in the TV Broadcast Bands, Additional Spectrum for Unlicensed Devices below 900 MHz and in the 3 GHz Band," Docket Number 10-174, September 23, 2010.
3. K. Harrison, S.M. Mishra, and A. Sahai, "How Much White-Space Capacity Is There?", IEEE symposium on The International Dynamic Spectrum Access Networks 2010, (DySPAN 2010), April, 2010.
4. Federal Communication Commission, FCC 07-138. http://hraunfoss.fcc.gov/edocs_public/attachmatch/FCC-07-138A1.pdf, August 2007.

5. Federal Communications Commission, "List of All Class A, LPTV, and TV Translator Stations," Federal Communications Commission, Tech. Rep., 2008. [Online]. Available: http://www.dtv.gov/MasterLowPowerList.xls.

6. T. S. Rappaport, *Wireless Communications: Principles and Practice*, 2nd Ed., Upper Saddle, NJ: Prentice Hall, 2002

7. Qualcomm Europe, "HNB and HNB-Macro Propagation Models," 3GPP TSG-RAN WG4, October 2007.

Chapter 5

TV White Spaces in Europe

Heikki Kokkinen, Jukka Henriksson, and Risto Wichman

Contents

5.1 Introduction

This chapter analyzes the potential for TV White Spaces adoption in Europe. We describe the TV digitalization background to understand the historical basis. We carry out two analyses: a capacity analysis based on signal propagation and an innovation diffusion estimate. By applying the country-specific capacity and parameter values in the technology diffusion theory, we create a forecast of the TV white spaces adoption in Europe. We give recommendations as to what actions could help the adoption of the TV white spaces in Europe both at European Union and country level.

5.2 Digital TV Development in Europe

In Europe, the use of white spaces is generally expected to take place on TV bands—mainly on the 470–862 MHz part of the ultrahigh frequency (UHF) band. This frequency range has varying degrees of utilization in different parts of the world. Europe is interesting because this band is fairly filled with TV transmissions especially in the central parts: United Kingdom, Germany, France, etc. If that band can be taken in the white spaces use, then it should also be possible elsewhere in the world.

In the beginning of 1990s, many European countries planned to upgrade their TV networks. First, there were activities to bring high-definition television (HDTV) to homes via high-definition multiplexed analog component (HD-MAC) satellites. This turned out to be unfeasible with the technology of those days and soon the efforts were concentrated toward a digital approach for TV systems; first for standard definition and later extension to HDTV. One motivation was the high utilization rate of the UHF TV spectrum that began early in Europe; while the United States had about 1700 high-power analog TV transmitters in the whole country, Europe had about that number already on one UHF channel—nearly 50,000 TV transmitters in total in the beginning of the 1990s [1]. A digital approach substantially increases spectrum efficiency—roughly four times the channel capacity of the analogue system is available in digital TV broadcasting.

In co-operation with European Telecommunications Standards Institute (ETSI), the industry-led Digital Video Broadcasting (DVB) consortium of over 250 organizations has published several successful digital TV standards: DVB-S for satellite reception, DVB-C for cable reception, and DVB-T (1997) for terrestrial reception [2], plus a large number of related, supporting standards. The standard for terrestrial mobile TV, DVB-H, was later released (2004). The new terrestrial standard DVB-T and its mobile version DVB-H increased the demand for broadcasting spectrum, and the old frequency allocation agreement (Stockholm 1961) was not capable of handling the new situation efficiently. A series of International Telecommunication Union (ITU) meetings were arranged, and the new agreement (Geneva 2006, GE06) was reached [3].

The focus in the GE06 plan is that ITU Region 1 including Europe will fully go to digital TV broadcasting, and the target date for the analog switch off (ASO) should be in 2015 at the latest (on UHF) or already in 2012 as set by the European Commission. The new plan took into account both analog and digital transmissions, but the main goal was to reach an efficient plan for digital services after the analog switch off. Generally, the GE06 plan gives seven nationwide coverages of DVB-T in Bands IV/V (8 MHz channel) and one DVB-T coverage in Band III (7 or 8 MHz channel) for the majority of the European countries within the European

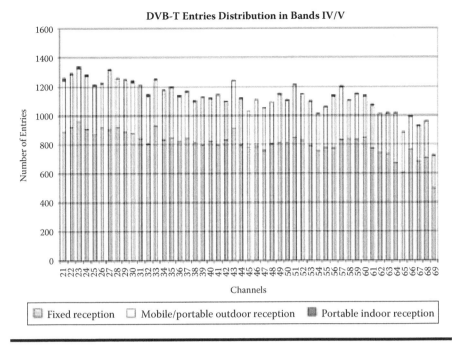

Figure 5.1   Indication of the channel usage in Bands IV/V for DVB-T and for the different reception modes, in the whole planning area [4].

Conference of Postal and Telecommunications Administrations (CEPT). In addition, there may be possibilities for local channels in some countries [4].

Figure 5.1 shows the channel reservations in Europe. They indicate a high utilization of the UHF band. There are over 1000 entries per UHF channel, with slightly fewer entries in channels above 65 [4]. Because that is the main interest in the planning area, a slightly different view is given by Figure 5.2, which shows the allocations (and coverage) on a geographical map for just one example channel (UHF channel 65). The availability of the channel in various locations in Europe gives a promise that there might be some room left for low-power transmissions in suitable geographical locations. Careful technical assessment is, nevertheless, needed before final conclusions [5]. The other point is that a few channel reservations will not be used all the time, and especially for a couple of years from 2012, there will be some room in the spectrum, while new services will be gradually set up. In the next section, we take a more detailed look on the availability of white space (WS) in Finland. This gives an example of the potential of WS in the part of Europe where the congestion of UHF band is not too severe.

Table 5.1 gives the schedule of launching digital terrestrial television (DTT) and the expected year of analog switch off for selected European countries [6–9]. Many of the countries have already implemented ASO, and many others are just about to do it. It is expected that a major part of Europe will be free of analog TV transmissions

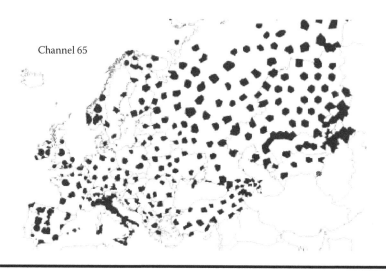

Channel 65

Figure 5.2   Map of DVB-T plan entries in UHF channel 65 in Europe and neighboring countries.

by 2012. It should be noted that, while the original target for ASO in Germany was 2010, the switchover was already achieved at the end of 2008. Some dates in the table are sophisticated guesses because not all the administrations have published confirmed plans, or they were not available for the authors at the time of writing.

When Geneva plan GE06 was made, the main digital TV systems to take into account were DVB-T and DVB-H. The planning principles, however, were made quite flexible for future developments. And indeed, such developments have already taken place. The new DTT standard, DVB-T2, was developed by DVB and published by the European Telecommunications Standards Institute (ETSI) in September 2009 [10].

The DVB-T2 standard uses the same 8 (or 7) MHz channel raster as DVB-T, but provides more efficiency with about 30% to 40% increase in capacity. The spectrum mask fits to that of DVB-T, and spectral planning requires no extra complexity. DVB-T2 has already been adopted and launched in the United Kingdom, and several other countries have shown interest in it. While most operators are considering adoption of DVB-T2 in order to provide HDTV services, the standard also lends itself for standard definition TV, radio, or data services.

The new situation with fixed digital TV via DVB-T2 has naturally also raised the question about mobile TV. Currently, it seems that DVB-T2 could form a very good basis for a new mobile TV standard as well. DVB has already started to work on this subject, and we will possibly see the new standard (next generation handheld, NGH) emerging in 2011/2012. This would allow a flexible use of DVB-T2 and NGH in the same planning area.

Similarly to the analog to digital switchover, the introduction of DVB-T2 transmissions will occupy more TV spectrum during the switchover period than

Table 5.1   Launch of Digital Terrestrial TV in Some European Countries and Expected Analog Switch-Off Dates

Country	DTT Launch	ASO Date	Estimated ASO
Netherlands	2004	2006	2006–2008
Germany	2004	2010	
Finland	2002	2007	
Sweden	1999	2008	2009–2012
Denmark	2006	2009	
Norway	2007	2009	
Switzerland	2005	2009	
Belgium	2004	2011	
Austria	2006	2010	
France	2005	2011	2012–2015
United Kingdom	1998	2012	
Spain	2000	2010	
Italy	2004	2012	
Czech Republic	2005	2012	
Hungary	2008	2011	
Ireland	2013?	2012	
Luxembourg	2006	2011	
Portugal	2009	2012	
Poland	2010?	2013	
Slovak Republic	2009	2012	
Greece	2009	2012/15	

DVB-T or DVB-T2 alone would. The future transition period from DVB-T to DVB-T2 will limit available spectrum for white space use in Europe.

Finally, we recognize that the digital TV field is very active in Europe, with several systems already on the air and new ones coming in. The UHF band in Europe is quite congested with current analog transmissions, DVB-T, and DVB-H. The situation may have temporary relief from the analog switch off during 2012–2015, but it is expected to saturate later on. Still, there might be geographical and temporal windows to use other systems on these frequencies with appropriate precautions taken into use.

The concept of "digital dividend" was launched several years ago to mean those frequencies and new allocations which might be available after the digital transition of TV transmissions. This has been quite a controversial issue because, understandably, the old TV operators do not want to give up their access to the current spectrum. On the contrary, mobile operators and also other parties with increasing need for spectrum look upon this change as a tempting opportunity to gain new access to spectrum. In Europe, there have been many activities starting from the upper EU level [11,12] together with various activities in CEPT, the European Committee for Electrotechnical Standardization (CENELEC), and ETSI. These efforts try to clarify various issues related to the coexistence of various services in the same frequency band. The general European Union level activities on digital dividend are reported on the European Union's Web page [13].

The relevant group trying to solve the technical details of adoption of white spaces in Europe is CEPT SE 43. So far it has published a couple of reports on the issue, most prominently reports, 24, 25, and 32 [14–16]. Currently the group SE 43 is working on an extensive report, "Technical and Operational Requirements for the Possible Operation of Cognitive Radio Systems in the White Spaces of the Frequency Band 470–790 MHz." The progress of this work can be followed on the Electronic Communications Committee Web pages (available to the public) [17]. In particular, WG SE (Spectrum engineering) and the subgroup SE 43 (Cognitive radio systems–White spaces (470–790 MHz)) are worth checking.

The European Telecommunications Standards Institute has been active in the area of reconfigurable radio systems (RRS) and software defined radio (SDR), lately also related to cognitive radio and its use in white spaces. The standardization work in this area is parallel and related to the activities in IEEE 802.19. ETSI group TC RRS is divided into four working groups. The system group WG1 is working on reports TR 102 907, "Operation in White Space Frequency Bands," and TS 102 908, "Coexistence Architecture for Cognitive Radio Networks on UHF White Space Frequency Bands," both due in the first half of 2011. The scope of the first report is "Use cases and system requirements for networks operating on a secondary basis in UHF White Spaces and Methods for protecting primary/incumbent users of UHF White Spaces." The scope of the technical specification (the latter document) is "System architecture (covering both centralized and distributed solutions) for spectrum sharing and coexistence of cognitive radio networks operating as secondary spectrum users in UHF White Space bands." Progress of these activities can be followed on the ETSI portal and related pages [18].

There are at least two significant new developments related to the use of white spaces in Europe. The first one is the adoption of technical rules on the allocation of radio frequencies in 790–862 MHz by the European Commission. The spectrum will be used for high-speed wireless Internet services by avoiding harmful interference. The EU countries can choose whether they free the frequencies, and if they do it, the new rules must be applied [19]. The technical rules allow the frequencies to be used for broadband wireless networks like 3GPP Long Term Evolution (LTE) or Worldwide Interoperability for Microwave Access (WiMAX). The other one is that wireless microphones that earlier used the band around 800 MHz and 820 MHz are now forced to other bands. One solution is to use white spaces from 470 MHz to 790 MHz. These are currently under evolution (e.g., in Denmark, Finland, and other countries).

5.3 White Space Case Study

In this section, we present a case study on the spectrum available as white space in Finland. Finland has 36 main DVB-T transmitter sites and over 200 relay sites broadcasting 4 DTT multiplexes on 42 8-Mhz channels between 198 MHz and 826 MHz. Thus, the channel usage is rather sparse, and one may anticipate that substantial amount of spectrum should be available even when the transmitters in the neighboring countries Estonia, Norway, Russia, and Sweden are taken into account in interference studies.

We used ITU propagation model [20] to calculate coverage areas of TV transmitters and signal strengths in 25 × 50 km grid points. We did not use geographical data in calculations. The main TV transmitter sites follow the assignment in the GE06 plan, while the relay sites give flexibility to fine-tune the broadcast network on demand. Main transmitters and relays are depicted in Figure 5.3. The parameters used in the WS study, adopted from [21], are outlined in Table 5.2. The received SNR corresponds to 64 QAM and 2/3 code rate for fixed reception.

The Finnish Ministry of Transport and Communications has made the channels in the 470–790 MHz range available in Finland for cognitive radio trials on the condition that the trial systems do not interfere with licensed radio transmission. Figure 5.4 shows the number of free channels within this range. A channel is classified as free if the grid point is outside of the DTT coverage area on the corresponding frequency band. This assumes that a white space device (WSD) does not employ spectrum sensing to detect transmit opportunities, because the sensing threshold should be much lower than the DTT signal strength at cell edge to protect the broadcast service. Thus, a WSD should use a geo-location database to be able to take the benefit from the available free channels in Figure 5.4. As anticipated, the number of potential channels is smaller in more densely populated

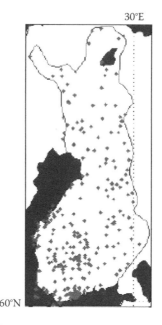

Figure 5.3 DTT transmitters in Finland.

Table 5.2 Parameters Used in the Study

Reception	Fixed
Thermal noise density	–174 dBm/Hz
Receiver noise figure	7 dB
Minimum SNR at cell edge	21 dB
Field strength exceeded in locations	95%
Field strength exceeded in time	50%
Receiver antenna height	10 m

areas, although the amount of available spectrum is still substantial. However, the available capacity in a grid point would depend on several factors, like heights of the transceiver antennas, and use cases, and we do not pursue the study further to avoid introducing additional parameters to the system.

The amount of spectrum given in this study may be overestimated because the protection of the adjacent channels used by DTT, and the protection of the program making and special event (PMSE) are not taken into account. Furthermore, only the main transmitters in the neighboring countries are included in the interference calculations, and the usage of UHF bands in Russia for radionavigation

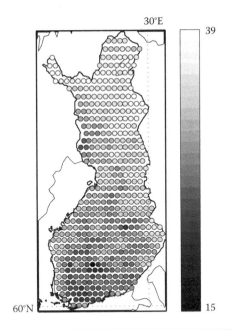

Figure 5.4 Number of free channels within 470–790 MHz in Finland. The mapping of grayscale to the number of channels is shown in the sidebar.

and military purposes is not considered. On the other hand, the study assumes that TV transmitters are continuously broadcasting, which may not always be true. In practice a WSD may have more opportunities to transmit near DTT coverage areas.

In the next section, Finland is estimated to be one of the fastest adopters of WS technology, and a lot of WS spectrum is available in the country. Thus in the future, WS technology may assume a significant role in Finland.

5.4 TV White Spaces Diffusion in Europe

Here we discuss the TV white spaces in the context of diffusion of innovations. We forecast the takeoff of the white spaces in Europe. The analyzed countries are limited to 23 European high-income Organization for Economic Co-operation and Development (OECD) members [22]. The theory is based on the Bass diffusion model [23]. We first estimate the Bass model coefficients by carrying out curve fitting to the World Bank Development Indicators data [24] of Internet users and mobile users. Then we apply the estimated coefficients to the Bass model again and draw the diffusion curve forecasts of the European white spaces diffusion.

A thorough discussion of the innovation diffusion with multiple examples can be found in the Rogers' textbook [25]. He describes the probability density function (pdf) of the adopters to have a bell shape in time. The cumulative distribution function (cdf) of the adoption follows an S-curve, respectively.

Stoneman [26,27], Kiiski and Pohjola [28], Pohjola [29], and Rouvinen [30] estimate the Information and Communication Technology (ICT) diffusion with the Gompertz [31] model. Further ICT diffusion modeling papers include Kros [32], Nurmilaakso [33], and Chandrasekaran and Tellis [34]. The cross-country and cultural ICT modeling has been discussed in Zhu et al. [35]. The geographic dimension has been taken into account in diffusion by Nyblom et al. [36] applying the theory of Moran's I. The impact of social relationships in diffusion have been studied by Young [37] and modeled as a network structure by Cowan and Jonard [38]. Lee et al. [39] summarized technology acceptance models (TAMs), which describe the technology diffusion and acceptance from an individual perspective.

5.4.1 Bass Model

Bass [23] included the separation of innovators and imitators, when he created the Bass model for the diffusion of innovations:

$$S(t) = m \frac{(p+q)^2}{p} \frac{e^{-(p+q)t}}{\left(1 + \frac{q}{p} e^{-(p+q)t}\right)^2} \tag{5.1}$$

where m is the market potential; p and q are the coefficients of innovation and imitation, respectively; t is time, and in this study the unit of time is a year. The coefficient p describes how strongly the consumers adopt through external influence like advertising. The coefficient q represents the adoption through internal influence like learning from the experience of the social peers. The Bass model has been applied in forecasting of diffusion, e.g., by Mahajan and Peterson [40] and Lawrence et al. [41].

5.4.2 Data and Model

The analyzed countries include 23 European high-income OECD members [22] listed in Table 5.3. For each country, we collect the data from World Bank Development Indicators data [24] of Internet users per 100 people (IT.NET.USER.P2) and mobile cellular subscriptions per 100 users (IT.CEL.SETS.P2) The used data begins in 1978 and continues till 2008. We estimate the Bass model coefficients by carrying out curve fitting to the data. An example of the curve fitting can be found in the Figure 5.5.

The mobile industry adoption describes the use of a strictly regulated licensed spectrum and shows how well the government and industry can together adopt new technology. The number of Internet subscribers depends less on regulation. It describes the interest of the inhabitants to try out new technology and the

Table 5.3 European Countries Included in the Study

Austria	AUS
Belgium	BEL
Switzerland	CHE
Czech Republic	CZE
Denmark	DEN
Germany	DEU
Spain	ESP
Finland	FIN
France	FRA
United Kingdom	GBR
Greece	GRC
Hungary	HUN
Iceland	ISL
Ireland	IRL
Italy	ITA
Luxembourg	LUX
Netherlands	NLD
Norway	NOR
Poland	POL
Portugal	PRT
Slovak Republic	SVK
Slovenia	SVN
Sweden	SWE

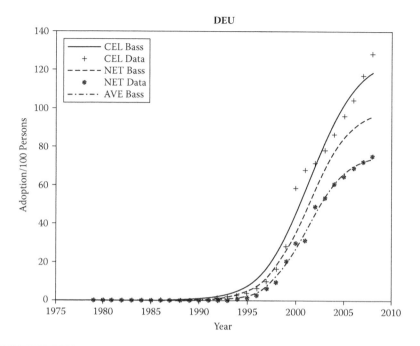

Figure 5.5 The Bass curve fitting of Internet (IT) and mobile (CELL) users and average value of Internet and mobile users (AVE) in Germany (DEU). Discrete values refer to the input data and solid lines to the model output.

reaction speed of the service industry for the demand. TV white space requires changes in the regulation, but if the regulation gives freedom for innovation (e.g., by allowing cognitive use of the radio spectrum without restricting the implementation too much), the Internet service-type development can be expected. Both these statistics represent technology adoption of end users. End user devices such as white space devices are only one way of utilizing the technology. The others include wireless microphones, machine-to-machine communication, and core network links. Some of those areas are more business-to-business applications, and the historical adoption patterns of the end users may not describe the adoption of those use cases well.

The curve fitting is carried out by having m, p, and q as variables in the basic Bass model 1.1. Additionally, the starting year of the curve fitting is optimized with one year steps, and it is marked as delay in the related figures. The curve fitting procedure minimizes the sum of squared differences of the data point value and the Bass curve at that point. The method is also called as least squares. Octave successive quadratic programming (sqp) is used for the optimization [42]. The country based coefficients are determined separately for the mobile cellular (CELL) and Internet (IT) users. The third Bass curve fitting is carried out for the average value

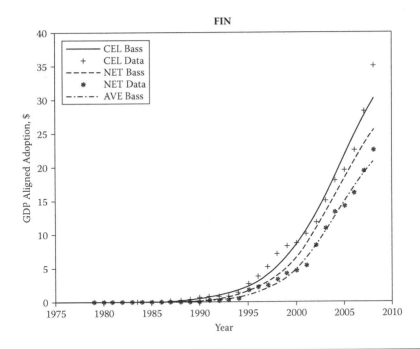

Figure 5.6   GDP-aligned Bass curves of Internet (IT) and mobile (CELL) users and average value of Internet and mobile users (AVE) in Finland (FIN). Discrete values refer to the input data and solid lines to the model output.

of the mobile and Internet users. For forecasting purposes, we multiply the mobile and Internet user numbers with the gross domestic product (GDP) in U.S. dollars (NY.GDP.MKTP.CD) [24]. An example of the curve fitting with GDP can be found in Figure 5.6 for Finland.

The studied countries are mapped in Figure 5.7 according to their innovation (p) and imitation (q) values. Figure 5.8 shows the market potential m is the Bass curve-fitted average saturation level of mobile and Internet users multiplied by the GDP and divided by 10^{12}. The delay (d) describes delay in years from the year 1979. The year 1979 represents the year 0 in the Bass curve equation.

5.5 TV White Spaces Diffusion Forecast

The TV white spaces diffusion forecast is created from the assumption that its diffusion curve is a combination of the mobile user and Internet user diffusion. The absolute forecast values are only estimates, but the curves are expected to be on a correct level relative to each other. The average diffusion forecast curves from the Bass model are grouped into Southern, Western, Northern, and Eastern European

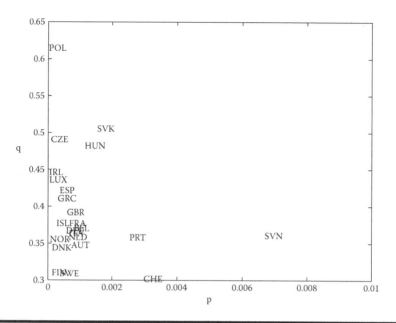

Figure 5.7 Bass coefficients innovation (*p*) and imitation (*q*) as the average of Internet and mobile users. See Table 5.3 for the names of the countries.

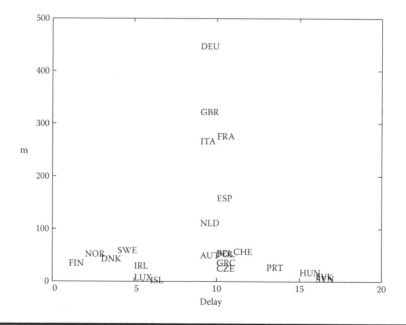

Figure 5.8 GDP-aligned market potential *m* and delay *d* as the average of mobile and Internet users. See Table 5.3 for the names of the countries.

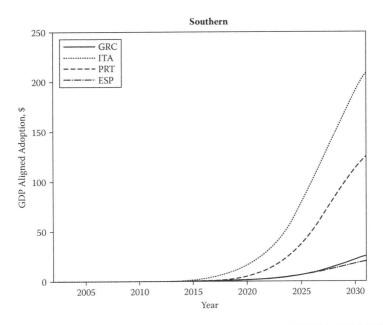

Figure 5.9   Southern Europe TV white spaces diffusion forecast.

countries according to the United Nations geographical classification [43]. The forecast diffusion results are shown for Southern, Northern, Western, and Eastern European countries in Figure 5.9, Figure 5.10, Figure 5.11, and Figure 5.12, respectively. The year 0 of the forecasts of the white space diffusion is 2002 in this study. It is comparable to the year 1979 of the mobile and Internet diffusion statistics.

In the p and q map, Nordic countries have very similar characteristics. Finland and Sweden are on the fastest adopting edge, and Iceland is closer to the Central European countries. The United Kingdom has characteristics of the Central European countries. Ireland is similar to the Eastern European countries. The Central European countries form a homogeneous group excluding Switzerland, which resembles Eastern European countries on the p and q map. The Southern European countries are quite homogeneous, as well. Italy has similarities with Central Europe and Portugal with Eastern Europe. The Eastern European countries are most diverse.

On the delay map (Figure 5.8), Central and Southern European countries, with Poland and the Czech Republic, form into a distinctive group. The Nordic countries, Luxembourg, and Ireland have the shortest delay, and the rest of Eastern European countries and Portugal have the longest delay. The market potential m is headed by Germany and followed by the United Kingdom, France, and Italy. The market potential of Spain and the Netherlands lies below the four countries with the highest market potential (DEU, GBR, FRA, and ITA) and above the rest of the European countries studied.

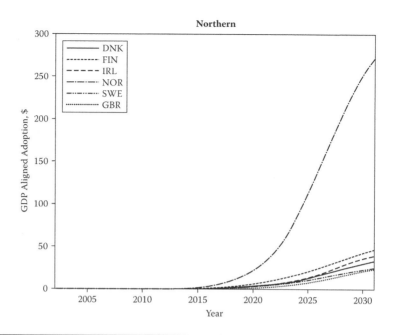

Figure 5.10 Northern European countries' TV white spaces diffusion forecast.

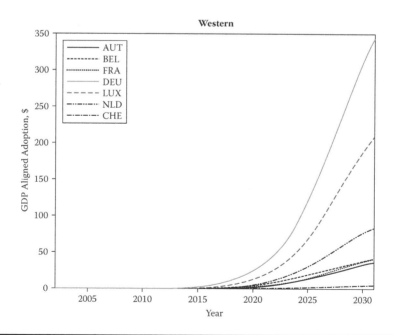

Figure 5.11 TV white spaces diffusion forecast of Western European countries.

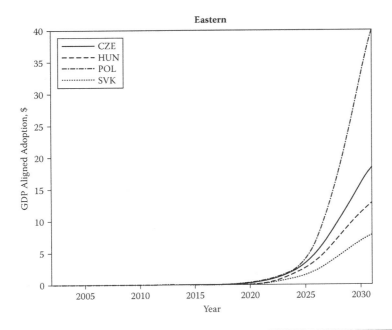

Figure 5.12   Eastern Europe TV white spaces diffusion forecast.

From the GDP-aligned adoption curves, we can observe that the adoption time has only a small impact on the early years of adoption (i.e., trialing and piloting phase). When the adoption begins in the large economies, Germany, the United Kingdom, France, and Italy, they soon shadow the first adopters on the diffusion curves.

The recommendation for the Nordic countries is to develop the white space technology together similar to the Nordic Mobile Telephone (NMT) network. That way they can reach a reasonable critical mass and have a leading edge in the white space technology development. Central Europe will most likely be well synchronized as a result of European Commission regulative work including European Conference of Postal and Telecommunications Administrations (CEPT) and The European Telecommunications Standards Institute (ETSI) standardization bodies. Southern and Eastern Europe should try to keep the coming timetable for white spaces adoption. In TV broadcasting, Germany is represented by its federal states rather than as a single country. This will delay the adoption of white space technology in Germany. A single representative for all federal states could make the adoption faster. Finally, the major investors in the European white spaces technology should follow the development in Germany and in the United Kingdom, and put the highest bets on the four largest European economies.

Europe has less potential white space capacity than the United States. Europe is administrationally very diverse, which potentially spreads the beginning of the adoption over a long period. Europe has also been slower in the regulation and white space trials than the United States. The Nordic countries are flexible as to regulation, and there is plenty of know-how in the communications and radio technologies. Due to a long common border with Russia, the differences between the radio spectrum planning in European Union and in Russia may delay the adoption of white space in Finland. Germany has a high demand of radio spectrum, but the TV spectrum administration in the federal states may slow down the spread of white space technology. Ofcom in the United Kingdom has been one of the most visible and active regulators in the white space area in Europe. The United Kingdom has also been a leading country in the adoption DVB-T2. If the industry is able to respond to the interest of Ofcom in the right way, white space adoption will begin from the United Kingdom into Europe.

5.6 Acknowledgments

The authors would like to express their gratitude to several persons and organizations who helped in providing necessary information and comments—to name a few: Ms. Margit Huhtala, Finnish Communications Regulatory Authority (FICORA) her colleague, Mr. Kari Heiska (Digita) and his colleague, Mr. Pekka Talmola (Nokia) and several others.

References

1. U. Reimers, "DVB, the Family of International Standards for Digital Broadcasting," 2nd edition, Berlin: Springer, 2005, 408 pp.
2. European Telecommunications Standards Institute, "Digital Video Broadcasting (DVB), Framing Structure, Channel Coding and Modulation for Digital Terrestrial Television," ETSI EN 300 744, March 1997.
3. International Telecommunication Union, "ITU: Final Acts of the RRC-06 and Associated Frequency Plans and List," ITU, Geneva June 16, 2006 Available in: http://www.itu.int/pub/R-ACT-RRC.14-2006/en.
4. T. OLeary, E. Puigrefagut, and W. Sami, "GE06—Overview of the Second Session (RRC-06) and the Main Features for Broadcasters," EBU *Technical Review,* October 2006, 20 pp.
5. J. Doeven, "Implementation of the Digital Dividend—Technical Constraints to Be Taken into Account," *EBU Technical Review,* January 2007, 10 pp.
6. E. Wilson, "Future Focus," *DVB-Scene,* Edition No. 21, March 2007, p. 7.
7. Cocom, "Information from Member States on Switchover to Digital TV," EC Report: COCOM09-01, Brussels, January 14, 2009, 12 pp.
8. Analysys Mason, "Report for the European Commission, 'Exploiting the Digital Dividend—A European Approach,'" Final report, August 14, 2009, http://www.analysysmason.com/EC_digital_dividend_study.
9. Digital Video Broadcasting Project, DVB Worldwide, http://www.dvb.org/about_dvb/dvb_worldwide/.

10. European Telecommunications Standards Institute, "Digital Video Broadcasting (DVB), Frame Structure Channel Coding and Modulation for a Second Generation Digital Terrestrial Television Broadcasting System (DVB-T2)," ETSI EN 302 755, September 2009.

11. Commission of the European Communities, Communication from the Commission, "Transforming the Digital Dividend into Social Benefits and Economic Growth," COM(2009) 586 final, Brussels, October 28, 2009.

12. Commission of the European Communities, Commission Recommendation of October 28, 2009: "Facilitating the release of the digital dividend in the European Union," (2009/848/EC).

13. European Commission, Information Society, "eCommunications: Radio Spectrum Policy," http://ec.europa.eu/information_society/policy/ecomm/radio_spectrum/topics/reorg/dividend/index_en.htm, accessed on November 11, 2010.

14. Conference on Postal and Telecommunications Administrations, "A Preliminary Assessment of the Feasibility of Fitting New/Future Applications/Services into Non-Harmonised Spectrum of the Digital Dividend (Namely the So-Called 'White Spaces' Between Allotments)," CEPT REPORT 24, July 1, 2008.

15. Conference on Postal and Telecommunications Administrations, "Technical Roadmap Proposing Relevant Technical Options and Scenarios to Optimise the Digital Dividend, Including Steps Required During the Transition Period Before Analogue Switch-Off," CEPT REPORT 25, July 1, 2008.

16. Conference on Postal and Telecommunications Administrations, "Recommendation on the Best Approach to Ensure the Continuation of Existing Program Making and Special Events (PMSE) Services Operating in the UHF (470–862 MHz), Including the Assessment of the Advantage of an EU-Level Approach," CEPT REPORT 32, October 30, 2009.

17. Electronic Communications Committee of the CEPT, http://www.ero.dk/, accessed on November 11, 2010.

18. European Telecommunication Standards Institute, http://www.etsi.org/website/technologies/RRS.aspx, accessed on November 11, 2010.

19. European Commission Decision 2010/267/EU on harmonised technical conditions of use in the 790–862 MHz frequency band for terrestrial systems capable of providing electronic communications services in the European Union.

20. International Telecommunications Union, "Method for Point-to-Area Prediction for Terrestrial Services in the Frequency Range 30 MHz to 3000 MHz," Rec. ITU-R P.1546-3, 2007.

21. Electronic Communications Committee (ECC) within the European Conference of Postal and Telecommunications Administrations (CEPT), Annex 3 to Doc. SE43(10)103, June 2010.

22. World Bank, "Country and Lending Groups," available in: http://data.worldbank.org/about/country-classifications/country-and-lending-groups#OECD_members, accessed on July 15, 2010.

23. F. M. Bass, "A New Product Growth for Model Consumer Durables," *Management Science,* vol. 15, No. 5, pp. 215–227, 1969.

24. World Bank, "World Development Indicators," available in: http://data.worldbank.org/indicator, accessed on July 15, 2010.

25. E. Rogers, *Diffusion of Innovations*, 5th Edition, New York: Free Press, 2003.

26. P. Stoneman, *The Economic Analysis of Technological Change*, New York: Oxford University Press, 1983.

27. P. Stoneman, *The Economics of Technological Diffusion*, Oxford: Blackwell Publishers Ltd., 2002.

28. S. Kiiski and M. Pohjola, "Cross-Country Diffusion of the Internet," *Information Economics and Policy*, vol. 14, No. 2, pp. 297–310, 2002.

29. M. Pohjola, "The Adoption and Diffusion of ICT across Countries: Patterns and Determinants," *The New Economy Handbook*, San Diego, CA, Academic Press, 2003.

30. P. Rouvinen, "Diffusion of Digital Mobile Telephony: Are Developing Countries Different?" *Telecommunications Policy*, vol. 30, No. 1, pp. 46–63, 2006.

31. B. Gompertz, "On the Nature of the Function Expressive of the Law of Human Mortality, and on a New Mode of Determining the Value of Life Contingencies," *Philos. Trans. R. Soc. London,* vol. 115, pp. 513–585, 1825.

32. J. F. Kros, "Forecasting with Innovation Diffusion Models: A Life Cycle Example in the Telecommunications Industry," in: K. D. Lawrence and M. D. Geurts, *Advances in Business and Management Forecasting,* vol. 5, Jai Press, 2008.

33. J.-M. Nurmilaakso, "Adoption of e-Business Functions and Migration from EDI-Based to XML-Based e-Business Frameworks in Supply Chain Integration," *International Journal of Production Economics,* vol. 113, No. 2, pp. 721–733, 2008.

34. D. Chandrasekaran and G. J. Tellis. "A Critical Review of Marketing Research on Diffusion of New Products," *Review of Marketing Research,* vol. 3. pp. 39–80, 2006.

35. K. Zhu, K. Kraemer, and S. Xu, "Electronic Business Adoption by European Firms: A Cross-Country Assessment of the Facilitators and Inhibitors," *European Journal of Information Systems,* vol. 12, No. 4, pp. 251–268, 2003.

36. J. Nyblom, S. Borgatti, J. Roslakka, and M. A. Salo. "Statistical Analysis of Network Data–An Application to Diffusion of Innovation," *Social Networks,* vol. 25, No. 2, May 2003, pp. 175–195, 2003.

37. P. H. Young, "Innovation Diffusion in Heterogeneous Populations: Contagion, Social Influence, and Social Learning," *The American Economic Review,* vol. 99, No. 5, pp. 1899–1924, 2009.

38. R. Cowan and N. Jonard, "Network Structure and the Diffusion of Knowledge," *Journal of Economic Dynamics and Control,* vol. 28, No. 8, pp. 1557–1575, 2004.

39. Y. Lee, K. A. Kozar, and K. R. T. Larsen, "The Technology Acceptance Model: Past, Present, and Future," *Communications of the Association for Information Systems,* vol. 12, No. 50, pp. 752–780, 2003.

40. V. Mahajan and R. A. Peterson, "Forecast Level of Diffusion: Models for Innovation Diffusion," Sage Publications, Inc. 1985.

41. K. D. Lawrence, D. R. Pai, and S. M. Lawrence, "Forecasting New Adoptions: A Comparative Evaluation of Three Techniques of Parameter Estimation," *Advances in Business and Management Forecasting,* vol. 6, pp. 81–91, 2009.

42. J. W. Eaton. "Octave Nonlinear Programming," available in: http://www.gnu.org/software/octave/doc/interpreter/Nonlinear-Programming.html, accessed on August 22, 2010.

43. United Nations, "Composition of Macro Geographical (Continental) Regions, Geographical Sub-Regions, and Selected Economic and Other Groups," Available in: http://unstats.un.org/unsd/methods/m49/m49regin.htm, accessed on September 2, 2010.

STANDARDS

Chapter 6

Standardization Activities Related to TV White Space: PHY/MAC Standards

Tuncer Baykas, Jianfeng Wang,
M. Azizur Rahman, and Zhou Lan

Contents

6.1 Introduction

The wireless communication industry started standardization activities in TV white spaces (TVWS) as early as 2004 to provide PHY/MAC layer solutions. In this chapter we will look at 4 of those activities as shown in Table 6.1. On the other hand, because TVWS bands will have larger coverage areas, interference between devices of different standards is an important issue. As a result, there are standardization bodies working on standards which enable coexistence and dynamic spectrum access, which are the topic of the next chapter.

Ecma International is an industry association dedicated to the standardization of information and communication systems [1]. In Ecma International, technical committees (TC) are responsible for developing standards. Ecma TC48-TG1 was responsible for developing a cognitive radio networking standard for personal/portable devices in TV white spaces. The standard comprises Physical (PHY) and Medium Access Control (MAC) layers including a protocol and mechanisms for intersystem coexistence. The group completed its activities, and the standard was published at the end of 2009 as Ecma 392 [2].

The IEEE 802 LAN/MAN Standards Committee (LMSC) develops and maintains the IEEE 802 standard family [3]. The committee began their operations in 1980 on wired communications. The focus of the LMSC is the Data Link and physical layers of the Open System Interconnection (OSI) reference model. Some of the most well-known standards from the LMSC are the Ethernet family (802.3), the Token Ring (802.4), the wireless LAN (802.11), and last but not least, the Wireless

Table 6.1 PHY/MAC Layer Standardization Activities Related to the TVWS

Standardization Organization	Responsible Group/Committee	Current Name of the Standard/Project	Type of the Standard	Status/Publication Date
Ecma International	Task Group 1 under Technical Committee 48	ECMA 392 MAC and PHY for Operation in TV White Space	PHY layer and MAC Layer	Published 2009
IEEE 802 LAN/MAN Standards Committee	802.22 Working Group	IEEE P802.22™ Draft Standard: IEEE 802.22/Cognitive Wireless RAN Medium Access Control (MAC) and Physical Layer (PHY) specifications: Policies and procedures for operation in the TV Bands	PHY Layer and MAC Layer	Published 2011
IEEE 802 LAN/MAN Standards Committee	802.11 Task Group af under 802.11 Working Group	IEEE P802.11af™/Draft Standard for Information Technology Telecommunications and information exchange between systems—Local and metropolitan area networks—Specific requirements—Part 11: Wireless LAN Medium Access Control (MAC) and Physical Layer (PHY) specifications Amendment 5: TV White Spaces Operation	PHY Layer and MAC Layer	Draft Standard 2012 (Expected)
IEEE Standards Coordinating Committee 41	Ad Hoc Group on TVWS Radio		PHY and Mac Layer	Submitted Project Authorization Request

MAN (802.16). The work of the LMSC is organized into working groups. A working group may delegate drafting of a standard to one of its task groups. As of 2010, there are 2 working groups under the LMSC with activities related to TVWS PHY/MAC layer design. The first working group is the 802.22, which provided a standard for rural broadband wireless access [4]. The 802.11 working group will bring, through its Task Group af, wireless local area networks into the TVWS spectrum [5]. IEEE Standards Coordinating Committee 41 (IEEE SCC41) is another standardization body of the IEEE which was established as the IEEE P1900 Standards Committee in 2005 [6]. The scope of SCC41 includes new techniques and methods of dynamic spectrum access, interference management, and coordination of wireless technologies. The SCC41 is considering working on a PHY/MAC standard to operate in TVWS band. For this purpose, an ad hoc group on TVWS Radio was created. The ad hoc group prepared a project authorization request, and SCC41 is seeking approval from the IEEE standards association [7]. The outline of the chapter follows the order in Table 6.1. The next section provides the details of Ecma 392 standard. Afterwards two IEEE 802 projects, P802.22 and P802.11af, will be explained. The last section of the chapter is reserved for SCC41's ad hoc group on TVWS.

6.2 Standard ECMA-392 MAC and PHY for Operation in TV White Space

The Ecma 392 standard specifies the PHY and MAC layers for cognitive radio networking for personal/portable devices operating in TV white spaces [8]. It supports flexible network formation with three types of devices: master devices, slave devices, and peer devices. A network can be formed as master–slave or peer-to-peer, as illustrated in Figure 6.1, or as a mesh-network. In a master–slave network, one device is designated as master and the other devices that are associated with the master are referred to as slaves. The master coordinates dynamic frequency selection (DFS), transmit power control (TPC), and channel measurement on behalf of slave devices. A peer-to-peer network is made up of peer devices. Peer devices coordinate DFS, TPC, and channel measurement in a distributed fashion. A peer device is able to communicate directly with any other peer device as long as it is within the range of that peer device. In other words, a peer-to-peer network can be ad hoc, self-organizing, and self-healing.

The interoperability of the three device types is built in due to the fact that all devices follow the same beaconing and channel access protocols. Two or more networks can share the same channel and are also able to communicate with each other. As a result, a number of networks may form a large-scale network such as a mesh-network or a cluster-tree network using a single channel or multiple channels. While not explicitly addressed by the standard, additional support from higher layers will allow multi-hop routing of messages from any device to any other device in the extended network.

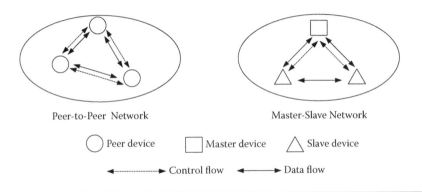

Figure 6.1   Basic network formation.

Regulations require the protection of incumbent users in order to operate in TV white spaces. These incumbent protection regulations may vary from one region to another. The standard specifies a number of incumbent protection mechanisms, including DFS, TPC, and spectrum sensing, that may be adapted based on the regulatory requirements of a particular region. While geo-location/database access is treated as a higher-layer function, and therefore out of the scope of the standard, the standard facilitates the use of information (e.g., available channel list) obtained by the devices to protect incumbents. As an example, for networks operating in the United States under the Federal Communication Commission (FCC) rules, a master device as defined in the Ecma 392 standard will meet the requirements of the FCC-defined Mode II device by including a geo-location function and periodically obtaining an available channels list from an authorized spectrum database via the Internet. All slave devices associated with such a master device will comply with the requirements of an FCC-defined Mode I device. A peer device without access to an authorized spectrum database can act as an FCC-defined sensing-only device. In addition, a peer device that includes the geo-location function and periodically obtains the available channels list from an authorized spectrum database can act as an FCC-defined Mode II device, also enabling other sensing capable devices as Mode I devices.

The Ecma 392 standard aims to serve the robust delivery of high definition video inside home and across multiple rooms, among a broad range of applications. Protocol efficiency and quality-of-service (QoS) provisioning are key design features of Ecma 392. For instance, to support one full high-definition television (HDTV) stream over a TV channel, the effective throughput at the MAC service access point (SAP) shall be at least 19.3 Mbps. Supposing that the maximum PHY layer data rate is 23.74 Mbps on a 6 MHz channel, the protocol overhead including PHY and MAC layer has to be about 19% or less. In addition to effective throughput, the delay jitter and packet loss rate have to be low for real-time video streaming.

Figure 6.2 illustrates in-home multimedia distribution via Ecma 392. In one scenario, the access point (AP) serves as the master device and facilitates the video streaming through triple-play gateway to TV or laptop via master–slave

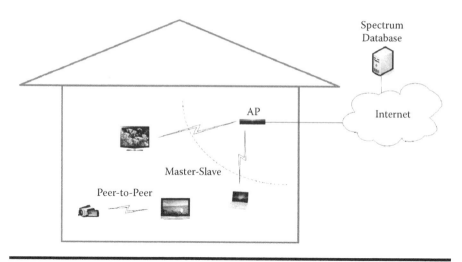

Figure 6.2   In-home multimedia distribution.

communication. The AP connects Internet and determines channel availability by querying spectrum database, for example, based on its home address. In another scenario, the camcorder streams recorded video or slideshow pictures to another TV set using peer-to-peer communication. The camcorder and TV set in the second scenario may determine the channel availability via sensing.

6.2.1 Physical Layer Design and Key Features

The PHY design is based on a 128-point FFT (fast Fourier transform) orthogonal frequency division multiplexing (OFDM) structure. This size was chosen as the best compromise between overhead and complexity. Table 6.2 shows the OFDM parameters and the corresponding values for each channel bandwidth. The various channel bandwidth options use the same 128-point FFT but use a different subcarrier spacing value as shown in Table 6.2. The subcarriers are classified according to their function as data, pilot, or NULL subcarriers.

In all OFDM symbols following the Physical Layer Convergence Protocol (PLCP) preamble, 98 subcarriers among the 102 used subcarriers are allocated for data transmission, and the remaining 4 subcarriers are allocated for pilot signals. The data subcarriers carry the complex constellation points. A group of complex constellations is sequentially mapped to the IFFT inputs from −51 to 51, excluding the inverse fast Fourier transform (IFFT) inputs for pilot and DC subcarriers. The location of the pilot subcarriers varies with symbol number, and the pattern repeats every 13 OFDM symbols. The pilot subcarriers facilitate coherent detection and provide robustness to the transmission system against frequency offsets and phase noise. Null subcarriers include the DC subcarrier and the guard subcarriers. No power is allocated to the null subcarriers. For each OFDM symbol, 25 subcarriers are allocated as guard

Table 6.2   OFDM Parameters

TV channel bandwidth (MHz)	6	7	8
Total number of subcarriers, N_{FFT}	128		
Number of guard subcarriers, N_G (L,DC,R)	26 (13,1,12)		
Number of used subcarriers, $N_T=N_D+N_P$	102		
Number of data subcarriers, N_D	98		
Number of pilot subcarriers, N_P	4		
Sampling frequency (MHz)	48/7	8	64/7
FFT period, T_{FFT}	18.667	16	14
Subcarrier spacing (KHz)	53.571	62.5	71.429
Signal bandwidth (MHz)	5.518	6.438	7.357

subcarriers. These guard subcarriers are located on either edge of the OFDM symbol, with 13 subcarriers on the left band edge and 12 subcarriers on the right band edge.

The standard draws heavily from well-known OFDM-based standards such as 802.11a. However, some key differentiators have been included in order to improve the performance. The enhancements include a forward error correction (*FEC*) scheme using the concatenation of a Reed-Solomon (RS) outer code and a convolutional inner code, two types of preamble structures, an improved retransmission scheme, and support for multiple antennae. In the following, we briefly describe these enhancements.

■ *Concatenated RS-Convolutional Code:* a systematic (245, 255) RS code over GF(256) has been included as an outer code in order to improve the packet error rate performance. The polynomial used is $p(x) = x^8 + x^4 + x^3 + x^2 + 1$. The same code is punctured and truncated to form a systematic (15, 23) code that is used to encode the PLCP header. Thus, a single RS decoder can be used to decode both the PLCP header and the data code blocks. Each RS block is convolutionally encoded with a mother code rate of R = 1/2. A total of 10 transmission modes are supported by combining 3 modulation types (quadrature phase shift keying [QPSK], 16-quadrature amplitude modulation [QAM], and 64-QAM) and 5 coding rates (1/2, 7/12, 2/3, 3/4, and 5/6). Table 6.3 shows the supported data rates in the single antenna mode. Using the 2×2 spatial multiplexing (SM) multiple antenna mode, the data rates could reach 47.48 Mbps.

■ *Two Types of Preamble Structures:* The PLCP preamble is used by the receiver for initial synchronization and channel estimation. Two types of PLCP preambles are defined: normal PLCP preamble and burst PLCP preamble. The

Table 6.3 Ecma-392 Modulation and Coding Schemes

Modulation	Outer Coding	Inner Coding Rate	Data Rate (Mb/s)	Spectral Efficiency (bit/s/Hz)
QPSK	(245,255,5)	1/2	4.75	0.79
QPSK	(245,255,5)	2/3	6.33	1.05
16-QAM	(245,255,5)	1/2	9.49	1.58
16-QAM	(245,255,5)	7/12	11.08	1.85
16-QAM	(245,255,5)	2/3	12.66	2.11
64-QAM	(245,255,5)	1/2	14.24	2.37
64-QAM	(245,255,5)	7/12	16.62	2.77
64-QAM	(245,255,5)	2/3	18.99	3.16
64-QAM	(245,255,5)	3/4	21.36	3.56
64-QAM	(245,255,5)	5/6	23.74	3.96

normal PLCP preamble is used for all the packets in normal mode transmission and for the first packet in burst mode transmission, while the burst PLCP preamble is shorter and used for the second and the subsequent packets in the burst mode transmission to improve efficiency. The PLCP preamble is modulated using binary phase shift keying (BPSK) and uses a cyclic prefix of length 1/8 of the FFT period.

■ *Retransmission Strategy:* to exploit the inherent frequency diversity in the channel, Ecma 392 supports using a different interleaver on the retransmission, with optional soft-combining at the receiver of the original and retransmitted packets. Simulation results show that soft-combining provides up to 7 dB of additional gain. If the receiver does not choose to implement soft-combining, the performance is no worse than that obtained by retransmitting with the same interleaver but with a different scrambler. Hence, this feature allows differentiation at the receiver.

■ *Multiple Antennae Support:* Though the use of multiple transmit antennae is optional, the standard includes "hooks" that would allow devices in the future to implement multiple antennae options (up to 2 antennae) without sacrificing throughput due to increased preamble and header length. This is accomplished as follows:

1. If a transmitter uses two transmit antennae, it transmits a defined short preamble sequence that is orthogonal to the one used when it transmits on only one antenna. All receivers shall be capable of detecting which short preamble was transmitted by correlation. Thus, receivers are capable

of distinguishing a single antenna transmission from a dual antenna one without any additional signaling.

2. When using two transmit antennae, a different long preamble is transmitted that is frequency interleaved over the two antennae (i.e., Antenna 1 transmits only over even subcarrier indices and Antenna 2 over odd subcarrier indices). Again, receivers shall be able to derive the channel estimate from such a preamble.

3. A transmit diversity scheme called frequency interleaved transmit diversity (FITD) is defined where, after coding, interleaving and modulation, the symbols are frequency interleaved over the two antennae as described above. Since receiving such a signal does not require additional complexity, unlike space time block coding (STBC), all receivers shall be able to receive such a signal. When a transmitter uses two antennae, the PLCP header is always transmitted using FITD so that all receivers are capable of receiving it.

4. Assigned bits in the PLCP header indicate the multiple antenna scheme used for the data as FITD, STBC, or SM (spatial multiplexing).

6.2.2 MAC Layer Design

The MAC timing structure is based on recurring superframe. One superframe is composed of 256 medium access slots (MASs). As shown in Figure 6.3, it consists of a beacon period (BP), data transfer period (DTP), and a contention signaling window (CSW). A reservation-based signaling window (RSW) could be appended right after the BP to support signaling between a master and slave devices in a master–slave network. RSW is not needed for a peer-to-peer network. The signaling windows and beacon period are used for sending and receiving critical information for management of network and channel.

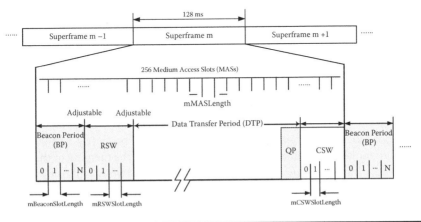

Figure 6.3   MAC superframe structure.

The BP length is adjustable and depends on the number of regular beaconing devices that participate in the same BP. A device is defined as a beaconing device if it owns a beacon slot in BP and regularly transmits beacons. A peer device or a master device is by default a beaconing device. A slave device is normally a nonbeaconing device unless promoted to be a beaconing device. A nonbeaconing device can be promoted to beaconing device to facilitate self-coexistence across neighboring networks. Multiple beacon periods, one from each neighbor network, can merge into one BP to enable efficient sharing of radio resources as well as establishment of an extended quiet zone across neighboring networks.

A beacon packet contains important information for network operation including device identification, beacon slot occupation, medium reservation, traffic indication map (TIM), quiet period (QP) schedule, and channel management. Periodical transmission of network and channel management information by using the scalable multidevice beaconing scheme described above enables easy device discovery, robust slot reservation, reliable channel measurement, and evacuation. Moreover, because the beaconing status of a slave device can be changed on demand, the BP length can be adjusted accordingly to minimize the overhead.

■ *Channel Access Methods:* The Ecma 392 standard supports both channel reservation access (CRA) and prioritized contention access (PCA) during DTP. To guarantee quality of service (QoS), the standard supports various channel reservation types such as hard reservation, soft reservation, and private reservation. A device may reserve MASs per stream by including channel reservation protocol (CRP) information element (IE) via beacon or control message. The reservation status of MASs is exchanged among devices via beacons regularly whereby reservation collisions can be either avoided, or discovered and resolved. All unreserved MASs in the DTP may be used for PCA. The PCA mechanism provides differentiated, distributed contention access to the medium for four access categories (ACs) of frames. The ACs are classified from low priority to high as, background (BC), best effort (BE), video (VO), and voice (VI).

The Ecma 392 standard specifies a number of frame processing mechanisms including frame aggregation, burst transmission, and block-acknowledgment (B-ACK) to support very high efficiency data transmission. These mechanisms, together with channel reservation, can provide the data rate measured at MAC SAP more than 20 Mbps which is sufficient for one HDTV streaming.

■ *Internetwork Self-Coexistence:* Two networks may be closely located or come into range due to mobility. The standard supports the merging of the two networks into one superframe if desired. Note that two neighboring networks can continue to operate as independent networks for channel management and security management. One network can freely move to another channel without disrupting the operation of the other network. Network association and device authentication are also controlled within each network.

6.2.3 Incumbent Protection Mechanisms

Ecma 392 supports sensing [9], database, and a combination of the two. The specific regulatory domain and market determine the features ("tools") required for certification, and the implementers can then make the optimal uses of those features, jointly or separately, to meet the certification requirements as well as the use-case requirements.

A key challenge of sensing is the requirement to detect incumbents reliably at very low signal levels (e.g., −114 dBm). This makes sensing highly susceptible to interference from other unlicensed TV band devices (TVBDs). To prevent such interference, quiet zone is set up such that all unlicensed devices can remain quiet while some of them perform sensing [10]. Ecma 392 allows the establishment of the extended quiet zone using over-the-air multidevice beaconing, network synchronization, and regular quiet period (QP) schedule. Given the QP protection, sensing algorithms still have to be very sensitive and robust. Specific algorithms to meet the regulation requirements are implementation dependent. For example, FFT-based pilot sensing algorithms (informative in the standard) may be used to perform ATSC (Advanced Television Systems Committee) sensing.

Based on sensing, or database, or a combination of the two, channels are classified into six categories: Disallowed, Operating, Backup, Candidate, Protected, and Unclassified. The classification algorithm itself is out of the scope of the standard. However, the classification results should be consistent across devices within a network. In a master–slave network, the master coordinates the channel measurement and classification, while in a peer–peer network, channel measurement and classification are coordinated in the distributed manner.

Ecma 392 specifies the TPC mechanisms to minimize the interference to incumbents. For example, when the incumbents are found in the adjacent channel(s), the transmission power limit will be reduced from 100 mW down to 40 mW. TPC mechanisms are also designed to achieve and maintain good link quality while minimizing the interference to neighboring networks for better self-coexistence. A sender is able to adjust transmission power as well as data rate based on the feedback from its receiver, such as the signal-to-noise-ratio, received signal strength, frame error ratio, or other parameters.

Another challenge for operation in TVWS is to maintain smooth operation during and after incumbent detection. Upon discovery of an incumbent, a device shall suspend data communication and transmit only management messages up to a certain time duration (e.g., 200 milliseconds as per the FCC rules) before evacuating the channel within a limited time (e.g., 2 seconds). On the other hand, the time for a group of devices to resume transmission in a new channel could be long, due to channel scan, device reassociation, and channel reservation reestablishment. To save time in identifying a new channel for operation, Ecma 392 proactively maintains at least one backup channel. The selected backup channels are checked regularly to make sure that they are available and ready for use as soon as needed.

To reduce time for device reassociation and network initialization, Ecma 392 supports copying of network settings such as beaconing status, channel reservation, and security establishment from the old channel to the new channel. Therefore, the devices do not have to go through the steps of joining beacon group, performing authentication, and establishing channel reservation again in the new channel. Certain conditions for using channel copy operation may apply. For example, the new channel should be empty in order to copy the same channel reservations from the old channel to the new channel. In a master–slave network, the master coordinates channel evacuation, while in a peer-to-peer network, any peer device may initiate channel evacuation with the new channel setup parameters preagreed. In the case that the incumbent signal is too strong to allow devices to exchange beacon/control message for evacuation, the devices shall move to the preagreed backup channel after a specified time-out period.

6.3 IEEE 802.22–2011™ Standard: Cognitive Wireless RAN Medium Access Control (MAC) and Physical Layer (PHY) Specifications: Policies and Procedures for Operation in the TV Bands

IEEE 802.22 Working Group started its activities in October 2004, after the FCC's Notice of Proposed Rule Making (NPRM) about TVWS in May 2004 [11]. The first project of the group 802.22 specifies medium access control layer (MAC) and physical layer (PHY) of point-to-multipoint wireless regional area networks. The networks include a professionally installed fixed base station (BS) with fixed and portable user terminals. Wireless rural area broadband service is one of the most suited use cases of TVWS spectrum. TV band channels, especially at the high end of UHF and low VHF spectrum, enjoy good propagation characteristics, reasonable antenna sizes, and relatively low levels of interference from electrical equipments [12]. Considering the cost of providing wired broadband access to large areas with low population density, a wireless solution is preferable. Another important advantage of the rural areas is the availability of vacant TV channels. In metropolitan areas most of the TV spectrum is occupied either by TV stations or by other incumbent devices such as wireless microphones, whereas in rural areas the possibility of finding a vacant channels is much higher. The standard was approved by IEEE NesCom in June 2011 [13].

The IEEE 802.22 Working Group has two other projects that are authorized under its umbrella. The 802.22.1 Standard to Enhance Harmful Interference Protection to Low Power Licensed Devices Operating in the TV Broadcast Bands is complete and issued as an official IEEE Standard [14]. The P802.22.2 Draft Standard on Recommended Practice for Installation and Deployment of 802.22 Systems is at the Working Group Letter Ballot stage.

In 802.22 networks are cellular type [15]. Typical radius of a cell varies between 10 km to 30 km, assuming outdoor directional antennas. Current MAC layer design allows propagation delays of up to 100 km range. Considered frequency range is between 54 MHz and 862 MHz. OFDM with a long cyclic prefix is used to be able to withstand multipath delays of up to 37 microseconds. The 802.22 PHY specifies the use of 2048 subcarriers for maximum utilization of the channel bandwidth, as well as to provide protection against highly frequency-selective channels with deep nulls that are encountered because of long channel impulse response and many multipath reflections. The maximum PHY layer data rate is 22.69 Mb/s, and the minimum rate is 4.54 Mb/s.

The 802.22 BS is capable of supporting up to 512 user terminals, which are also called the Customer Premises Equipment (CPEs). The number of users supported depends upon the bandwidth requirements of each user.. One can easily envision a single 802.22 BS supporting up to 512 CPEs that may be associated, for example, with the power meters of a home or an industrial complex for monitoring and control in the smart grid type of applications, where the bandwidth requirements are small and unsustained. However, given that a TV channel is only 6, 7, or 8 MHz wide, and in most regulatory domains, channel bonding is not allowed, the network manager of a wireless Internet service provider (WISP) needs to determine the number of users that can be supported over a given coverage area to provide the required quality of service (QoS) and adequate user experience.

6.3.1 802.22 Reference Architecture and System Basics

Figure 6.4 illustrates the reference architecture of a 802.22 BS. The 802.22 CPE reference architecture is similar to the 802.22 BS, except that it does not contain an entity called the Spectrum Manager (SM). Traditional reference architectures include data, control, and management planes. The unique feature of the IEEE 802.22 reference architecture is the addition of the cognitive plane, which contains cognitive functions that enable the cognitive radio operation of the 802.22 device. Cognitive plane includes the optional spectrum sensing function (SSF), the geo-location (GL) function, the Spectrum Manager/Spectrum Sensing Automaton (SM/SSA) and a dedicated Security Sublayer 2. SSF implements spectrum sensing algorithms whereas the GL function provides location information of the 802.22 device. The most important entity of the cognitive plane is the Spectrum Manager (SM). The functions of the SM are listed below:

- Accessing the database service
- Scheduling quiet periods for spectrum sensing
- CPE registration and location tracking
- Information fusion of the location dependent available channel list from the incumbent database and the situational awareness gathered as a result of spectrum sensing

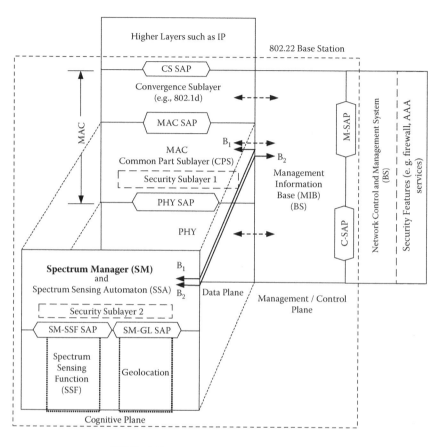

PHY: Physical Layer
MAC: Medium Access Control Layer
CS: Convergence Sublayer
MIB: Management Information Base
PHY SAP: Physical Layer Service Access Point
MAC SAP: Medium Access Control Service Access Point
CS SAP: Convergence Sublayer Service Access Point
SSF: Spectrum Sensing Function
SM: Spectrum Manager
SSA: Spectrum Sensing Automation
SM-SSF SAP: Spectrum Manager, Spectrum Sensing Function Service Access Point
SM-GL SAP: Spectrum Manager, Geolocation Service Access Point
Security Sublayer 1: Security functions for the Data/Control Plane
Security Sublayer 2: Security functions for the Cognitive Plane
NCMS: Network Control and Management System
AAA: Authentication, Authorization and Accounting
M-SAP: Management Service Access Point
C-SAP: Control Service Access Point

Figure 6.4 802.22 WRAN base station reference architecture (© IEEE).

- Channel classification, selection, and channel set management
- Association control
- Enforcing 802.22 and regulatory domain policies
- Making channel move decisions for one or more CPEs or the entire cell
- Self-coexistence with other WRANs

Another unique feature of the IEEE 802.22 Standard is the addition of a Security Sublayer 2 in the cognitive plane that is used to provide protection for the 802.22 systems against denial of service (DoS) attacks for the primary users of the spectrum, as well as the opportunistic users such as the 802.22 systems.

As mentioned above, the reference architecture for the CPE is similar to the one at the BS, the major difference being that the CPE does not contain a Spectrum Manager. SSA exists in the cognitive plane of both CPEs and BSs. They are responsible for running procedures related to sensing and geo-location. They act independently at the initialization of the BS and if a CPE is not registered to a BS. Afterward they are controlled by base station SM. There are two explicit connections at any CPE. The first one is used to communicate with the SM through the data plane, whereas the second one provides connection to local interface via management information bases (MIB). The reference architecture also includes the traditional components such as the PHY layer, the MAC layer, and interface to higher layers through a convergence sublayer. Interfaces to network control management system are provided via MIBs.

Data plane is responsible for carrying information input from the upper layers and also information related to management and control. It includes the PHY layer, the MAC layer, and the convergence sublayer (CS).

The management/control plane includes MIB, which is used for system configuration and monitoring statistics.

6.3.2 802.22 Physical Layer

As mentioned in the introduction, IEEE 802.22 WRANs are designed to provide broadband access in data rates similar to ADSL. The main features of 802.22 PHY are shown in Table 6.4.

Frequency range and channel bandwidth depend on the regulatory domain in which the 802.22 system is operating. Over the world, 6 MHz, 7 MHz, and 8 MHz channelizations are being used. A 802.22 system shall fully occupy one and only one TV channel in a regulatory domain while operating. For the case of the United States, each TV channel has 6 MHz bandwidth; therefore the 802.22 system should use 6 MHz channelization. The system uses orthogonal frequency division multiplexing (OFDM) for uplink and downlink and orthogonal frequency division multiple access (OFDMA) for uplink if multiple CPEs are transmitting to the BS at the same time.

Table 6.4   PHY Layer System Parameters for WRAN

Parameters	Specification
Frequency range	54~862 MHz
	Depends on the regulations
TV channel bandwidth	6, 7, or 8 MHz
	Depends on the regulations
Data rate	4.54 to 22.69 Mbit/s
Multiple access	OFDMA
FFT size (total number of subcarriers)	2048
Number of data subcarriers	1440
Number of pilot subcarriers	240
Number of guard subcarriers	368
Signal bandwidth	5.6 MHz, 6.5 MHz, 7.4 MHz,
	Depends on the regulations
Payload modulation	QPSK, 16-QAM, 64-QAM
Payload FEC	Binary Convolutional Code (Mandatory)
	Convolutional Turbo Code
	Shortened Block Turbo Code
	LDPC
Duplex	Time Division Duplex
Cyclic prefix length	$1/4$, $1/8$, $1/16$, $1/32$ of the symbol duration
Frame size	10 ms
Superframe size	16 frames
Support of multiple antennas	None

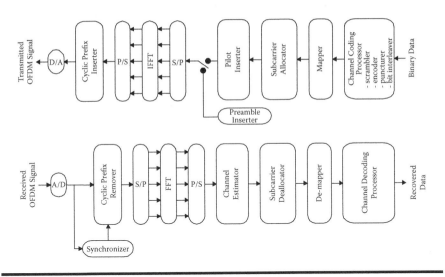

Figure 6.5 **Transmitter and receiver block diagram for the OFDMA PHY.**

The transmitter and receiver block diagram of the PHY layer is shown in Figure 6.5. Transmitted OFDM signal can be written as:

$$s(t) = \mathrm{Re}\left\{ e^{j2\pi f_c t} \sum_{\substack{k=-N_T/2 \\ k\neq 0}}^{N_T/2} c_k e^{j2\pi k\Delta f(t-T_{CP})} \right\}$$

where

t is the time elapsed since the beginning of the current symbol, with $0<t<T_{\text{SYM}}$.

T_{SYM} is the symbol duration, including cyclic prefix duration.

Re(.) is the real part of the signal.

f_c is the carrier frequency.

c_k is a complex number; the data to be transmitted on the subcarrier whose frequency offset index is k, during the current symbol. It specifies a point in a QAM constellation.

Δf is the subcarrier frequency spacing.

T_{CP} is the time duration of cyclic prefix.

N_T is the number of used subcarriers (not including DC subcarrier).

To each OFDM symbol, a cyclic prefix (CP) is added. As shown in Table 6.4, there are four options for CP, $T_{\text{FFT}}/4$, $T_{\text{FFT}}/8$, $T_{\text{FFT}}/16$, and $T_{\text{FFT}}/32$, where T_{FFT} is the time duration of the IFFT output signal.

The IEEE 802.22 defines 16 different PHY modes, in which 12 of them can be used for data communication, as shown in Table 6.5. Modes 1 to 4 are used for

Table 6.5   PHY Modes of IEEE 802.22

PHY Mode	Modulation	Coding Rate	Data Rate (Mb/s) ($T_{CP} = T_{FFT}/16$)
1	BPSK	Uncoded	—
2	QPSK	$1/2$ repetition: 4	—
3	QPSK	$1/2$ repetition: 3	—
4	QPSK	$1/2$ repetition: 2	—
5	QPSK	$1/2$	4.54
6	QPSK	$2/3$	6.05
7	QPSK	$3/4$	6.81
8	QPSK	$5/6$	7.56
9	16-QAM	$1/2$	9.08
10	16-QAM	$2/3$	12.10
11	16-QAM	$3/4$	13.61
12	16-QAM	$5/6$	15.13
13	64-QAM	$1/2$	13.61
14	64-QAM	$2/3$	18.15
15	64-QAM	$3/4$	20.42
16	64-QAM	$5/6$	22.69

transmission of control signals. Modes 5 to 16 are for data transmission and are obtained by combinations of 3 modulation schemes and 4 coding rates. Since there are four different FEC options for each coding rate, 802.22 gives system implementers flexibility to choose best data rate with acceptable complexity and desired robustness against channel conditions including interference.

Among FEC options, binary convolutional coding (BCC) is mandatory. Implementation of convolutional turbo code, shortened block turbo code, and low-density parity check (LDPC) are optional. Main encoder of the BCC operates at a rate of ½ with constrained length 7. The generator polynomials are 171 and 133. Using different puncturing schemes, all data rates are obtained.

There are 3 types of preambles defined in the 802.22 system: the superframe preamble, which is used for synchronization of CPEs to the BS, the frame preamble, which is used for maintaining synchronization, channel estimation, and frequency offset estimation, and the coexistence beacon protocol (CBP) preamble, which is used for CBP detection. All preambles are based on a long training sequence (LTS) and a short training sequence (STS), and their total length is an OFDM symbol with a CP length of $T_{FFT}/4$. LTS is half the length of an OFDM symbol without CP, whereas STS is ¼ of an OFDM symbol without CP. Superframe preamble uses 5 STS, whereas frame preamble uses 2 LTS with a CP consisting of the second half of the LTS. CBP preamble uses 5 STS similar to superframe preamble. The STS used by CMP preamble has low correlation with STS used by superframe preamble to differentiate them. Although preambles provide initial channel estimation, pilot tones are necessary to track changes in the channels. For this purpose, one pilot tone is positioned for every 6 data tones. Positions of the pilot tones are changed from one OFDM symbol to another to enable channel estimation for each subcarrier.

6.3.3 802.22 MAC Layer

The IEEE 802.22 is designed to be a point to multipoint standard; therefore in the MAC layer the BS controls all CPEs and all activities inside the WRAN. All BSs and CPEs have a 48-bit universal MAC address to uniquely define each device and a 12-bit connection identifier that indicates 4096 possible connections within an 802.22 network. Downstream medium access is based on time division multiplexing, whereas upstream is based on demand-assigned multiple access (DAMA) based on orthogonal frequency division multiple access. Timing in 802.22 WRANs is based on superframes, which consist of 16 frames of length 10 ms as shown in Figure 6.6.

In each superframe, frames can be assigned to only one BS (normal mode) or different BSs (coexistence mode). Each BS shall send a superframe preamble, a frame

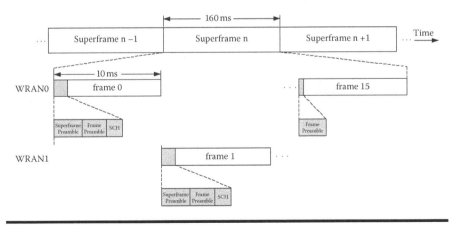

Figure 6.6 General superframe structure in a coexistence mode.

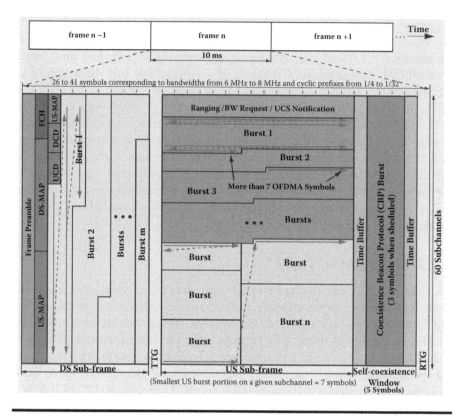

Figure 6.7 MAC frame structure of 802.22.

preamble, and a superframe control header (SCH) at the beginning of the first frame assigned to itself. Afterward the frame header and data payload are sent.

A frame is divided into 3 parts, downstream sub-frame, upstream sub-frame and self-coexistence window as shown in Figure 6.7. The length of the downstream sub-frame depends on the data transmitted by the BS. In any case, BS shall send the frame preamble, upstream and downstream MAPs, which indicate the structure of the upstream data and downstream data, respectively, frame control header, and upstream and downstream channel descriptors. For both upstream and downstream subcarrier allocation is based on subchannels, with 24 data subcarriers and 4 pilot tones, and therefore 60 subchannels are available per OFDM symbol. In downstream, to enable fast channel estimation and frequency diversity gain, pilot subcarriers are set first, and then interleaved data subcarriers are added. Between the downstream and upstream there is a transmit/receive transition gap (TTG). The default value is set to 210 μs to allow propagation delays up to 30 km. In downstream, 2 subchannels are allocated for ranging, bandwidth request, CPE association, CPE link synchronization, power control, geo-location, and urgent coexistence situation (UCS) identification. The rest of them can be used by CPEs for their upstream bursts; each upstream burst shall be at least 7 OFDM symbols in length. In the upstream pilot tones are interleaved with data subcarriers to ensure reliable channel estimation for each CPE burst. After the upstream sub-frame, there could be a 5 OFDM symbol length of self-coexistence window (SCW), in which coexistence beacon protocol bursts are sent from CPEs. A coexistence beacon protocol burst includes information about the backup channel sets, sensing times, geo-location information, and device identification data. One important feature of the self-coexistence window is that BS and CPEs can listen to CBP bursts from other 802.22 systems at different TV bands. Each WRAN announces known SCW patterns so that other WRANs can synchronize and obtain coexistence-related information. After the WCS there is a mandatory receive/transmit transition gap (RTG). RTG is used to align frames at 10 ms and allow propagation delays.

6.3.4 802.22 Cognitive Radio Capabilities

IEEE 802.22 standard defines a comprehensive set of cognitive functions. The cognitive capabilities include mechanisms for incumbent protection, self-coexistence among overlapping 802.22 BSs and SM and SSA operation. There are three major mechanisms for incumbent protection, namely, sensing, database access and specially designed beacon. The standard defines the interfaces for sensing devices, database access and geolocation devices, as well as, lists a good number of sensing algorithms in informative annex. Some of the sensing schemes can sense incumbent signal level up to –120 dBm/8MHz [16]. The standard also defines a number of methods for determining geolocation. The specially designed beacon was developed by IEEE 802.22.1 standard. This beacon can be used by incumbent users. 802.22 devices will be able to receive the beacon and take appropriate action for protection.

To support sensing in the operating channel, 802.22 BS can schedule network wide quiet period (QP). QP in 802.22 can be flexibly scheduled by using two modes: explicit mode where QP can be scheduled by using MAC management message and implicit mode where the schedule is done by using fields in SCH. QP duration can also be flexibly managed by following any of the two available options. Inter-frame sensing is performed where the QP is less than a frame duration and initiated between two frames. Conversely, in inter frame sensing, the QP is longer than a frame duration but shorter than a superframe duration and initiated inside a frame by splitting it in to two pieces. In self-coexistence mode where there could be a number of 802.22 BSs with overlapped coverage area, the QPs of the BSs are synchronized.

In IEEE 802.22 standard, channels are classified in a number of groups which is somewhat identical to what was discussed in the previous section of ECMA 392. More importantly, the standard defines mechanisms for channel management and treats actions related to channel management with high priority. Both BS and CPEs support all channel management actions that could be done either in embedded or explicit manner. Embedded management is done through downstream channel description (DCD) that provides advantage in better spectrum usage and easier control of all CPEs simultaneously. By contrast, explicit management is done through MAC messages and provides flexibility of unicast, multicast or broadcast. Termination of using a channel, channel switching, QP scheduling etc. can be achieved through these messages.

Two unique design features of 802.22 are that the OFDM frame contains urgent coexistence situation (UCS) slots and a self-coexistence window (SCW). The UCS slots are useful in the sense that if an incumbent is detected by a CPE, which didn't receive any upstream allocation from the BS can still inform the BS about the incumbent by transmitting generic MAC header during the UCS slots. The SCW provides opportunity for multiple 802.22 BSs to talk with each other to resolve self-coexistence issues if their coverage areas are over lapped. To facilitate self-coexistence, IEEE 802.22 standard defines coexistence beacon protocol (CBP) that is composed of spectrum etiquette and on demand frame contention (ODFC). Spectrum etiquette is the courteous method of selecting non-interfering operating channels by neighboring 802.22 BSs. ODFC is the method of sharing frame duration within a superframe by two or more synchronized 802.22 BSs operating in the same channel.

As mentioned earlier, the SM is the 802.22 BS controls all the cognitive functions. The 802.22 standard defines details policies and procedures for operation of the SM, as well as, SSA. The standard supports transmit power control on a link-by-link basis. Techniques such as shutting down certain CPEs due to interference issues can be easily implemented.

6.4 IEEE P802.11af™/Draft Standard for Information Technology Telecommunications and Information Exchange between Systems—Local and Metropolitan Area Networks—Specific Requirements—Part 11: Wireless LAN Medium Access Control (MAC) and Physical Layer (PHY) Specifications Amendment 5: TV White Spaces Operation

The 802.11 standard is one of the most successful standards under the 802 LAN/MAN standardization committee. When the initial standard was published in 1997, as the first wireless network standard from IEEE 802, the main design goal was realizing wireless local area networks up to 100 meters in range [17]. In less than a decade, 802.11 enjoyed a remarkable commercial success, and 802.11 chipsets are part of not only computers but also smart phones and cameras.

The initial 802.11 standard was supporting 3 separate PHY layers. The first one was based on infrared communication. The other two radio frequency PHY layers were at the 2.4 GHz band, one based on frequency hopping spread spectrum and the other based on direct sequence spread spectrum. All of them were providing a mandatory data rate of 1 Mb/s and an optional data rate of 2 Mb/s. Afterward, new amendments to the main standard added new properties such as security, transmit power control, operation in new spectrum bands, new regulatory domains, and at higher data rates [17]. In 2007, 802.11 Working Group published IEEE 802.11-2007, which combines and supersedes all previous amendments [18]. After 2007 the group continued to expand and improve the 802.11 standard. In 2010 a project authorization request with the scope "An amendment that defines modifications to both the 802.11 physical layers (PHY) and the 802.11 Medium Access Control Layer (MAC), to meet the legal requirements for channel access and coexistence in the TV White Space" put 802.11 in the TVWS world [5]. During its first meeting, the group approved TGaf purpose, principles, and vision/outcome [19], and decided to use the OFDM PHYs of 802.11 to specify the basis for a system that the regulators can approve for operation in the TVWS bands. The important additions to 802.11 for operation in TVWS are:

1. Channelization
2. Channel availability check

Channelization is required to enable OFDM PHYs operation in TVWS. The MAC layer introduces mechanisms to allow the client device to obtain available channels and operate in TVWS under the regulatory requirements. To get into the details of enablement, we first provide system basics of 802.11 and its reference architecture.

6.4.1 System Basics and Reference Architecture

An 802.11 LAN has a cellular structure which is divided into basic service sets [18]. There are two modes of basic service sets. The first one is the independent basic service set, in which 802.11 wireless stations (STAs) communicate to each other as an ad hoc network. The second one is the infrastructure mode in which access points (APs) enable connection to a distribution system. An AP acts as a master to control the STA within a basic service set (BSS). A distribution system can be wired or wireless. Using the distribution system, multiple BSSs can be interconnected to create an extended service set. An extended service set is seen by upper layers as a single 802 network and enables STAs to roam between different APs. Beyond allowing roaming, APs may provide beaconing, power management, and data traffic management. Normal 802.11 traffic flows through APs even if it is between two STAs of the same network.

The 802.11 reference model is shown in Figure 6.8 [18]. IEEE 802.11 systems address the physical (PHY) layer and the medium access control (MAC) sublayer

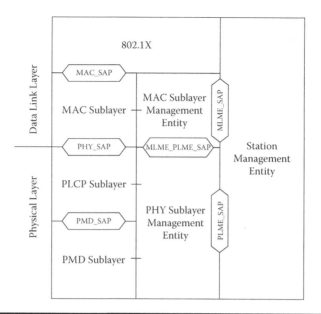

Figure 6.8 IEEE 802.11 reference model.

of the data link layer. The PHY layer consists of physical layer convergence protocol (PLCP) sublayer and physical medium-dependent (PMD) sublayer. Because 802.11 is based on one common MAC layer and multiple PHY layers, PLCP layer is introduced to enable IEEE 802.11 MAC to operate with minimum dependence on the PMD sublayer. The PMD sublayer provides a transmission interface used to send and receive data between two or more LAN devices. There are sublayer management entities, which enable communication with the station management entity. MAC_SAP provides the link between 802.11 MAC and 802.1X, which is the IEEE standard for wired and wireless LAN access control. It provides authentication and authorization devices attached to a LAN.

6.4.2 802.11 MAC Layer Basics

The 802.11 MAC defines two modes of medium access, mandatory distributed coordination function (DCF) and optional point coordination function (PCF). DCF is mandatory mode and based on CSMA/CA (carrier sense multiple access with collision avoidance) protocol. STAs contend for access and attempt to send frames when there is no other station transmitting. If another station is sending a frame, stations wait until the channel is free. After the transmission an acknowledgment frame is sent by the receiver to indicate success. The problem with this method is, although the transmitter can sense its environment, it cannot sense the whole area of the BSS, and packets from different transmitters can collide and reduce efficiency. To reduce this probability, RTS and CTS frames are used. An

RTS (request to send) frame is sent by the transmitting STA. The destination STA replies with a CTS frame (clear to send). RTS and CTS frames include information about the duration of the communication. Upon receiving either RTS or CTS, remaining devices defer accessing the medium. In point coordination function, APs allow individual STAs to communicate. Some other MAC layer functions include scanning for APs, authentication process, and data fragmentation.

6.4.3 802.11 Physical Layers Which Will Be Supported by 802.11af

6.4.3.1 OFDM PHY

The OFDM PHY is first designed to operate at 5 GHz band, and it supports three operation modes which are 20 MHz, 10 MHz, and 5 MHz channel spacing. It can provide data rates of up to 54 Mb/s at 20 MHz channel spacing. The support of transmitting and receiving at data rates of 6, 12, and 24 Mb/s is mandatory. As shown in Table 6.6, OFDM PHY is based on FTT length of 64. Among 64, 52 of

Table 6.6 PHY Layer System Parameters for OFDM PHY

Parameters	Specification		
FFT size (total number of subcarriers)	64		
Number of data subcarriers per OFDM symbol	48		
Number of pilot subcarriers	4		
Number of guard subcarriers	12		
Payload modulation	QPSK, 16-QAM, 64-QAM		
Payload FEC	Convolutional coding		
	Coding rates $1/2$, $2/3$, $3/4$		
Channel spacing	20 MHz	10 MHz	5 MHz
Subcarrier frequency spacing	0.3125 MHz (= 20 MHz/64)	0.15625 MHz (= 10 MHz/64)	0.078125 MHz (= 5 MHz/64)
FFT/IFFT period	3.2 µs	6.4 µs	12.8 µs
Guard interval length	0.8 µs	1.6 µs	3.2 µs
OFDM symbol duration	4 µs	8 µs	16 µs
Support of multiple antennas	No		

Table 6.7   PHY Modes of 802.11 OFDM PHY Layer

PHY Mode	Modulation	Coding Rate	Data Rate (Mb/s) (20 MHz)	Data Rate (Mb/s) (10 MHz)	Data Rate (Mb/s) (5 MHz)
1	BPSK	$1/2$	6	3	1.5
2	BPSK	$3/4$	9	4.5	2.25
3	QPSK	$1/2$	12	6	3
4	QPSK	$3/4$	18	9	4.5
5	16-QAM	$1/2$	24	12	6
6	16-QAM	$3/4$	36	18	9
7	64-QAM	$2/3$	48	24	12
8	64-QAM	$3/4$	54	27	13.5

the subcarriers are used for data and as pilot tones. Modulation schemes of OFDM PHY are BPSK, QPSK, 16-QAM, or 64-QAM. Forward error correction coding is based on convolutional coding with coding rates of $1/2$, $2/3$, or $3/4$. Different data rates supported by OFDM PHY are shown in Table 6.7.

The convolutional encoder shall use the industry-standard generator polynomials, g0 = 1338 and g1 = 1718, of rate R = $1/2$. Using different puncturing methods, rates of $2/3$ and $3/4$ are obtained.

Figure 6.9 shows the frame format of the OFDM PHY. It contains PLCP preamble which is 4 OFDM symbols long. It is followed by a SIGNAL field of an OFDM symbol duration and a DATA field which has a variable length.

The preamble in OFDM PHY consists of 10 short preamble symbols, which are used for detection of the 802.11 packets, automatic gain control, synchronization, and coarse frequency offset estimation. There are also 2 long symbols, which are used for channel and fine frequency offset estimation. The signal field includes information about the data rate and packet length. It is always modulated with BPSK and rate ½ coding. Data field includes a service field which gives information

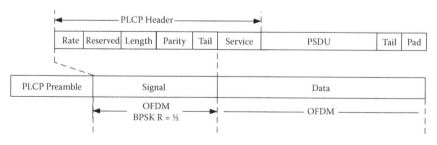

Figure 6.9   Frame format of OFDM PHY.

about the scrambling used and the transmitted information data, which is called a PSDU (PLCP Service Data Unit). Because convolutional coding is used in 802.11 OFDM PHY, both signal field and data field include tail bits to put the encoder to zero state. Data field may include pad bits to make the size of the data a multiple of OFDM symbols.

6.4.3.2 High Throughput PHY (802.11n)

After the initial success of the 802.11 standards, in 2002 802.11 WG started discussion on improving the data rates [20]. The High Throughput (HT) study group was formed by the 802.11 WG. HT SG prepared a project authorization request, which states the new MAC/PHY layer standard shall enable a MAC SAP throughput rate of 100 Mbps [21]. Subsequently the 802.11n Task Group was formed, and development of the standard began. After 7 years of development, in 2009, the 802.11n standard was approved [22]. The PHY layer of 802.11n is named High Throughput (HT) PHY [21]. It is based on the OFDM PHY defined in the previous clause. Main extensions to OFDM PHY are operation in 40 MHz bandwidth and support of multiple antenna technologies. As a result, at PHY layer data rates up to 600 Mb/s are available. We will look at the details of the HT OFDM PHY, starting with the PLCP layer. There are 3 formats that PLCP can use to format PPDU:

1. *Non-HT format* (**NON_HT**): This format has full compatibility with legacy 802.11 devices. When operating in this mode, the frames are all in legacy format. The device cannot use multiantenna modes and 40 MHz channels. This format is exactly the same as OFDM PHY. FFT length is 64. Total number of data and pilot subcarriers is 52. Multiantenna transmission is not available. Support of this mode is mandatory.
2. *HT-mixed format* (**HT_MF**): This format is for networks with an HT PHY access point and a mixed environment of HT PHY devices as well as legacy devices. There is a full legacy preamble, followed by the option to be either HT or legacy format. The preamble allows legacy clients to detect the transmission, and acquire the carrier frequency and timing synchronization. A legacy signal field (L-SIG), which is the same as the signal filed in OFDM PHY, allows them to estimate the length of the transmission. Operation can be in 20 MHz bandwidth or 40 MHz bandwidth. During **HT_MF** operation in 20 MHz channels, FFT length is 64. In this case 56 subcarriers are used. The 40 MHz bandwidth is created by using two adjacent 20 MHz channels. The broadcast and other control frames are sent in each legacy 20 MHz channel to allow the legacy devices to interoperate. In 40 MHz operation of the **HT_MF** case the FFT size is 128. Total number of data and pilot subcarriers is 102.
3. *HT-greenfield format* (**HT_GF**): This format does not contain non-HT-compatible parts. It operates in 40 MHz mode. This mode enables highest data

rate because it only operates in 40 MHz bandwidth and assigns 114 subcarrier to data and pilot tones.

Basic system parameters related to OFDM symbol for different modes of HT PHY are given in Table 6.8.

The data and pilot subcariers of HT PHY are modulated using the same modulation schemes of OFDM PHY: BPSK, QPSK, 16-QAM, or 64-QAM. In forward error control coding, on top of $^1/_2$, $^2/_3$, and $^3/_4$ rate binary convolutional codes of OFDM PHY a new rate ($^5/_6$) is added. For some applications more robust performance is necessary, and an optional LDPC encoder supports all data rates supported by convolutional codes. Before giving a detailed explanation of the support of multiantenna systems, we would like to provide PHY modes available with single antenna transmission in Table 6.9. MCS 0–7 are created by using HT_MF format and HT_GF formats for 40 MHz operation. MCS32 is the most robust MCS mode which operates in 40 MHz bandwidth, by repeating the same data subcarrier in upper and lower 20 MHz bands.

802.11 HT PHY uses three different multiantenna technologies: MIMO, transmit beamforming, and space time block codes.

6.4.3.3 MIMO Technology in HT PHY

HT PHY uses MIMO (multiple input, multiple output) technology to achieve its highest data rates in PHY. In MIMO, the transmitting and receiving stations each have multiple antennas and RF chains. HT PHY enables up to 4 transmit and receive antennas. From each transmit antenna a spatial stream can be transmitted. The term *spatial stream* is defined in the standard as one of several bit streams that are transmitted over multiple spatial dimensions created by the use of multiple antennas at both ends of a communications link. As a result, PHY data rates shown in Table 6.9 can be quadrupled. However, due to complexity and cost issues, only support for 2 antennas at APs is mandatory in HT PHY.

6.4.3.4 Transmit Beamforming

Transmit beamforming is an optional technique used in HT PHY to improve the received signal-to-noise ratio at the reception end. Such an increase may support higher data rates. It is achieved by carefully controlling the phase (creating a channel steering matrix) of the signal transmitted from multiple antennas. For such control, a transmitter should know the channel state between each transmitter and receiver antenna. Implicit and explicit feedback mechanisms are used to characterize the channel. In implicit feedback, it is assumed that the channel from the beamformer to beamformee is reciprocal to the channel from beamformer to beamformee. The beamformer uses the channel training symbols from the beamformee to calculate transmit steering matrix. In explicit beamforming, the beamformer first sends training symbols to the

Table 6.8   OFDM Related System Parameters for HT-PHY

Parameters	*Non-HT Format for 20 MHz Operation*	*HT_MF Format for 20 MHz Operation*	*HT_GF Format for 40 MHz Operation*	*Non-HT Duplicate and MCS32**
FFT size (total number of subcarriers)	64	64	128	128
Number of data subcarriers per OFDM symbol	48	52	114	48 (replicated)
Number of pilot subcarriers	4	4	6	4 (replicated)
Number of guard subcarriers	12	8	8	12 (replicated)
Subcarrier frequency spacing	0.3125 MHz (= 20 MHz/64)	0.3125 MHz	0.3125 MHz (= 40 MHz/128)	0.3125 MHz
FFT/IFFT period	3.2 μs	3.2 μs	3.2 μs	3.2 μs
Guard interval length (GI)	0.8 μs	0.8 μs/0.4 μs (optional)	0.8 μs/0.4 μs (optional)	0.8 μs
OFDM symbol duration	4 μs	4 μs/3.6 μs (optional)	4 μs/3.6 μs (optional)	4 μs
Support of multiple antennas	No	Yes	Yes	No

* MCS 32 and non-HT duplicate are special formats, in which data numbers and pilot values are replicated in upper and lower 20 MHz portions of 40 MHz signal to make a total of 104 subcarriers.

Table 6.9   PHY Modes of 802.11 HT PHY Layer for Single Antenna Transmission

MCS Index	Modulation	Coding Rate	Data Rate (Mb/s) (20 MHz)		Data Rate (Mb/s) (40 MHz)	
			0.8 μs Guard Interval	0.4 μs Guard Interval	0.8 μs Guard Interval	0.4 μs Guard Interval
0	BPSK	$1/2$	6.5	7.2	13.5	15
1	QPSK	$1/2$	13	14.4	27	30
2	QPSK	$3/4$	19.5	21.7	40.5	45
3	16-QAM	$1/2$	26	28.9	54	60
4	16-QAM	$3/4$	39	43.3	81	90
5	64-QAM	$2/3$	52	57.8	108	120
6	64-QAM	$3/4$	58.5	65	121.5	135
7	64-QAM	$5/6$	65	72.2	135	150
32	BPSK	$1/2$	—	6	6.7	

beamformee. The beamformee estimates the channel and sends it back to beamformer. The beamformer calculates steering matrix according to received information.

6.4.3.5 Space Time Block Coding

Space time block coding is another optional technique in HT PHY, which is used to improve robustness of the link between tranmitter and receiver. A spatial stream is sent from two transmit antennas using a special coding method. At the receiver side, signals from two transmit antennas can be combined to obtain a higher received signal-to-noise ratio.

All these MIMO systems are part of the PHY layer; however, they require extensive addition to MAC layer. We mentioned that the scope of the 802.11n task group stated the MAC throughput should be over 100 Mbps. During the standardization, it is shown that basic 802.11 MAC cannot reach that level even if the PHY layer provides 600 Mbps. The reason is the overhead from MAC and PHY layers become dominant compared to diminishing the packet lengths.

6.4.3.6 MAC Layer Enhancements

802.11n Task Group approved several important MAC changes to break the 100 Mbps barrier. Two forms of aggregation were added to the standard: MAC application protocol data unit (APDU) aggregation and MAC service data unit aggregation (A-MSDU). In order to accommodate both forms, the maximum length accepted by the PHY is increased from 4095 bytes in previous standards to 65535 bytes.

In the A-MSDU, multiple packets from higher layers are combined by the MAC layer. Each original packet is converted to a sub-frame within the aggregated MAC frame, including a sub-header containing source and destination addresses and length. Thus this method can be used for packets with differing source and destination addresses.

The second method, *Aggregated* Mac Protocol Data Unit (*A-MPDU*), allows concatenation of MPDUs into an aggregate MAC frame. Each individual MPDU is encrypted and decrypted separately. A header which includes length, cyclic redundancy check (CRC), and signature fields is prepended to each MPDU. This allows the receiver to skip corrupted MPDUs if necessary. On the other hand, because MPDUs are packed together, this method cannot use 802.11's legacy per-MPDU acknowledgment mechanisms. For A-MPDU a new block acknowledgment function is developed. The transmitter can send a burst of frames either aggregated or separately, and the receiver sends a Block Ack, which includes a bit-map to acknowledge each outstanding frame. This allows selective retransmission. Block Ack also enables reduced interframe spacing because switching from transmitter to receiver is not needed to obtain ACK frames. This ends the overview of PHYs which will be supported by P802.11af. Now we will look at special features in P802.11af for TVWS operation.

6.4.4 Channelization for 802.11af

802.11 TG af decided to focus first on channelization in North America, in which the TV band channels are 6 MHz wide. The task group decided to support 5 MHz, 10 MHz, 20 MHz, and 40 MHz channelization [23]. 5 MHz channels are placed at the center of TV bands, and 10 MHz channels are set at the center of two contiguous TV bands. This configuration is selected to enable 802.11af operation in urban areas in which availability is limited. Figure 6.10 shows examples of both cases.

For rural areas in which most of the TVWS is available, the TG focuses on HT PHY and its 40 MHz channels. HT PHY has the channelization strategy in which

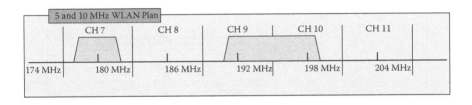

Figure 6.10   **Optimized channelization for 5 MHz and 10 MHz bands.**

40 MHz channels do not overlap. 802.11af selected 3 frequency bands: 520–560 MHz, 560–600 MHz, and 640–680 MHz. The 600–640 MHz band cannot be used because the 608–614 MHz band is reserved for radio astronomy, and TVWS operation is not allowed. Those bands are divided into two to create 20 MHz channels,. and each 20 MHz channel is divided into two to create 10 MHz channels. As a result not only HT PHY but also OFDM PHY systems can operate in those bands.

6.4.5 Channel Availability Check

For P802.11af, traditional 802.11 MAC has to be enhanced regarding the management methods to guarantee the operation of 802.11 systems will not affect the primary services. The management method, namely channel availability check, allows 802.11 systems to obtain a list of available channels that are not occupied by the primary services according to their current locations (Figure 6.11).

According to the rules, a personal portable Mode II device is capable of accessing the database and obtaining the list of the available channels. Another function of a personal portable Mode II is to provide an enabling signal that initiates the network. As shown in the figure, an 802.11 AP may serve in the role of a personal portable Mode II device. The geo-location information of an AP is known to the TVWS database by either an administrative method or GPS. The TVWS database provides the AP with the list of available channels to operate based on the geo-location information. AP as a master device sends out beacon as an enabling signal to initiate the operation of the network. 802.11 stations are personal portable Mode I devices

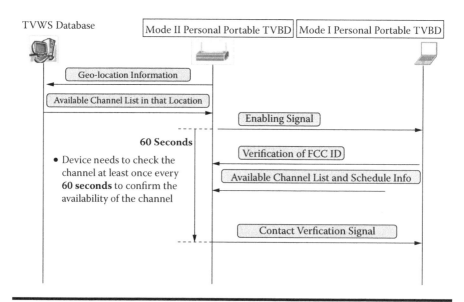

Figure 6.11   Channel availability check procedure.

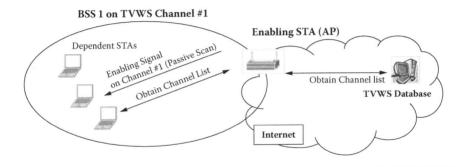

Figure 6.12   Deployment scenario 1.

according to the rules. They cannot release energy unless they are allowed to do so by a personal portable Mode II device. Upon receiving an enabling signal, the 802.11 stations may transmit frames for management under the maximum power constraints indicated in the enabling signal. They can perform data transmission only after obtaining a list of available channel from a personal portable Mode II or a fixed device.

In this work we will look into the details of 4 deployment scenarios for 802.11 systems in TVWS. In deployment scenario 1, the STAs and the enabling STA, which is an AP, are operating on the same TVWS channel, which is illustrated in Figure 6.12. The AP directly accesses the TVWS database to obtain the list of available channels with the means specified by the TVWS database implementer. The AP transmits the enabling signal required by the regulation on a TVWS channel to initiate the network. The STAs perform passive scans of all the supported TVWS channels to be able to receive the enabling signal. Upon receiving the enabling signal, the STAs may attempt to obtain the list of available channels from the enabling STA on the TVWS channel where the enabling signal is received. The frame transmission and reception shall conform to the maximum transmission power indicated in the enabling signal.

In deployment scenario 2, the AP does not have direct access to the database. It becomes a dependent AP in this case (Figure 6.13). It needs to access a TV bands (TVB) enabler to obtain the list of available channels. TVB enabler is instantiated on a server provided by the operator. The dependent AP accesses the TVB enabler in out-bands. FCC rules allow a Mode II or fixed device to obtain the list of available channels through another Mode II or fixed device.

Deployment scenario 3 considers that the BSS may operate on a TVWS channel that is different from the TVWS channel that the enabling STA is operating on (Figure 6.14). The AP that initiates the BSS is a dependent AP, because its operation needs to be under the control of the enabling STA. Both the enabling STA and the dependent AP are personal portable Mode II devices and are allowed to transmit enabling signals. STAs need to passively scan all the supported TVWS channels to receive an enabling signal. The enabling signal can be received on either the TVWS

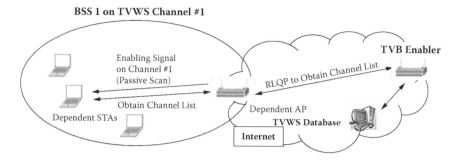

Figure 6.13   Deployment scenario 2.

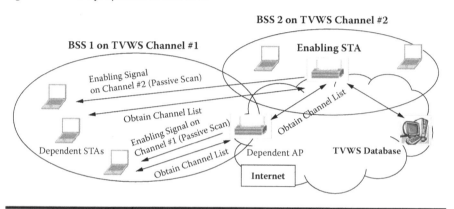

Figure 6.14   Deployment scenario 3.

channel of the dependent AP or on the TVWS channel of the enabling STA. The STAs may attempt to obtain the list of available channels from either the dependent AP or the enabling STA, depending on from which device the enabling signal is received. The frames shall be exchanged on the TVWS channel where the enabling signal is received. The transmission power lever shall be constrained by the maximum transmission power indicated in the enabling signal. Because the dependent AP does not have direct access to the database, it shall rely on the enabling STA to obtain the list of available channels from the TVWS database. The geo-location of the dependent AP shall be known to the enabling STA. It is possible that the AP on the other TVWS channel is a dependent AP that does not have direct access to the TVWS database. It shall rely on a TVB enabler provided by the operator for TVWS database access. The geo-location of the dependent AP shall be known to the TVB enabler.

In the last deployment scenario for enablement (Figure 6.15), the STAs and enabling STA have 2.4 GHz or 5 GHz air interface that allows the enabling signal to be transmitted on 2.4 GHz or 5 GHz bands. The BSS, however, operates on the TVWS bands. The AP that initiates the BSS on the TVWS bands is a dependent AP. STAs may receive enabling signal from TVWS channel or 2.4/5 GHz bands

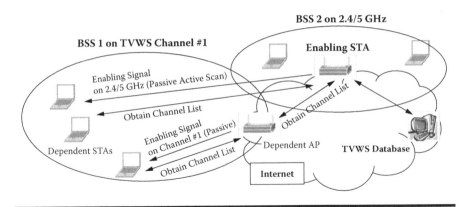

Figure 6.15 Deployment scenario 4.

after scanning the supported TVWS channels and 2.4/5 GHz channels. Both passive and active scan are allowed in 2.4 GHz bands, while only passive scan is allowed in TVWS channels and 5 GHz bands. STAs may attempt to obtain the list of available channels from the enabling STA on 2.4/5 GHz if the enabling signal is received on 2.4/5 GHz bands. STAs may attempt to obtain the list of available channels from the dependent AP on TVWS bands if the enabling signal is received on TVWS bands. The dependent AP relies on the enabling STA for TVWS database access. The geo-location of the dependent AP shall be known to the enabling STA. It is possible that the AP on the 2.4/5 GHz is a dependent AP that does not have direct access to the database. It shall rely on a TVB enabler provided by the operator for TVWS database access. The geo-location of the dependent AP shall be known to the TVB enabler.

6.5 The IEEE SCC41 Ad Hoc Group on TVWS Radio

On March 8, 2010 the ad hoc group on White Space Radio was created within IEEE SCC41 to consider interest in, feasibility of, and necessity of developing standard defining radio interface (MAC and PHY layers) for white space communication systems. Figure 6.16 illustrates the approach considered in the SCC41 WS Radio group. While a new radio interface and new radio interface control and management will be developed, for white space management existing standards of IEEE SCC41 will be used. Device management, sensing, and geo-location are considered out of scope by the group.

The group discussed the following topics:

- Usage models [24]
- System requirements [25]
- Collaboration with WS networking/management standards
- Potential frequency bands for WS Radio [26]

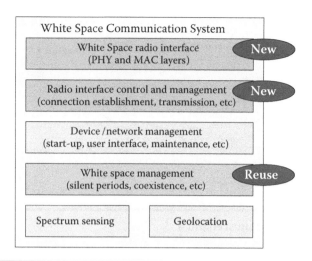

Figure 6.16   Approach in SCC41 WS Radio.

Major usage models for the standard are: "Wide Area Connectivity," "Transportation Logistics," "Land Mobile Connectivity," "High Speed Vehicle Broadband Access," and "Maritime Connectivity." To support the above usage models several require-ments are determined. For example, the system should minimize complexity of the architecture and protocols and avoid excessive system complexity. Both cellular and point-to multipoint topologies with mesh and relay extensions should be supported. For high-power device networks the system shall be optimized up to 5 km, whereas low-power device networks shall have optimized range of 100 m. Handover and other mobility-related procedures should be supported. Systems shall be able to achieve a mandatory spectral efficiency of at least 1 bit/s/Hz. Mechanisms to achieve power-efficient communications such as power control, power saving mode, and sleep modes shall be supported.

Because the system will operate in a white space environment, a means to protect incumbent users shall be provided. Channelization of the system shall be able to support operation in different regulatory domains. Because the standard is from SCC41, it should be able to support other SCC41 standards, IEEE P1900.4a mechanisms for white space management, IEEE 1900.5 mechanisms for policy languages, and IEEE 1900.6 mechanisms to obtain and exchange sensing related information (spectrum sensing and geo-location information).

As of December 2010, the group submitted its project authorization request to the new standards committee of the IEEE Standards Association [7]. According to the project authorization request, the resulting standard will support fixed, nomadic, and mobile operation in frequency bands, such as TV bands, public safety bands, and wireless medical telemetry bands, and facilitate a variety of applications, including both low-power and high-power, short-, medium-, and long-range, and a variety of network topologies.

References

1. Ecma International, "Ecma Organization," available: http://www.ecma-international. org/memento/org.htm.
2. Ecma International, "Standard ECMA-392 MAC and PHY for Operation in TV White Space," available: http://www.ecma-international.org/publications/standards/ Ecma-392.htm.
3. IEEE, "IEEE 802 Detailed Overview," available: http://www.ieee802.org/.
4. A. Moody and G. Choinard, "Overview of IEEE 802.22 Standard on Wireless Regional Area Networks and Core Technologies," June 2010, available: https://mentor.ieee. org/802.22/dcn/10/22-10-0073-02-0000-ieee-802-22-overview-and-core-technolo- gies.pdf.
5. R. Kennedy, "P802.11af Project Authorization Request," available: https://develop- ment.standards.ieee.org/get-file/P802.11af.pdf?t=41492700024.
6. IEEE, "IEEE Standards Coordinating Committee 41 (Dynamic Spectrum Access Networks) Background," available: http://grouper.ieee.org/groups/scc41/.
7. S. Filin, "IEEE SCC41 ad-hoc on White Space Radio PAR," available: http://grouper. ieee.org/groups/scc41/adhoc-wsr/contrib/scc41-ws-radio-10-0029-10.pdf.
8. J. Wang, M. S. Song, S. Santhiveeran, K. Lim, G. Ko, K. Kim, S. H. Hwang, M. Ghosh, V. Gaddam, and K. Challapali, "First Cognitive Radio Networking Standard for Personal/Portable Devices in TV White Spaces," *New Frontiers in Dynamic Spectrum, 2010 IEEE Symposium on New Frontiers in Dynamic Spectrum*, 2010.
9. M. Ghosh, V. Gaddam, G. Turkenich, and K. Challapali, "Spectrum Sensing Prototype for Sensing ATSC and Wireless Microphone Signals," *Cognitive Radio Oriented Wireless Networks and Communications, 2008. CrownCom 2008. 3rd International Conference on Cognitive Radio Oriented Wireless Networks and Communications*, 2008.
10. J. Wang and V. Gaddam, "Feasibility study of sensing TV whitespace with local quiet zone," *Systems, Man and Cybernetics, 2009. SMC 2009. IEEE International Conference on Systems, Man and Cybernetics*, 2009.
11. Federal Communication Commission, "Notice of Proposed Rule Making: ET Docket no. 04-113," May 2004.
12. C. R. Stevenson, G. Chouinard, Z. Lei, W. Hu, S. J. Shellhammer, and W. Caldwell, "IEEE 802.22: The First Cognitive Radio Wireless Regional Area Network Standard," *Communications Magazine, IEEE*, vol.47, no.1, pp.130–138, January 2009.
13. IEEE Standard 802.22-2011, "IEEE Standard for Information Technology Telecommunications and Information Exchange between Systems Local and Metropolitan Area Networks Specific Requirements – Part 22: Cognitive Wireless RAN Medium Access Control (MAC) and Physical Layer (PHY) Specifications: Policies and Procedures for Operation in the TV Bands", July 2011.
14. IEEE Std 802.22.1-2010, "IEEE Standard for Information Technology— Telecommunications and Information Exchange Between Systems—Local and Metropolitan Area Networks—Specific Requirements Part 22.1: Standard to Enhance Harmful Interference Protection for Low-Power Licensed Devices Operating in TV Broadcast Bands," pp. 1–145, November 1, 2010.
15. G. Chouinard, "802.22 Overview Presentation to the 802 Whitespaces Study Group," available: https://mentor.ieee.org/802-sg-whitespace/dcn/09/sg-whitespace-09-0058- 00-0000-802-22-presentation-to-ecsg.ppt.

16. M. A. Rahman, C. Song and H. Harada, "Development of a TV white space cognitive radio prototype and its spectrum sensing performance," Proc. IEEE CrownCom 2011, June 1–3, 2011.

17. G. Hiertz, D. Denteneer, L. Stibor, Y. Zang, X. P. Costa, and B. Walke, "The IEEE 802.11 Universe," *Communications Magazine, IEEE* , vol. 48, no.1, pp. 62–70, January 2010.

18. IEEE P802.11-2007, "IEEE Standard for Information Technology—Telecommunications and Information Exchange between Systems—Local and Metropolitan Area Networks—Specific Requirements—Part 11: Wireless Medium Access Control (MAC) and Physical Layer (PHY) Specifications," June 2007.

19. R. Kennedy, "TGaf Call for Proposals," available: https://mentor.ieee.org/802.11/dcn/10/11-10-0218-01-00af-call-for-proposals.doc.

20. V. K. Jones, R. De Vegt, and J. Terry, "Interest for HDR Extension to 802.11a," IEEE 802.11-02/081r0, January 2002.

21. E. Perahia, "IEEE 802.11n Development: History, Process, and Technology," *Communications Magazine, IEEE* , vol. 46, no. 7, pp. 48–55, July 2008.

22. IEEE P802.11n, "Amendment to Standard for Information Technology—Telecommunications and Information Exchange between Systems—Local and Metropolitan Networks—Specific Requirements—Part 11: Wireless LAN Medium Access Control (MAC) and Physical Layer (PHY). Amendment 4: Enhancements for Higher Throughput," September 2009.

23. P. Ecclestine and T. Baykas, "11af Channel Numbering Comment Resolutions," IEEE 802.11-10/1033r2, September 2010. Available: https://mentor.ieee.org/802.11/dcn/10/11-10-1033-03-00af-11af-channel-numbering-comment-resolutions.doc.

24. C. S. Sum et al., "IEEE SCC41 Ad Hoc on WS Radio Usage Models," IEEE doc. Scc41-ws-radio-10/31r0, 04/2010. Available: http://grouper.ieee.org/groups/scc41/adhoc-wsr/contrib/.

25. J. Wang et al., "IEEE SCC41 Ad Hoc on WS Radio, System Requirements," IEEE doc. Scc41-ws-radio-10/35r0, 09/2010. Available: http://grouper.ieee.org/groups/scc41/adhoc-wsr/contrib/.

26. T. Baykas et al., "IEEE SCC41 Ad Hoc on WS Radio, WS Frequency bands," IEEE doc. Scc41-ws-radio-10/15r0, Requirements, 07/2010. Available: http://grouper.ieee.org/groups/scc41/adhoc-wsr/contrib/.

Chapter 7

Standardization Activities Related to TV White Space: Coexistence and Dynamic Spectrum Access Standards

Tuncer Baykas, Markus Muck, Stanislav Filin, Mika Kasslin, Paivi Ruuska, and Junyi Wang

Contents

7.1 Introduction

While regulators are opening up new spectrum by allowing special access to unused TV channels in VHF/UHF bands, they are concentrating on means for protecting incumbents such as TV stations and wireless microphones. Neither the Federal Communications Commission (FCC) nor other regulators currently address the problem of coexistence of multiple technologies and service providers in the same TV white space (TVWS) spectrum or optimize spectrum usage with dynamic spectrum access techniques. From one perspective it is believed that the situation is no different from today's Industrial Scientific and Medical (ISM) bands where, as an example, Wi-Fi and Bluetooth coexist without common coexistence mechanisms. Propagation characteristics in the VHF/UHF bands are, however, so much different from the ones in the ISM bands that the TVWS coexistence problem is both more complex and much more severe. In this chapter we will look at activities from three standardization bodies which focus on providing coexistence mechanisms and dynamic spectrum access methods as shown in Table 7.1.

Under the IEEE 802 LAN/MAN Standards Committee (LMSC), 802.19 working group is developing a coexistence standard between heterogeneous networks

Table 7.1 Standardization Activities Related to Coexistence and Dynamic Spectrum Access Standards

Standardization Organization	Responsible Group/Committee	Current Name of the Standard/Project	Type of the Standard	Status/Publication Date
IEEE 802 LAN/MAN Standards Committee	802.19 Task Group 1 under 802.19 Working Group	Draft Standard for Information Technology—Telecommunications and Information Exchange Between Systems—Local and Metropolitan Area Networks—Specific Requirements—Part 19: TV White Space Coexistence Methods	Coexistence between TVBD networks	In progress/2013 (Expected)
IEEE Standards Coordinating Committee 41	1900.4 Working Group	IEEE P 1900.4a Architecture and Interfaces for Dynamic Spectrum Access Networks in White Space Frequency Bands	Dynamic Spectrum Access	In Progress
ETSI	ETSI TC RRS	Technical Specifications on "System Requirements for Operation in UHF TV Band White Spaces" and "Coexistence Architecture for Cognitive Radio Networks on UHF White Space Frequency Bands" Technical Reports on "Use Cases for Operation in White Space Frequency Bands" and "Feasibility Study on Radio Frequency (RF) Performances for Cognitive Radio Systems Operating in UHF TV Band White Spaces"	TVWS operation and coexistence	In Progress

in the TVWS [1]. IEEE Standards Coordinating Committee 41 focuses mainly on dynamic spectrum access, and its 1900.4 working group is developing standards specifying management architectures enabling distributed decision making for optimized radio resource usage in heterogeneous wireless access networks [2]. IEEE standard 1900.4 on "Architectural Building Blocks Enabling Network-Device Distributed Decision Making for Optimized Radio Resource Usage in Heterogeneous Wireless Access Networks" was published in February 2009 [3]. Currently, the 1900.4 working group (WG) is developing two new standards: P1900.4.1 and P1900.4a. Draft standard P1900.4.1 on "Interfaces and Protocols Enabling Distributed Decision Making for Optimized Radio Resource Usage in Heterogeneous Wireless Networks" is a direct extension of IEEE standard 1900.4. Draft standard P1900.4a on "Architecture and Interfaces for Dynamic Spectrum Access Networks in White Space Frequency Bands" is an amendment to IEEE standard 1900.4 focusing on white space access [4,5].

ETSI (The European Telecommunications Standards Institute) was created by CEPT (the European Conference of Postal and Telecommunications Administrations) in 1988 for standardization of information and communication technologies. ETSI TC RRS is responsible within ETSI for the standardization of reconfigurable radio systems (RRS), including software defined radio (SDR) and cognitive radio systems (CRS). In this framework, ETSI RRS considers the operation in the TVWS [6].

The outline of the chapter is as follows. First we will look at the coexistence need in TVWS in detail. Subsequently, we will focus on mentioned standard activities starting with IEEE 802.19 and IEEE P1900.4a. The last section of the chapter is reserved for ETSI RRS activities.

7.2 Need for Coexistence in TVWS

There are a number of wireless technologies that are likely to be deployed in the TVWS as specified in the previous chapter. In IEEE 802 both the IEEE 802.22 working group and the IEEE 802.11 working group are working on wireless standards for TVWS. The former working group is developing a standard for Wireless Regional Area Networks (WRAN) in the TVWS. The latter one has a task group, TGaf, developing an amendment to the 802.11 Wireless Local Area Network (WLAN) standard for the TVWS operations. In addition the ECMA 392 standard is already available. Due to the unlicensed nature of TVWS which is for secondary wireless systems, it is very likely that other wireless technologies will also be deployed in the TVWS.

As a result, a diverse set of wireless technologies will try to operate in the same piece of spectrum. This is expected to lead to interference issues between TVWS users in geographic locations with a limited number of TVWS channels. The number of TVWS channels varies from location to location due to the number of TV stations operating in any given area. The number of TVWS channels can be reduced even further due to the use of wireless microphones, since channels occupied by these wireless microphones are not to be used by white space devices (WSDs). Also, because the usage patterns of the wireless microphones can change from day to day or even hour to hour, the number of the channels can vary with time.

Even if there were a lot of channels available for the WSDs, when different wireless technologies, like an 802.22 WRAN and an 802.11 WLAN, are deployed in a common region, there is a potential for interference. If the WSDs share the same TVWS channel, it is possible for either of the networks to cause interference to the other network. In this example case it is very probable that the WLAN deployed in a home causes interference to a WRAN customer premise equipment (CPE) which is connected to the WRAN base station (BS) on the channel used by the WLAN. The case is extremely severe if and when the CPE is transmitting using a directional antenna directed toward the BS. The WLAN may not even be aware of the interference it is causing to the WRAN CPE if the WLAN devices are outside the coverage area of the directional link between the BS and the CPE. Consequently the WLAN devices do not detect much if any transmissions from the WRAN CPE and consider the channel to be free. In some other deployments the WLAN is the victim of the interference generated by the WRAN. For example, if the WLAN network happens to be directly between the WRAN CPE and the BS, the WRAN transmissions are so high in power due to the antenna gain of the CPE from the WLAN perspective that the channel is sensed as occupied for long periods of time.

All that can be imagined to be true also in the unlicensed bands at 2.4 GHz and 5 GHz. Why bother with intersystem interference and coexistence in TVWS if we have been able to utilize an enormous amount of diverse devices utilizing different technologies in those bands? There is neither coexistence system nor protocol that these devices can use to coordinate usage of the unlicensed band. They do it autonomously. What is so different in TVWS that this same approach would not be enough in TVWS as well?

First, the coexistence situation in the widely used unlicensed bands at 2.4 GHz and 5 GHz is not as good as one could imagine. The bands and especially the 2.4 GHz ISM band have become so popular, thanks to the global availability as a common low-barrier resource, that there are actually no resources available for anybody. Everybody tries to use the resources imprudently and autonomously, which leads to a situation in which no device gets any resources because there is no collaboration on coexistence.

The resource exhaustion problem is even more severe in TVWS due to differences in propagation characteristics, compared to the ones in the 2.4 GHz and especially in the 5 GHz. Signals in those bands lose power much faster than in the TVWS band. They are absorbed by the environment much more than in the TVWS band. These propagation characteristics severely exacerbate the coexistence problem by bringing devices within each other's transmission range. The range difference is even further increased by the significantly higher dynamic range of

Figure 7.1 **TV channel availability in the United States based on method in [7].**

allowed power in TVWS compared to the situation in today's unlicensed bands. All the devices in the 2.4 GHz and 5 GHz bands can be easily classified as low-power devices. This is not true in TVWS where the maximum power limits cover the range from 40 mW up to 4 watts.

Second, a problem specific to the TVWS is that where there are users there are few available channels. In the 2.4 GHz and 5 GHz we have quite the same amount of spectrum and channels available globally from one location to another. In TVWS the channels are available only when they are not used by the incumbents. In areas with high population density this spectrum is, therefore, likely to be scarce. Several studies on TV channel availability for unlicensed operation in the United States have been recently conducted; Figure 7.1 below is taken from one of these [7]. While most of the United States has plenty of TVWS spectrum, most people in the United States live in areas where such spectrum is very scarce. If this spectrum is to serve the largest number of people (as opposed to covering the most area), then the various players in the spectrum need to find a way to share a very precious resource.

All these reasons together show that, to ensure efficient and fair use of the available spectrum resources, the secondary users in TVWS should be able to:

■ Discover the information of the spectrum use, including the discovery of licensed operation
■ Discover the neighbor networks with which coexistence is needed
■ Discover the characteristics and resource needs of the networks
■ Share the available resources fairly between the neighboring networks

Since current PHY/MAC standards do not support most of the above functions, coexistence standards are needed.

7.3 802.19 Task Group 1 Wireless Coexistence in the TV White Space

The IEEE 802.19 working group has taken actions to work on the coexistence in TVWS and a task group (TG), 802.19 TG1, was formed in early 2010 [1]. This TG has been chartered with the specific task of developing a standard for TVWS coexistence methods. Because the standard development is still ongoing, it is not possible to present the coexistence solution. However, the group has defined the logical architecture in the System Design Document [8] which serves as the high-level guideline for standard contributions. In the next section we will provide details of system architecture followed by a few deployment scenarios.

7.3.1 IEEE 802.19.1 System Architecture

The coexistence system architecture (Figure 7.2) in System Design Document defines the system entities, which are the core of providing the coexistence for the TV band networks, and external entities which provide information to the coexistence system.

The coexistence system architecture comprises three logical interfaces and six related interfaces. The logical elements are:

- Coexistence Enabler (CE)
- Coexistence Manager (CM)
- Coexistence Discovery and Information Server (CDIS)

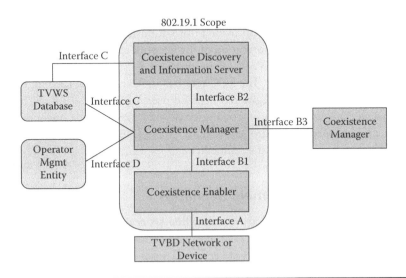

Figure 7.2   802.19.1 system architecture.

The coexistence system is expected to interface with three external logical elements that are:

- TVBD network or device
- TVWS database
- Operator management entity

The key functions of the CE are to obtain information required for coexistence from the TV band device (TVBD), and to reconfigure TVBD operation according to the coexistence decisions. The collected information covers the capabilities and the resource needs of the TVBD network, and the characteristics of the radio environment. The CE may reside in a TVBD (e.g., in an access point, base station, or mesh point), but it may also be a SW entity in a network management system. Typical deployment is such that a network is represented by a CE.

The CM is the decision maker of the coexistence system. It discovers and solves the coexistence conflicts of the networks operating in the same area. A CM serves one or more networks. Depending on the deployment, it resides either in a TVBD, or in backbone network, or in Internet as public service. The CM discovers the interfering networks and their CMs and shares information with other CMs. Based on the collected information, it reconfigures the operation of its own network(s), but also performs resource reallocation for the whole neighborhood as needed.

The CDIS assists the CM in the neighbor discovery. It keeps a record of the registered CMs and location of the networks they serve, and provides a list of candidate neighbors for a CM adding or updating location of a network. CDIS may also store some other information relevant for coexistence (e.g., statistics of spectrum use).

The system external entities provide information needed for the spectrum use decisions. TVBDs provide information of the network capabilities, resource needs, and radio environment. TVWS database provides a list of available channels for the secondary use in the area. The FCC requires the existence of such an entity to ensure the operation of the registered licensed users of the TV band, such as TV broadcasters. Operator Management Entity (OME) may provide operator-related policies for the operator controlled systems. There is no such entity existing or required by the authorities at the moment.

7.3.2 Example Deployment Scenarios

Unlike the licensed radio spectrum, the white spaces may be used by heterogeneous networks with different type, range, utilization, and mobility. The devices which operate in the white spaces are expected, even required, to be well aware of the spectrum availability at their location. Thus, locally and dynamically managed spectrum sharing between the networks which are within interference range to each other ensures the most efficient use of local spectrum resources.

The basic assumptions in these deployment scenarios are:

- Each network is served by only one Coexistence Manager.
- A Coexistence Manager serves one or more networks.
- There is no hierarchy between the Coexistence Managers.

A Coexistence Manager can connect to any other Coexistence Manager to solve the coexistence conflicts between the networks they are serving. The 802.19.1 architecture is not limited to the presented scenarios at the moment, and the cases discussed in the following are just example cases.

The networks may be divided in two groups based on how they are managed: coordinated networks and independent networks. A set of coordinated networks are set up and managed by an operator. For example, the IT department may plan the whole wireless connectivity in the office building and set up and manage the networks accordingly. The set of coordinated networks is served by a shared CM entity residing, for example, in the network management system or in Internet. Independent networks are stand-alone networks set up and configured typically by the user of the network. An independent network is served by a local CM residing, for example, in the access point. Any networks should be able to coexist with other networks within interference range, independent of their type, to ensure efficient use of available spectrum.

The networks using the same spectrum resources may have different ranges. In that case the longer-range network may only provide information to the shorter-range networks, not actually negotiate on the spectrum sharing. This is because otherwise it may need to negotiate with all the short-range networks within in its range, and that causes a lot of signaling.

7.3.2.1 Coexistence of Coordinated Networks

In this scenario there is only a set of coordinated networks within the interference range. All the networks are coordinated by the same operator and served by the same Coexistence Manager. The scenario is presented in Figure 7.3. First a network (e.g., a NW6) connects to the CM. The CM discovers, with the help of CDIS, if there are any networks served by other CMs within the interference range of the network. Since in this case there are not, the CM coordinates the coexistence of the network with other networks which it is also serving, and which are within the interference range of the network (NW3 and NW5). Depending on the spectrum availability, and the type and resource needs of the networks, the CMs may agree on operation on different frequency channels and/or time-sharing.

7.3.2.2 Coexistence of Independent Networks

Figure 7.4 presents a scenario in which all the networks within the interference range are independent networks and have local Coexistence Managers. This scenario may

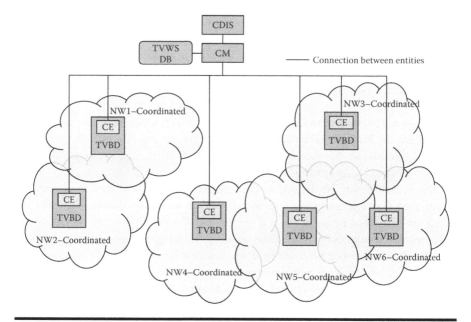

Figure 7.3 Centralized coexistence coordination.

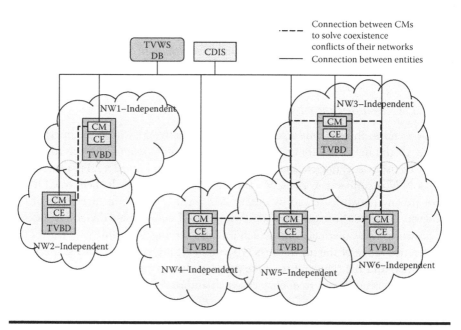

Figure 7.4 Distributed coordination of independent networks.

be valid, for example, in the residential area where each house/apartment owner has set up their own network. When a network (e.g., NW6) is activated, its CM discovers, with the help of CDIS, which networks are within interference range. Then the CM connects to the CMs of those networks (NW3 and NW6) and starts coordinating the coexistence of the networks with them.

7.3.2.3 Coexistence of Coordinated and Independent Networks

In this deployment scenario there are independent networks with local CMs and coordinated networks with shared CMs. A network is able to participate in coexistence coordination with any other network within the interference range, independent of whether it is an independent or coordinated network. Also, the coordinated networks managed by different operators are able to coexist. Figure 7.5 presents a scenario where an independent network (NW3) is within interference range of coordinated networks (NW5 and NW6). First the CM of a network (e.g., NW3) discovers with which networks the coexistence is needed and connects to the CM of those networks (NW5 and NW6). The CMs start coordinating coexistence of those networks.

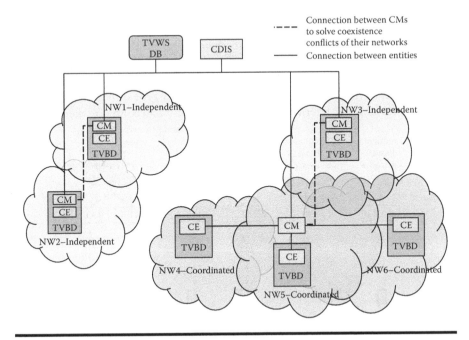

Figure 7.5 **Distributed coordination of independent and coordinated networks.**

7.4 IEEE P1900.4a Architecture and Interfaces for Dynamic Spectrum Access Networks in White Space Frequency Bands

Among the working groups of SCC41, IEEE 1900.4 WG is developing standards specifying management architectures enabling distributed decision making for optimized radio resource usage in heterogeneous wireless access networks. P1900.4a focuses on operation in white space frequency bands. Before going into details of P1900.4a, the details of its baseline standard 1900.4 are given. Afterward, system overview and deployment scenarios of P1900.4a are provided.

7.4.1 Baseline Standard 1900.4

The IEEE standard 1900.4, "Architectural Building Blocks Enabling Network-Device Distributed Decision Making for Optimized Radio Resource Usage in Heterogeneous Wireless Access Networks," was published in February 2009. It considers three use cases: distributed radio resource usage optimization, dynamic spectrum assignment, and dynamic spectrum sharing (as shown in Figure 7.6) [3].

In the **distributed radio resource usage optimization** use case, one or several operators operate several radio access networks (RAN) using the same or different radio access technologies (RAT). Frequency bands assigned to these RANs are fixed. Also, reconfiguration of radio equipment on the network side is not considered. These RANs provide services to their users, which use different terminals. One type of terminal are legacy terminals, designed to use a particular RAT. Such terminals can connect to one particular operator or to other operators having roaming agreements with the home operator. Another type of terminals is reconfigurable terminals. Such terminals have the capability to reconfigure themselves to

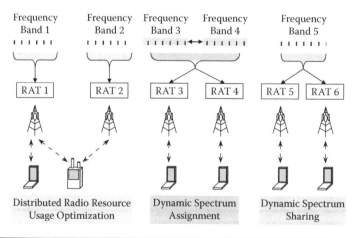

Figure 7.6   IEEE 1900.4 use cases.

use different RATs. Correspondingly, such terminals can hand over between different RANs using different RATs and operated by different operators. Optionally, reconfigurable terminals can support multiple simultaneous links with RANs.

In the **dynamic spectrum assignment** use case, one or several operators operate several RANs using the same or different RATs in different frequency bands. To improve radio resource usage, configuration of these frequency bands can be dynamically changed. One example application of this use case is joint management of several RANs within one operator, where spectrum allocated to this operator is dynamically redistributed between its RANs. Another example is spectrum trading between different operators.

In the **dynamic spectrum sharing** use case, several RANs using same or different RATs can share the same frequency band. One example of this use case is when several RANs operate in unlicensed spectrum. In this example IEEE 1900.4 can enable coexistence of such systems. Another example is when a secondary system operates TVWS. In such example, IEEE 1900.4 should provide protection of primary service (TV broadcast) and coexistence between secondary systems. To accomplish this, development of P1900.4a is started which enables mobile wireless access service in white space frequency bands without any limitation on used radio interface by defining additional components of the IEEE 1900.4 system.

7.4.2 P1900.4a System Overview

P1900.4a uses IEEE standard 1900.4 as a baseline. It concentrates on dynamic spectrum sharing use case of IEEE standard 1900.4. P1900.4a defines additional entities and interfaces to enable efficient operation of white space wireless systems. Figure 7.7 introduces the P1900.4a system [9].

White Space Manager (WSM) enables collaboration between P1900.4a system and IEEE 1900.4 system, provides regulatory context information to the Cognitive Base Station Manager (CBSRM), and enables communication between the CBSRM and white space database.

The Cognitive Base Station Reconfiguration Manager (CBSRM) is the entity that manages Cognitive Base Station (CBS) and terminals for network–terminal distributed optimization of spectrum usage. The key functions of the CBSRM, which are specific to dynamic spectrum access in white space frequency bands are management of spectrum sensing (e.g., by coordinating silent periods for measurements), classification of white space resources, and coordination of white space resource usage with CBSRMs of the same RAN for radio resource management and with CBSRMs of other RANs for coexistence.

7.4.2.1 P1900.4a System Architecture

P1900.4a system architecture is shown in Figure 7.8. Among entities defined in the network side, the Operator Spectrum Manager (OSM) is the entity that enables an

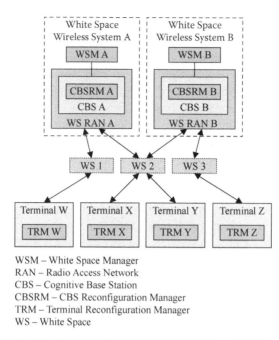

WSM – White Space Manager
RAN – Radio Access Network
CBS – Cognitive Base Station
CBSRM – CBS Reconfiguration Manager
TRM – Terminal Reconfiguration Manager
WS – White Space

Figure 7.7   IEEE P1900.4a system overview.

operator to control Network Reconfiguration Manager (NRM) dynamic spectrum assignment decisions. For this purpose, the draft standard defines *spectrum assignment policies*. These policies express regulatory framework, defining spectrum usage rules for the frequency bands available to this operator. Also, these policies express operator objectives in radio resource usage optimization related to dynamic spectrum assignment. Spectrum assignment policies are sent from the OSM to the NRM.

The RAN Measurement Collector (RMC) is the entity that collects RAN context information (RAN radio resource optimization objectives, RAN configuration, and RAN measurements and operator capabilities) and provides it to NRM. RMC may be implemented in a distributed manner.

The Network Reconfiguration Manager (NRM) is the entity that manages network and terminals for distributed optimization of spectrum usage. On the network side, the NRM makes RAN reconfiguration decisions and sends RAN reconfiguration requests to the RRC. In other words, the NRM directly decides on reconfiguration of RANs.

For managing terminal reconfiguration by the NRM, the draft standard defines *NRM radio resource selection policies*. These policies are sent from the NRM to the TRMs under its management and create the framework within which the TRMs will make terminal reconfiguration decisions.

The RAN Reconfiguration Controller (RRC) is the entity that controls reconfiguration of RANs based on requests from NRM.

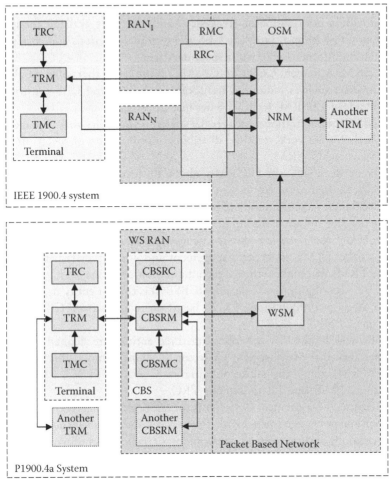

RAN – Radio Access Network
WS RAN – White Space RAN
CBS – Cognitive Base Station
OSM – Operator Spectrum Manager
NRM – Network Reconfiguration Manager
TRM – Terminal Reconfiguration Manager
CBSRM – CBS Reconfiguration Manager

RMC – RAN Measurement Collector
TMC – Terminal Measurement Collector
CBSMC – CBS Measurement Collector
RRC – RAN Reconfiguration Controller
TRC – Terminal Reconfiguration Controller
CBSRC – CBS Reconfiguration Controller
WSM – White Space Manager

Figure 7.8   P1900.4a architecture.

The White Space Manager (WSM) is the entity that enables collaboration between P1900.4a system and IEEE 1900.4 system, provides regulatory context information to CBSRM, and enables communication between CBSRM and white space database. Communication between CBSRM and white space database may include providing information on frequency bands currently available for secondary usage to CBSRM and providing CBS and terminal locations and corresponding registration information to white space database.

The CBS Measurement Collector (CBSMC) is the entity that collects CBS context information such as radio and transport capabilities and CBS confirmation and provides it to CBSRM. Each CBS has one CBSMC.

The CBS Reconfiguration Manager (CBSRM) is the entity that manages CBS and terminals for network–terminal distributed optimization of spectrum usage. Each CBS has one CBSRM.

On the network side, the CBSRM makes CBS reconfiguration decisions and sends CBS reconfiguration requests to the CBSRC. In other words, the CBSRM directly decides on reconfiguration of its CBS.

For managing terminal reconfiguration by the CBSRM, the draft standard defines *CBSRM radio resource selection policies.* These policies are sent from the CBSRM to the TRMs under its management and create the framework within which the TRMs will make terminal reconfiguration decisions.

The CBS Reconfiguration Controller (CBSRC) is the last entity in the network side. It controls reconfiguration of CBS based on requests from CBSRM. Each CBS has one CBSRC.

As shown in Figure 7.7, the following three entities are defined on the terminal side: Terminal Measurement Collector (TMC), Terminal Reconfiguration Manager (TRM), and Terminal Reconfiguration Controller (TRC). Each terminal shall have one TMC, one TRM, and one TRC.

The Terminal Measurement Collector is the entity that collects terminal context information and provides it to the TRM. The *terminal context information* may include user subscriptions and preferences; application parameters including QoS requirements, measurement capabilities, and measurements; terminal parameters including capabilities, configuration location information, and registration information.

The Terminal Reconfiguration Manager is the entity that manages its terminal for network–terminal distributed optimization of spectrum usage. This management is performed within the framework defined by the NRM radio resource selection policies and the CBSRM radio resource selection policies and in a manner consistent with user preferences and available context information.

The Terminal Reconfiguration Controller is the entity that controls reconfiguration of terminal based on requests from the TRM.

The OSM, RMC, NRM, RRC, TMC, TRM, and TRC entities have been defined in the IEEE standard 1900.4. The IEEE draft standard P1900.4a additionally defines WSM, CBSMC, CBSRM, and CBSRC entities, which perform management functions specific to dynamic spectrum access in white space frequency bands.

7.4.2.2 P1900.4a Deployment Scenarios

Management entities defined in IEEE draft standard P1900.4a can be deployed in different ways. The informative part of P1900.4a describes several deployment examples.

The basic deployment scenario is shown in Figure 7.9. CBSRM, CBSMC, and CBSRC are deployed inside CBS. TRM, TMC, and TRC are deployed inside the terminal. WSM deployed, for example, in Internet provides interface between CBSRM and WS database.

Coexistence deployment scenario is shown in Figure 7.10. Two WS radio systems can exchange information required for coexistence directly via interface between CBSRMs or via WSM. Figure 7.11 shows an example reference model of CBS/terminal showing deployment of P1900.4a entities inside CBS/terminal.

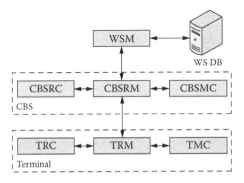

Figure 7.9   Basic deployment scenario of P1900.4a.

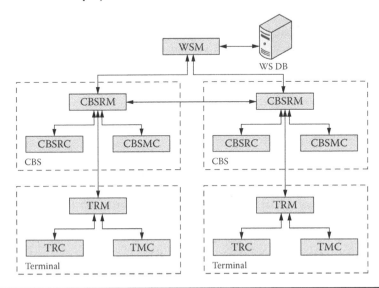

Figure 7.10   Coexistence deployment scenario of P1900.4a.

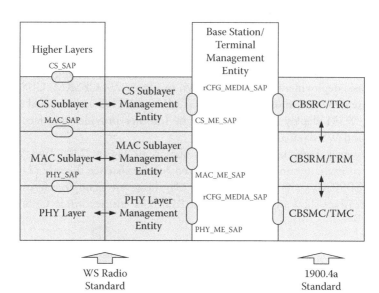

Figure 7.11 Example reference model of CBS/terminal with P1900.4a entities.

Finally, Figure 7.12 shows a cognitive femtocell deployment scenario which illustrates collaboration between 1900.4 system and P1900.4a system. The femto CBS is capable of operating in different frequency bands including white space frequency bands. Also this CBS is a part of a heterogeneous wireless network.

In such a deployment scenario, there are two types of management tasks:

■ Management of heterogeneous wireless network
■ Management of each component of heterogeneous wireless network

Management of heterogeneous wireless network is performed as defined in the IEEE standard 1900.4. For this purpose such entities as OSM, NRM, RMC, RRC, TRM, TMC, and TRC are used.

Examples of such management are:

■ Reconfiguration of the femto CBS to switch between white space frequency bands and operator frequency bands
■ Reconfiguration of terminals to connect to femto CBS in white space frequency bands and to the RAN in operator frequency bands
■ Using femto CBS to implement interface between NRM and TRM for terminals operating in white space frequency bands only

In this example, the heterogeneous wireless network has two components: RAN and femtocell. Management of RAN is performed by management entities described

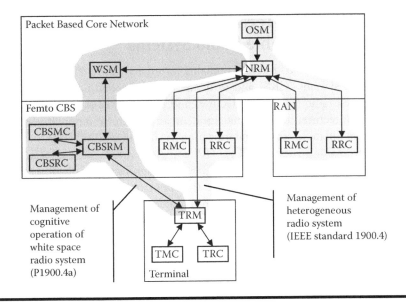

Figure 7.12 Cognitive femtocell deployment scenario.

in standards corresponding to the operating RAT of this RAN, for example, as described in 3GPP. Management of femto CBS is performed as defined in P1900.4a. For this purpose such entities as WSM, CBSRM, CBSMC, CBSRC, TRM, TMC, and TRC are used. Examples of such management are:

- Management of spectrum sensing performed by the femto CBS and terminals
- Management of white space resource usage of the femto CBS performed for the purpose of coexistence with other secondary systems
- Management of switching the femto CBS and its terminals to another white space resource in case primary radio system starts its operation in the femto CBS operating white space resource

In this deployment scenario reconfiguration of CBS is managed by 1900.4 system and P1900.4a system. The tasks of these two management systems are well distinguished so that there is no contradiction in their operation. 1900.4 system manages reconfiguration of CBS to its operation between white space frequency bands and RAN frequency bands. P1900.4a system manages reconfiguration of CBS when it operates in white space frequency bands. Reconfiguration requests from 1900.4 system may have priority over reconfiguration requests from P1900.4a system, but in this deployment scenario this should be defined by the manufacturer of the CBS.

7.5 ETSI TC RRS—Technical Committee on Reconfigurable Radio Systems

ETSI RRS addresses four different domains in order to drive the introduction of cognitive radio systems (CRS) and to define the corresponding enablers, such as software defined radio (SDR) architectures and related APIs: system aspects, radio equipment architecture, cognitive management, and control and public safety. ETSI RRS gained mandate to start standardization in the end of 2009 [10].

Building on this structure, ETSI RRS complements ongoing efforts in other bodies, such as IEEE standardization bodies, by proposing technological solutions. Furthermore, ETSI TC RRS has a role in supporting the European regulations work related to CRS and TVWS, and defining specifications required for operation in Europe. For example,

1. The Radio and Telecommunications Terminal Equipment (R&TTE) Directive regime in force in Europe is based on declaration of conformity and includes neither type approval nor registration of the equipment nor equipment identifier (in the United States, type approval is still necessary). This self-declaration is preferably a reference to a harmonized standard, which may be developed by ETSI RRS.
2. Protection of operation TV bands: In Europe, DVB-T does not show a residual carrier as is the case in the United States (the possibility for detection of the U.S. ATSC signal below noise (i.e., at −114 dBm) is made possible thanks to the residual carrier which is present in the ATSC signal). A corresponding adaptation of sensing based standards needs to be defined for Europe.
3. Broadcasting, wireless microphones, and assignment to radio stations are managed in Europe at the national level. Any sharing scheme based on a database will require some level of integration of the national data.

In order to address the above and other European regulatory aspects, the Electronic Communications Committee (ECC) within the European Conference of Postal and Telecommunications Administrations (CEPT) has set up the SE43 group working on "Technical and operational requirements for the operation of cognitive radio systems in the 'white spaces' of the frequency band 470–790 MHz." ETSI RRS is the competence center within ETSI to implement those regulatory requirements.

Specifically related to the TV white spaces, ETSI RRS has in scope to define technical reports and specification on the use cases, system requirements, system architectures, protocols and interfaces for the operation of the cognitive radio systems in the TV white spaces. Examples of the use cases are given in this section. In addition, ETSI RRS defines feasibility studies on specific topics, such Radio Frequency performances for cognitive radio systems operating in TV white spaces.

7.5.1 ETSI TC RRS Use Cases for Operation in TV White Space Frequency Bands

ETSI RRS has in scope to define use cases for TVWS, and also other possible frequency bands which enable cognitive features. In this section, examples of the use cases for TVWS are described.

7.5.1.1 Mid/Long Range Wireless Access over White Space Frequency Bands

Internet access is provided from a base station to the end users by utilizing white space frequency bands over ranges similar to today's cellular systems, e.g., in the range of 0 to10 km (Figure 7.13).

This use case can be divided into three scenarios dependent on the mobility of the end-user devices: no mobility, low mobility (walk speed), and high mobility (car or train speed). This differentiation is made because the constraints for detecting primary users or other secondary users, as well as on retrieving the geographical position, may differ dependent on the mobility of the users.

In this use case, Multimode user terminals (i.e., terminals that support multi-RAT in licensed spectra, for instance High Speed Packet Access [HSPA] and Long Term Evolution [LTE]) are also provided with the capability of accessing TV white space spectrum bands in order to provide wireless broadband access (e.g., Time Division LTE [TD-LTE]), for instance, in rural areas where high-data-rate connections are commonly not available. This use case takes the benefit of

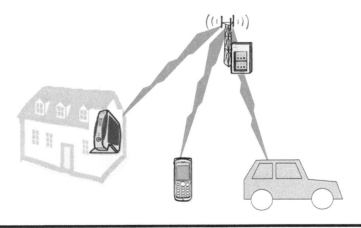

Figure 7.13 **Mid-/long-range wireless access over white space frequency bands.**

the excellent propagation performance of a radio network operating in TV white space frequency bands.

TDD can be considered more suitable for a secondary/overlay spectrum access compared with FDD. TDD only needs one frequency band, so it is simpler to find one single suitable white space frequency band. For FDD a pair of separated frequency bands—uplink (UL) and downlink (DL)—is required with strict separation bandwidth requirements that make candidates' frequency bands more difficult to find. With two frequency bands used by FDD, there are more chances to interfere with incumbent users on any of the 2 bands than TDD and its single band—furthermore interference on any of the 2 frequency bands will result in a handover or a break of the link both DL and UL. It could be simpler to detect incumbent users on one single frequency band (TDD) than a pair of frequency bands (FDD). TDD—allowing asymmetric DL/UL data connection on a single frequency band—may fit well a dynamic spectrum assignment with optimized/dynamic channel bandwidth.

This use case is considered in further detail in the following scenarios.

7.5.1.1.1 Mid/Long Range, No Mobility

In this scenario, wireless access is provided from a base station toward fixed devices (e.g., a fixed mounted home base station/access point, with the range of 0 to 10 km). The geo-location of the base station as well as of the fixed device are well-known via GPS or professional installation (Figure 7.14).

7.5.1.1.2 Mid/Long Range, Low Mobility

In this scenario, wireless access is provided in the range of 0 to 10 km from a base station toward mobile devices where the users have low mobility of 0 to 20 km/h (e.g., they are staying at their location or walking). In that respect, sensing results for primary users retrieved for the current location are not getting invalid due to the mobility of the user. The geo-location from the base station is well known. The geo-location from the mobile device must be determined during operation, via GPS or cellular positioning systems (Figure 7.15).

Figure 7.14   Mid-/long-range wireless access, no mobility.

Figure 7.15   Mid-/long-range wireless access, low mobility.

7.5.1.1.3 Mid/Long Range, High Mobility

In this scenario, wireless access is provided in the range of 0 to 10 km from a base station toward mobile devices, and the mobile devices may move fast with the speed of 0 to 250 km/h (e.g., because a user is in a car or a train). In that respect, sensing results for primary users retrieved for the current location may get invalid quickly due to the mobility of the user. Thus, this use case sets high constraints for the detection of primary users.

The geo-location from the base station is well known. The geo-location from the mobile device must be determined during operation, via GPS or cellular positioning systems (Figure 7.16).

7.5.1.2 Network-Centric Management of Terminals in TV White Spaces

This scenario considers a network centric solution for allocating available TVWS for the user terminal.

Available TVWS frequency band is considered based on location; it is assumed that TVWS would be largely available in rural area. However, the availability of the bands cannot be excluded and thus shall be taken into account by the system.

In the case of a network-centric solution, the terminal can get the required information from its current connectivity and its current RAT (i.e., TD-LTE operating in TVWS), or from another RAT (e.g., HSPA in 3G bands).

Figure 7.16   Mid-/long-range wireless access, high mobility.

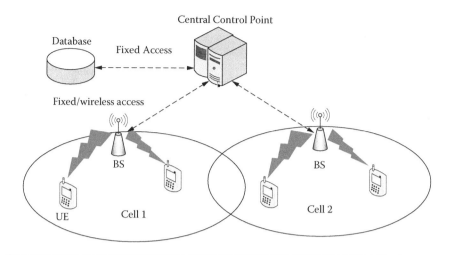

Figure 7.17   Access to TVWS, centralized mode.

Once the terminal accesses the network, it can be left under the control of the network. High-layer signaling can be used for this purpose (e.g., handover command to hand-off to a new frequency), or system broadcast messages can be used to notify terminals about the change of the frequency.

Figure 7.17 illustrates a central control point deployed to manage the access of the TD-LTE system to the TV white space. It can be either an enhanced radio base station or a stand-alone node.

7.5.1.3 Short-Range Wireless Access over White Space Frequency Bands

In this scenario, Internet access is provided via short-range wireless communication (e.g., in the range of 0 to 50 m) from an access point or base station to the end users by utilizing TV white space frequency bands (Figure 7.18).

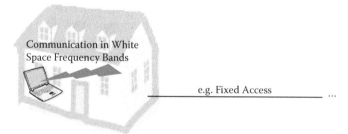

Figure 7.18   Short-range wireless access over white space frequency bands.

This use case is divided into three scenarios dependent on the coordination of the networks: uncoordinated networks, coordinated networks, and hybrid of uncoordinated and coordinated networks.

7.5.1.3.1 Uncoordinated Networks

In this scenario multiple networks access white space frequency bands. The different networks are independent and the backbone connectivity is provided by different network operators. The scenario may be valid e.g. in an apartment house where residents have their own local area access points operating in TV white space frequency band. The access points are operated and maintained by the residents or the ISPs. This scenario is illustrated in Figure 7.19.

For efficient operation, this case requires effective coexistence mechanisms between the secondary TV white space users. As the networks are set up in an independent manner, there needs to be means for the access points or base stations discover the other networks operating in the same geo-location area. A functionality called coexistence service provider is needed to facilitate the coexistence of the neighboring white space networks. It enables networks to discover each other and to exchange information related to the spectrum use. Due to the independent nature of the networks, in this scenario the coexistence service provider functionality is distributed.

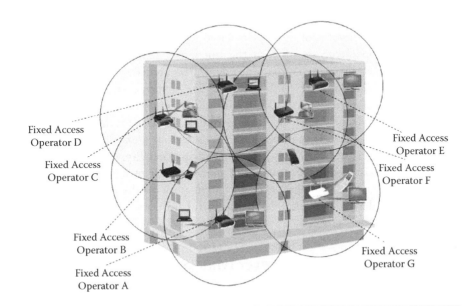

Figure 7.19 **Short-range wireless access, uncoordinated networks.**

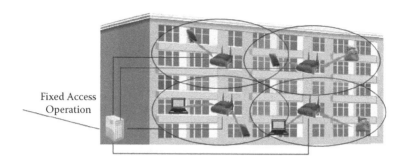

Fixed Access
Operation

Figure 7.20 Short-range wireless access, coordinated networks.

7.5.1.3.2 Coordinated Networks

In this case all the networks in the same geo-location area are operated in a coordinated manner by a network operator. This scenario is illustrated in Figure 7.20. Examples of this kind of usage can be small scale corporate networks, and networks for academic institutions. Also in this scenario there is a functionality called coexistence service provider which is responsible for the coexistence of the networks, but it is centralized functionality and may reside in the server of the operator managing the networks.

7.5.1.3.3 Hybrid of Uncoordinated and Coordinated Networks

In this case, the networks of the previous two scenarios are overlapping, i.e. the overall deployment consists of both the uncoordinated and coordinated networks. Examples where such situations could arise are campus areas, shopping malls. Figure 7.21 illustrates this scenario. The coverage area of coordinated networks is blue, coverage area of uncoordinated networks is red. For efficient operation, effective coexistence methods need to be in place for this scenario case. The overall coexistence management in this scenario is distributed. Therefore the coexistence service provider is a distributed functionality. The networks with centralized coexistence management will manage their own share of resources in centralized fashion, but they need to cooperate with the other neighboring networks in order to facilitate overall successful utilization of the white space frequency bands.

7.5.1.4 Ad Hoc Networking over White Space Frequency Bands

In this use case, the devices (user devices and other devices like access points) communicate directly with each other to share information, or to run joint applications or services. For the communication, an ad hoc network operating on TVWS is formed (Figure 7.22). There may be two or more devices in the ad hoc network. The

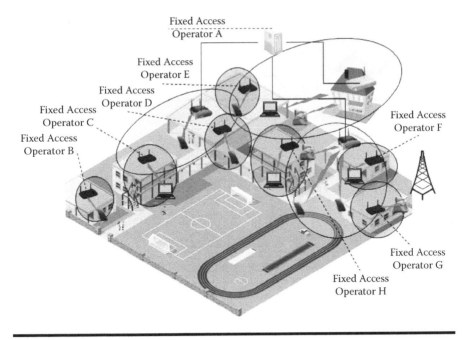

Figure 7.21 Short-range wireless access, hybrid of uncoordinated and coordinated networks.

Figure 7.22 Short ad hoc networking over white space frequency bands.

geo-location of at least one node should be known. For the other devices in close distance to that node, it may not be necessary to derive an own geo-location.

This use case can be divided into two scenarios dependent on the coordination of the networks: device-to-device connectivity and ad hoc networking.

7.5.1.4.1 Device-to-Device Connectivity

In this case, two devices connect in a peer-to-peer manner to exchange information between each other (Figure 7.23). The information can be, e.g., multimedia content,

Figure 7.23 Device-to-device connectivity.

Figure 7.24 Ad hoc networking.

or control information like measurement results shared between the devices. The capabilities of the devices may be similar (like two mobile devices), or different (like a mobile device and external printer or display).

7.5.1.4.2 Ad Hoc Networking

In this case, the devices form an ad hoc network to communicate and collaborate with other devices in the same geo-location area (Figure 7.24). As an example an ad hoc network may be formed to share a localized social networking or various sensing information.

7.5.1.5 Opportunistic Access to TVWS by Cellular Systems during Absence of Primary User

In this use case, TVWS slots are only available sporadically for secondary users. The multi-mode user terminals are able to operate in multiple networks and bands. Thus, the cellular network operating in licensed band may also use the TVWS when it is available.

The time-limited switch of a base station to operate in TVWS spectrum is leading to a number of advantages which are further detailed in the scenario descriptions detailed in the following subsections (note: similar approaches are discussed for opportunistic relaying in [11]).

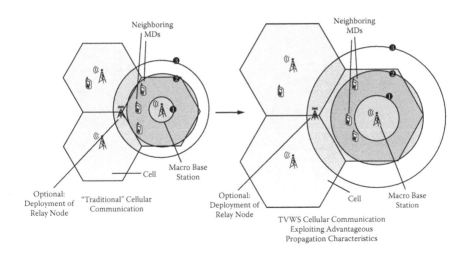

Figure 7.25 **Reduced propagation loss in TVWS and resulting improved coverage.**

7.5.1.5.1 Lighter Infrastructure Deployment through Larger Cell Sizes

Due to the improved propagation characteristics in the TVWS bands compared to typical licensed bands, a large cell size is chosen which will lead to a lighter infrastructure deployment and thus to an overall reduced capital expenditure to operational expenditure ratio, as illustrated by Figure 7.25.

In the analysis given in [12] for a typical example using the deployment of a relay node, it is observed that it is possible to transmit over a wireless link at 470 MHz with d=200 m by getting the same performance as a legacy wireless link at 2 GHz with d=100 m. In this setup, the communication distance can be increased twofold by keeping the same QoS.

In the case that the primary user reappears, the base station accessing to TVWS typically has to switch to a different TVWS frequency band (if available) or switch to the licensed spectrum (for example, in the 2 GHz band). In the latter case, the coverage is reduced. In order to maintain the coverage of the cell, either a lower QoS has to be tolerated at the cell edges, or relay nodes may be deployed in order to attenuate the corresponding effects.

7.5.1.5.2 Increased Spectral Efficiency through Reduced Propagation Loss

Due to the improved propagation characteristics in the TVWS bands compared to the typical licensed bands, a possible deployment choice is to keep a cell size as it is for the case of licensed band deployment, since a higher QoS can be achieved within the given cell (Figure 7.26). However, those propagation characteristics

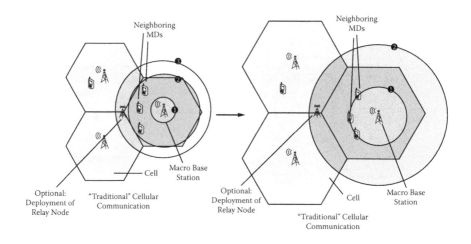

Figure 7.26 Reduced propagation loss in TVWS and thus improved QoS.

may also increase interference issues which require an adequate management (e.g., suitable frequency reuse-factor for TVWS, power management, etc.); due to the improved propagation characteristics in the TVWS bands, an identical QoS is achieved within the given cell at a lower RBS/MD output power level. The inherent power consumption can be reduced. A hybrid solution is also possible (i.e., a moderate reduction of the RBS output power levels combined with a moderate improvement of the QoS).

7.5.1.5.3 Increased Spectral Efficiency through Extended Macro-Diversity

Another possible deployment choice is to keep a cell size (or to increase it only slightly) as in the licensed band deployment. Then, the joint operation of neighboring RBS can be exploited in order to achieve a higher macro-diversity gain in the UL (multiple RBS are jointly decoding the received signals) or in the DL (multiple RBS are contributing to jointly optimized transmission) as illustrated in Figure 7.27.

7.5.1.6 TVWS Band-Switch in the Case That Primary User Reenters

The TVWS usage rules vary over the geographical regions. Typically, when a primary system arrives, the corresponding band is no longer available for opportunistic spectrum access by secondary systems. Also, a guard band is introduced as is illustrated in Figure 7.28 for the example of an FCC TVWS usage context.

In this scenario, it is suggested that the secondary user switches to another TVWS channel that is still available for the opportunistic access—if such a channel

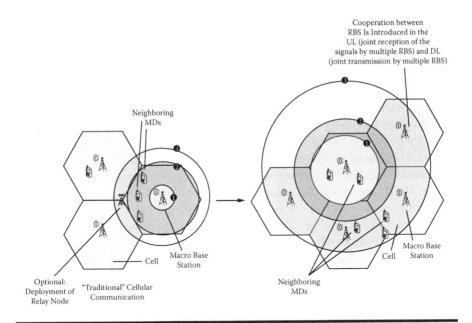

Figure 7.27 Reduced propagation loss in TVWS and thus improved Macro-Diversity.

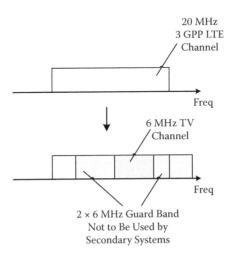

Figure 7.28 Arrival of primary user in TV white space framework.

is available. Otherwise, the secondary user is assumed to switch to operate in a suitable licensed band.

7.5.1.7 Carrier Aggregation between IMT and TVWS Bands

In this scenario, TVWS radio resources are used as a component carrier (CC) pool and can offer temporary extra radio resources to, e.g., cellular systems operating primarily in the IMT bands. Carrier Aggregation is one of the most distinct features of 4G systems such as 3GPP LTE Advanced. It allows the support of wider transmission bandwidths: up to 100 MHz by aggregating two or more component carriers (CC), up to 5 CCs. A Long Term Evolution (LTE) advanced terminal may simultaneously receive one or multiple component carriers, depending on its capabilities, and it is possible to aggregate a different number of component carriers of possibly different bandwidths in the UL and the DL. Both intra- and interband carrier aggregation are considered by LTE advanced as potential scenarios. They are illustrated in Figure 7.29. It covers both contiguous component carrier and noncontiguous component carrier aggregation in UL and DL.

Opportunistic spectrum aggregation use case is going one step beyond, by considering, e.g., an LTE system, able to manage the aggregation of the licensed bands and opportunistic spectrum in the TV white spaces. The main principle is to consider noncontiguous Carrier Aggregation between CCs from IMT bands and from UHF WS band. TVWS can be used for improving user peak rate, network traffic offload, and/or, in specific geo-graphical areas for coverage extension. Carrier Aggregation, as a network controlled function, may be considered in the downlink only (as for DC-HSPA) to fit asymmetric services or in both downlink and uplink. As an example, in FDD radio technologies the Carrier Aggregation in TVWS may

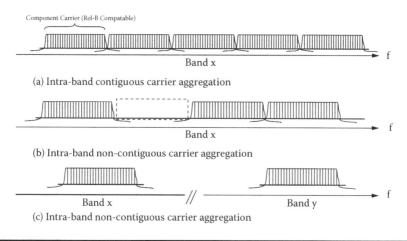

(a) Intra-band contiguous carrier aggregation

(b) Intra-band non-contiguous carrier aggregation

(c) Intra-band non-contiguous carrier aggregation

Figure 7.29 Carrier aggregation types.

improve asymmetric services by enabling extra DL resources. An operator which operates in both IMT and TVWS can control which part of the spectrum should be accessible to the terminals.

7.5.1.8 System Requirements

ETSI RRS has in scope to define the system requirements for the operation in TVWS. The initial system requirements are derived from the use cases:

1. Base station based sensing:
   - The market may be sensitive to the terminal price, so it would be beneficial that using white space would not add too much complexity on terminals. Cognitive capability may not be required for terminals; in this case the radio base stations are able to detect the primary signal (geo-location database and sensing) without assistance from mobiles.
2. Geo-location database management
   - The spectrum sensing has not been considered to be a reliable method to detect the available TVWS frequency resources. Therefore, geo-location database has been considered as mandatory. It shall be possible to get information on incumbent user(s) and optionally also on secondary user(s) of the TVWS frequency bands from a reliable and secure geo-location database. The usage of TVWS frequency bands may be registered to the geo-location database. The operation of the geo-location database can be done by a third party service provider.
3. Primary user protection
   - The interference to TV receiving signal should be managed below an acceptable level. The reliability of primary user protection determines the possibility whether the spectrum owners are willing to share their white space or not.
   - There are two phases for primary user protection: (1) secondary user has the capability to discover TVWS which is not occupied by primary user; (2) secondary user has the capability to free the spectrum in a reasonable time after primary user appears.
4. Centralized spectrum allocation
   - In order to avoid the interference between different cells, a central control point is provided with the capability to allocate available TVWS spectrum to the cells. All involved radio base stations operating in TVWS has to connect to the central control point to negotiate available TVWS frequency resource before using it. The central control point has the capability to reduce the interference between neighbor cells by allocating different TVWS frequency resources to them or configuring the radio base stations with reasonable transmission power.

5. Coexistence between secondary spectrum users
   - A network operating in the TVWS needs to have capabilities to co-exist with other networks/users in the same band. These capabilities include (1) a capability to discover other secondary users of the spectrum including networks and nodes, (2) a capability to negotiate and agree on spectrum usage including configuration of the network operation parameters, and (3) a capability to identify changes in the resources including configuring network to perform measurements.

6. Fast initial access to network
   - Network provides the terminal with the information about the available TVWS frequency bands. For example, a terminal accessing to a certain RAT such as HSPA operating in licensed band to obtain the information through in-band CPC. As a result, there is no need for the terminal to scan large frequency bands to find available TVWS resource.

7. Seamless handover within TVWS spectrum
   - Since TVWS may be used to solve the coverage problem of future 3G/4G networks, it is important to ensure a stable connection for the secondary user when the primary signal emerges. Seamless handover within TVWS spectrum is needed, and the real-time services should be able to be handed over to the new frequency within a reasonable disruption time.

8. TDD operation
   - The system operating in TVWS shall address heterogeneous network made of various radio technologies, for instance, the system shall not prevent TDD radio mode of operation.

9. Base station management of TVWS access
   - It may be assumed that a base station accessing TVWS will apply a combination of (distributed/centralized) sensing and geo-location database access (contain information of intended spectrum usage by primary users) in order to determine available and suitable TVWS bands for the operation of cellular systems. A corresponding "TVWS Access Management" function will deal with the acquisition of the required information and the corresponding configuration of the concerned mobile devices.

Acknowledgments

The authors would like to thank Kari Kalliojarvi, Antti Piipponen, Jens Gebert, Gianmarco Baldini, Pierre-Jean Muller, and Stanislav Filin for their contributions

References

1. S. Shellhammer, "P802.19.1 Project Authorization Request," available: https://development.standards.ieee.org/get-file/P802.19.1.pdf?t=41492900024.
2. IEEE Standards Association, "IEEE Standards Coordinating Committee 41 (Dynamic Spectrum Access Networks) Background," available: http://grouper.ieee.org/groups/scc41/.
3. IEEE Standards Association, "IEEE Standard for Architectural Building Blocks Enabling Network-Device Distributed Decision Making for Optimized Radio Resource Usage in Heterogeneous Wireless Access Networks," *IEEE Std 1900.4-2009*, pp. C1–119, February 27, 2009.
4. S. Filin, H. Harada, H. Murakami, K. Ishizu, and G. Miyamoto, "IEEE Draft Standards P1900.4.1 and P1900.4a for Heterogeneous Type and Spectrum Sharing Type Cognitive Radio Systems," *Personal, Indoor and Mobile Radio Communications, 2009 IEEE 20th International Symposium on Personal, Indoor, and Mobile Radio Communications,* 2009.
5. S. Filin, K. Ishizu, and H. Harada, "IEEE draft standard P1900.4a for architecture and interfaces for dynamic spectrum access networks in white space frequency bands: Technical overview and feasibility study," *Personal, Indoor and Mobile Radio Communications Workshops (PIMRC Workshops), 2010 IEEE 21st International Symposium on Personal, Indoor, and Mobile Radio Communications,* 2009.
6. Markus Mueck et al., "ETSI Reconfigurable Radio Systems: Status and Future Directions on Software Defined Radio and Cognitive Radio Standards," *Communications Magazine, IEEE* , vol. 48, no. 9, pp. 78–86, September 2010.
7. K. Harrison, S. M. Mishra, and A. Sahai, "How Much White-Space Capacity Is There?" *New Frontiers in Dynamic Spectrum, 2010 IEEE Symposium on New Frontiers in Dynamic Spectrum,* 2010.
8. T. Baykas, M. Kasslin, and S. Shellhammer, "System Design Document," doc.: IEEE 802.19-10/0055r3, March 2010. Available: https://mentor.ieee.org/802.19/dcn/10/19-10-0055-03-0001-system-design-document.pdf.
9. Draft Standard for Architectural Building Blocks Enabling Network-Device Distributed Decision Making for Optimized Radio Resource Usage in Heterogeneous Wireless Access Networks—Amendment: Architecture and Interfaces for Dynamic Spectrum Access Networks in White Space Frequency Bands, IEEE P1900.4a D0.4, March 2010.
10. M. Mueck et al. "ETSI RRS—The Standardization Path to Next Generation Cognitive Radio Systems," *Personal, Indoor and Mobile Radio Communications, IEEE 21st International Symposium on New Frontiers in Dynamic Spectrum,* 2010.
11. Federal Communications Commission, "Authorization and Use of Software Defined Radio: First Report and Order," Federal Communications Commission: Washington, D.C., September 2001.
12. M. Mueck, M. Di Renzo, and M. Debbah, "Opportunistic Relaying for Cognitive Radio Enhanced Cellular Networks: Infrastructure and Initial Results," *Wireless Pervasive Computing (ISWPC), 2010 5th IEEE International Symposium on Wireless Pervasive Computing (ISWPC),* 2010.

Chapter 8

System Level Analysis of OFDMA-Based Networks in TV White Spaces: IEEE 802.22 Case Study

Przemysław Pawełczak, Jihoon Park,
Danijela Čabrić, and Pål Grønsund

Contents

8.1 Introduction

Wireless networks equipped with opportunistic spectrum access (OSA) capabilities [1,2] allow the network subscriber, often referred to as secondary user (SU), to access frequency bands temporarily and spatially unused by the primary user (PU) of the licensed spectrum. Such spectrum, called (radio frequency) white space, could be a temporarily vacated paging band or an unused TV channel. Through its flexible operation, OSA is believed to improve spectrum utilization significantly and to solve the spectrum scarcity problem [3]. To be able to gain from spectrum access opportunities, OSA network usually operates on multiple channels. However, the complexity of network implementation for multichannel OSA networks operating with multiple users is increased. Therefore, a centralized network, in which a single node (usually denoted as base station (BS)) manages the network, is considered as a practical OSA network implementation. This is due to the fact that BS has far more processing power than SU nodes; thus all network management and control functions can be performed directly by such a central entity.

For a centralized network, orthogonal frequency division multiple access (OFDMA) is a popular multiple access technique where orthogonally divided frequency subcarriers are assigned to individual users of the network. For OFDMA, subcarrier assignment is usually performed by the BS and can be based on the quality of service requirements of the individual users. Because of high spectral efficiency, as well as robustness against intersymbol interference and time synchronization errors, OFDMA has been a design choice of such recent wireless networking standards as IEEE 802.16 [4], IEEE 802.20 [5], and 3GPP Long Term Evolution [6]. In the OSA context, OFDMA has also been chosen as a multiple access technology for IEEE 802.22 draft standard [7–9]. Therefore, the study of OFDMA in the context of OSA network is essential. In this chapter, we develop a throughput evaluation framework for the OFDMA-based centralized OSA system. Specifically, we focus on IEEE 802.22 as a case study, because it is the first wireless networking standard operating based on the OSA principle.

We present an analytical model that enables an evaluation of a specific design option of Opportunistic Spectrum Orthogonal Frequency Division Multiple Access (OS-OFDMA). Considered OS-OFDMA design mimics in many ways the structure of IEEE 802.22. The proposed model takes into account continuous subchannel allocation algorithms, as well as the ability to bond separate noncontinuous frequency bands. Different user priorities and channel dwell times, for the SUs and the PUs, are investigated. Finally, the model accounts for different types of traffic. The model and the results presented in this part of the chapter are drawn from [10]. Extension of the analysis presented in this chapter are also available in [11].

We also present results of IEEE 802.22 evaluation based on the detailed implementation of complete protocol stack in Network Simulator-2 (ns-2) [12]. We describe the simulator and compare features of IEEE 802.22 with its predecessor: IEEE 802.16. Further, we describe the detailed implementation of physical layer, MAC layer, cognitive functions, and spectrum management of IEEE 802.22. Based on the developed simulator, we present results showing the impact of temporal and/or spatial wireless microphone operation on IEEE802.22 performance. The ns-2 implementation allowed us to evaluate components of the network that were difficult to assess analytically.

The rest of the chapter is organized as follows. Work related to the scope of this chapter is presented in Section 8.2. A system model for OS-OFDMA network design considered is presented in Section 8.3. An analytical model enabling performance evaluation of the system is presented in Section 8.4. Numerical and simulation results are presented in Section 8.5. Finally, the chapter is concluded in Section 8.6.

8.2 Related Work

The first paper that has introduced the idea of OS-OFDMA was [13], referred to there as Spectrum Pooling, where OFDM subcarriers assigned to individual SU of the spectrum have been deactivated whenever PU of the radio frequency band reappeared. Many recent papers have proposed optimal OFDMA resource allocation in an OSA scenario; for example, refer to a recent work of [14]. However, it usually has been assumed that information about channels occupied by PUs was known a priori, and traffic characteristics and priority levels of SUs were not considered.

In [15] a general framework of IEEE 802.16 with OSA capabilities has been proposed. The focus of the paper was mostly on propagation calculations, including coverage, interference, and protection distances. However, a very simplified networking model, based on Erlang-B formula, was proposed [15, Sec. V-A]. In

[16] IEEE 802.22-like network was evaluated using only simulations, focusing on a very limited set of scenarios (e.g., not considering different priorities and traffic characteristics of individual users). Lastly, authors in [17] analyzed OSA-based OFDMA network with focus on resource allocation, with one class of OSA traffic and static licensed user.

Presumably the first formal proposal of using OFDMA approach in an OSA scenario was IEEE 802.22 draft standard on Wireless Regional Access Networks [7] (for a general overview of the standard please refer to [8,9].) In the application domain IEEE 802.22 has been proposed to bridge remote wireless sensor networks with the command center [18] or support connectivity in the rural areas [19]. Note that IEEE 802.22 is not the only networking standard that operates in the TV white spaces. The other is recently published ECMA TC48-TG1 standard [20] focusing on porting local area networks to TV white spaces, and the recently initiated IEEE 802.11af [21], similar in scope to the aforementioned ECMA activity.

While many papers analyzed a certain functionality of IEEE 802.22 network, we are not aware of any work that has performed cohesive analysis of IEEE 802.22 taking into account its intrinsic features: multiple classes of traffic, different levels of PUs, their temporal activity, and OFDMA subchannel allocation process. Such types of analysis have been comprehensively performed in [22] to evaluate IEEE 802.16 system, but obviously this work cannot be used directly to evaluate OSA-based OFDMA system due to the lack of spectrum sensing and incumbent users activity analysis. IEEE 802.22 features that were discussed in isolation from the rest of the network stack include efficient spectrum sensing algorithm design [23,24], circuit design for dedicated spectrum sensing [25], MIMO (multiple input, multiple output) extensions for IEEE 802.22 physical layer [26], game theoretic analysis of the IEEE 802.22 networks coexistence [27–32] (with joint resource allocation), duplexing schemes [33] (frequency hopping operation), [19] (time division duplex design), and mesh establishment [34].

8.3 System Model

The considered OS-OFDMA system, mimicking IEEE 802.22 operation, was explored via analysis and simulations. For the purpose of analysis we relaxed some of the assumptions and simplified the properties of the network. Our aim was to investigate traffic characteristics of the OS-OFDMA system (i.e., average throughput and blocking probability); thus detail physical layer representation of the network was not required.

Physical layer properties (i.e., average throughput obtained by OS-OFDMA network as a function of primary user detection rate and signal characteristics) were evaluated using the detail simulation model developed in ns-2 [12]. This allowed the exact implementations of the higher layer protocols at the link, network, transport, and application layers of the Open Systems Interconnection (OSI) stack.

Because IEEE 802.22 is very similar to IEEE 802.16e [35], we have adapted an extensive IEEE 802.16e implementation for ns-2 [36], developed by the WiMAX Forum [37], to represent IEEE 802.22 exactly as specified by the standard [38]. The features that are different from IEEE 802.16 were implemented and conform to their best extent to the functional requirements as specified in the draft v.7.0 of the IEEE 802.22 standard. The most relevant parameters for the two systems were compared in [8]. This comparison has been extended in Table 8.1. The last column does not list all the profiles implemented in the ns-2 simulator, but only the actual profile that is used for the simulations of IEEE 802.22 throughout the rest of this chapter. This will be explained further in the following sections. In the subsequent sections we present the general system model, differentiating between assumptions needed for analytical evaluation and simulations, where necessary.

8.3.1 Network Structure

For the evaluation we consider an OS-OFDMA system limited to 1 BS and 4 Customer Premises Equipment (CPEs) (in case of simulations) or 1 CPE (in case of analysis). The CPEs are fixed at their locations, which are in accordance with the IEEE 802.22 draft standard. The network setup is illustrated in Figure 8.1 where the CPEs are IEEE 802.22 nodes, and the PUs denote primary users (both wideband primary user (WPU) and narrowband primary user (NPU), whose properties will be described in Section 8.3.2). The small oval illustrates the coverage area for "PU 1a," and the big oval the coverage area for the "802.22 BS." In the simulations, because we consider multiple secondary nodes, CPEs other than "CPE 1" can also detect the signals from "PU 1a" if the received signal strength is above the sensing threshold. The number of nodes able to detect the PU will depend on the distance between the BS and CPE, d_{cpe}, and also the distance between the CPE and PU, d_{wm}. Note that the BS covers the entire cell and that all CPEs are placed equidistantly and uniformly around the BS.

NPUs and WPUs are the greatest challenge for IEEE 802.22 operation because their channel occupancy is more dynamic than that of TV broadcasting and their presence cannot certainly be downloaded from a database. Also, they transmit with lower power than TV broadcasters and are therefore more challenging to detect. NPU and WPU activity will be detected only by sensing techniques with no help from beacon protocol (i.e., IEEE 802.22.1). The PUs operate in the UHF band and are configured with transmit power of 200 mW, unless otherwise stated.

8.3.2 User Class Division

In the considered OS-OFDMA system we assume PUs and SUs are being divided into classes. That is, for PUs we consider low-activity wideband users, denoted as WPUs, and high-activity narrowband PUs, denoted as NPUs. Note that such a setup can mimic many scenarios. In a white space context WPUs and NPUs can

Table 8.1   Parameters Used in the ns-2 Implementation of IEEE 802.22, Compared against IEEE 802.16 [35,39] and IEEE 802.22 [8,38] Standards

Parameter	IEEE 802.16	IEEE 802.22	ns-2
Bandwidth	10MHz	{6, 7, 8}MHz	6MHz
FFT size	1024	2048	2048
Frequency (f)/ channels	2.5–2.69 GHz	5–13, 14–51 MHz	575–593 MHz
Frame size	5 ms	10 ms	10 ms
Duplexing method	TDD	TDD	TDD
Tx/Rx Transit Gap (TTG)	105.7µs	83.33/190/270 µs	83.33 µs
Rx/Tx Transit Gap (RTG)	60 µs	210 µs	210 µs
Modulation types	{16,64}-QAM, QPSK	{16,64}-QAM, QPSK	{16,64}-QAM, QPSK
Coding rates	$^1/_2$, $^2/_3$, $^3/_4$, $^5/_6$	$^1/_2$, $^2/_3$, $^3/_4$, $^5/_6$	$^1/_2$, $^2/_3$, $^3/_4$
Error correction coding	CC, CTC, LDPC	CTC/BTC	No
Max power	BS: 43, CPE: 23 dBm	BS/CPE: 36 dBm	BS/CPE: 36 dBm
Assumed noise figure	BS: 4, CPE: 7 dB	BS/CPE: 10 dB	BS/CPE: 10 dB
QoS classes	UGS, rtPS, ErtPS, nrtPS, BE	UGS, rtPS, ErtPS, nrtPS, BE	UGS, BE
Cyclic prefix mode	$^1/_4$, $^1/_8$, $^1/_{16}$, $^1/_{32}$	$^1/_4$	$^1/_4$
OFDM mapping	Rectangular	DL: vert. UL: horiz.	vert.
Error protection	HARQ	ARQ	ARQ
Subcarrier. spacing	10.94 kHz	3.348/3.906/4.464 kHz	3.348 kHz
Useful symbol length	91.4 µs	298.7/256/224 µs	298.7 µs

Table 8.1 (*Continued*) Parameters Used in the ns-2 Implementation of IEEE 802.22, Compared against IEEE 802.16 [35,39] and IEEE 802.22 [8,38] Standards

Parameter	IEEE 802.16	IEEE 802.22	ns-2
Guard time	11.4 μs	37.34/32/28 μs	37.34 μs
Symbol duration	102.9 μs	373.3/320/280 μs	373.3 μs
Sampling frequency	11.2MHz	6.857/8/9.145MHz	6.857MHz
Sampling period	0.18 ms	0.299/0.256/0.224 ms	0.299 ms
Symbols per frame	48	26/30/34	26
Used subcarriers	840	1680	1680
Guard and null subcarriers	184	368	368
Pilot subcarriers	DL: 120, UL: 280	DL/UL: 240	DL/UL: 240
Data subcarriers	DL: 720, UL: 560	DL/UL: 1440	DL/UL: 1440
Subcarriers/ subchannel	DL: 24, UL: 28	DL/UL: 24	DL/UL: 24
Subchannels	DL: 30, UL: 35	DL/UL: 60	DL/UL: 60
Pilot location	Distributed	Distributed	Distributed
Power control	Distributed	Distributed	No
Subcarrier allocation	FUSC, PUSC, Cont.	PUSC	PUSC
Sensing strategy			Two stage sensing
Coarse sensing duration		—	0.5 ms
Fine sensing duration		—	30 ms
Fine sensing interval		—	∞ (event based)
Cooperative sensing		—	"OR" rule

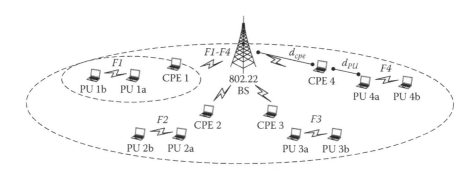

Figure 8.1   Illustration of network scenario considered for the ns-2-based evaluation of OS-OFDMA system (note that in the analysis only one CPE is considered).

resemble wideband wireless assist video devices and wireless microphones (WMs) [40], respectively. For SUs, we consider nonelastic constant bit rate (CBR, also called unsolicited grant service (UGS)) and elastic, variable bit rate (VBR, also called best effort (BE)) class [4,22]. We assume that at most $U_{w, \max}$, $U_{n, \max}$, $U_{c, \max}$, and $U_{v, \max}$ of, respectively, WPUs, NPUs, CBR, and VBR connections, can be active at the same time in the considered bandwidth [22, Sec. III-B].

8.3.3 Traffic Model

For the data traffic of SUs and PUs, for analytical tractability reasons [41], we assume that all users generate new connections according to the negative exponential distribution for the inter-arrival time and burst departure time. The average inter-arrival time and the average departure time are, respectively, $1/\lambda_w$ and $1/\mu_w$ for WPU, $1/\lambda_n$ and $1/\mu_n$ for NPU, $1/\lambda_c$ and $1/\mu_c$ for CBR, and $1/\lambda_v$ and $1/\mu_v$ for VBR. Also we assume that connection of each class except VBR occupies a fixed number of subchannels. We denote the number of subchannels utilized by a user as l_w for WPU, l_n for NPU, l_c for CBR, and l_v for VBR. Note that the number of subchannels assigned to a VBR connection can be a real number, which stems from the fact that one data frame consists of a group of OFDMA symbols, and the symbols in the frame can be assigned to multiple VBR connections. Also, for VBR connections, the burst departure time depends on the number of subchannels used by VBR; thus $1/\mu_v$ is an average departure time when one subchannel is assigned to the VBR connection.

Note that when NPU and WPU appear, the traffic pattern is characterized by a 100% duty cycle until the NPU or WPU disappears. In reality, NPUs and WPUs appear very infrequently and do not operate for very long durations (i.e., in the order of hours). In order to be able to simulate the activity of PUs in the ns-2 simulator, as well as in the simulations used to verify analytical results, we scale the realistic $1/\lambda_{x=w,s}$ and $1/\mu_{x=w,s}$ to values that result in short simulation time.

8.3.3.1 Traffic Scheduling

In the analysis we do not consider subcarrier scheduling. However, the simulation model implements subcarrier allocation procedure. The medium access control (MAC) layer of IEEE 802.22 uses linear scheduling to allocate OFDMA slots to traffic from the upper layers in both the downlink and uplink subframes, as opposed to a rectangular scheduling in IEEE 802.16. An OFDMA slot can be characterized as a {subchannel, symbol}-tuple in the frequency and time domain. Vertical striping is used for both DL and UL subframes in the simulator, which means that OFDMA slots are allocated in frequency first and then in time (i.e., the first symbol is filled with data before the next symbol). For more information please refer to [42].

The transmit/receive transition gap (TTG) is set to 210 μs, which supports a 30 km distance between BS and CPE. A dynamic TTG is needed for greater distances; however this is not implemented in the ns-2, because the simulation scenarios considered involve only small distances. The receive/transmit transition gap (RTG) is set to 88.33 μs. There are a total of 26 symbols, each of 373.33 μs duration, where the ratio is set to 2:1. Thus 16 and 9 symbols are used for the downlink and uplink subframe, respectively. Both the downlink and uplink subframe have a total of 60 subchannels, where each subchannel consists of 28 subcarriers out of which there are 24 data and 4 pilot subcarriers. Finally, note that the self coexistence protocol is not used in the simulations because a single-cell scenario is considered.

8.3.3.2 Traffic Prioritization

Further, we assume that the system has hierarchical structure such that the WPU has the highest priority in accessing bandwidth, and NPU is second in the access hierarchy, followed by CBR and VBR. In other words, if different user classes access the same subchannel, the lower priority class must evacuate in order for the higher class to utilize the subchannel. The evacuated CBR switches to the other idle subchannels or drops the connection if there is no idle subchannel available. For VBR, if the PU is detected, active VBR connection squeezes the bandwidth [22, Sec. III-B] excluding the subchannel occupied by PU, and if there is no subchannel detected as idle, it buffers data until PU disappears. Note that the behavior of VBR promises to obtain the highest possible throughput, as demonstrated in [43, Sec. V-B and Fig. 6], assuming there is no switching overhead in the event of squeezing the bandwidth, while CBR does not consider buffering due to the excessive delay that this class might experience while waiting for WPU or NPU to vacate the bandwidth.

Both CBR traffic and VBR traffic will be simulated in ns-2 over the network layer protocol UDP. A computer connected via Ethernet to the IEEE 802.22 BS will establish a link to the IEEE 802.22 users in order to measure the throughput. VBR traffic will mostly be used to measure maximum throughput. Maximum traffic rates will be simulated by constantly transmitting VBR traffic with bit rates higher than the channel capacity. Also, the packet size is specified to 1500 bytes. CBR will not be used to measure throughput because CBR is mostly used in voice and video applications where it is more interesting to measure the packet loss and call drop rate.

8.3.4 Channelization Structure

The OS-OFDMA system operates on X channels and consists of Y subchannels. In general, each subchannel consists of a set of subcarriers. The total number of subchannels which can be utilized by the SU is then $M = XY$. Moreover, in the analysis, we assume that OS-OFDMA system is capable of deactivating subcarriers so that the fractional bandwidth of a subchannel can be utilized [17,44]. The channel capacity used in the analysis is assumed to be constant and denoted as C, while in the simulations it will be a function of channel conditions and physical layer attributes.

Guard bands are not considered in the allocation of these bands, but are considered by the IEEE 802.22 system itself by nulling out subcarriers at both edges of the band. The IEEE 802.22 network can operate on any channel if it is not used by the primary user. Channel bonding of scattered available channels will not be considered; therefore only one of the available frequencies (i.e., $X = 1$) will be used by the IEEE 802.22 system at any time. As noted earlier, fractional frequency reuse, where some subcarriers occupied by PUs in one frequency band are notched out and the remaining are utilized by IEEE 802.22, will be considered in the analysis but will not be considered in the ns-2 simulations. The approach taken in ns-2 simulations follows the FCC rules for opportunistic spectrum use, while the analytical model provides more flexible access to spectrum based on fractional spectrum reuse.

In the IEEE 802.22 ns-2 implementation only the 6 MHz profile is used. The channels 31 to 34 in the UHF band are selected for operation, where only NPUs are the incumbent services. These channels correspond to the center frequency of 575, 581, 587, and 593 MHz, respectively.

8.3.5 Slot Structure

SU transmission in the analyzed OS-OFDMA system is divided into time slots of t_f seconds, equal to frame size transmitted by SU. At the beginning of each frame each SU senses all channels to detect sudden appearance of each PU during t_s seconds. We assume that PU is stationary for the duration of the sensing time, noting however that the arrivals of PUs do not necessarily have to be synchronized with OS-OFDMA network.

8.3.6 Propagation Model

As noted, the analysis considers perfect channel conditions (excluding errors during primary user detection). Therefore, in ns-2 we had to consider channel properties.

The propagation model used in the ns-2 simulations is the COST Hata [45] path-loss model which is intended for large cells, with BS being placed higher than the surrounding rooftops and designed for frequencies between 150 and 1500 MHz, such that propagation loss is calculated as

$$P_L = 69.55 + 26.16\log_{10}(f_c) - 13.82\log_{10}(h_{BS}) - C_H$$
$$+ [44.9 - 6.55\log_{10}(h_{BS})]\log_{10}(d) \tag{8.1}$$

where f_c denotes the carrier frequency, d denotes the distance between BS and CPE (given in km), and h_{BS} denotes the height of the BS antenna. A correction factor $C_H > 0$ for the CPE antenna height is used for small to medium cities and large cities. In suburban scenarios $C_H = 0$, which will be used in the simulations that consider suburban areas. Both the BS and CPEs are configured with 36 dBm transmit power. Channel gains for both the BS and CPE are set to 1 dB in the ns-2 simulations. Dynamic transmit power adaptation is not implemented.

Further, the COST Hata model is combined with a Clarke-Gans [46] implementation of Rayleigh fading, which has been extended to support 2048 subcarriers in IEEE 802.22. Three International Telecommunication Union (ITU) power delay spread models [47] are implemented: Pedestrian A, Pedestrian B, and Vehicular A. The Vehicular A model is best suited for the considered IEEE 802.22 scenario with larger cells in suburban areas and a high BS antenna and will therefore be used in the subsequent simulations.

Interference modeling is done at the subcarrier level by capturing packets from all transmitters in the system, whether from IEEE 802.22 nodes or primary users. Once the received signal to interference and noise ratio (SINR) on each subcarrier used by a packet is obtained, a decision is made to further process or drop the packet. This is done by first finding the exponential effective signal-to-interference ratio (SIR) mapping [48] to get the effective SINR, and then the block error rate (BLER) from SINR, modulation, and coding scheme and block size. Based on the BLER value a decision is made whether to drop the packet or not.* It is assumed for simplicity that the locations of all TV stations are obtained from the incumbent database, and interference from TV stations is therefore not considered.

8.3.7 Primary User Detection Process

Detection process is assumed to be imperfect, and we denote the probability of false alarm as p_f and the probability of detection as p_d. We assume that $p_d = p_{d,min}$, where $p_{d,min}$ is a required detectability level imposed by the spectrum regulator. For details

---

* Please refer to [36] for a detailed description of of the OFDMA physical layer implementation and interference modeling. Note that the propagation model, operating frequency range, and system profiles are reimplemented to fit the UHF bands, IEEE 802.22, and suburban scenarios.

of derivation of detection process parameters in the network performance analysis context, the reader is referred to [43]. In the analysis we assume simple single stage sensing without cooperation based on energy detection. The simulations in ns-2, on the other hand, consider more complex spectrum sensing implementation.

The main cognitive functions of IEEE 802.22 are the spectrum sensing function (SSF), the geo-location function (GEO), the spectrum manager (SM) in the BS and the spectrum automaton (SA) in the CPE. All functions are implemented in ns-2, except for GEO and transmit power control limits, which might be imposed by the radio regulator. For the GEO function, it is assumed that the list of occupied TV channels is obtained from a database. Because the transmit power control limits are not implemented, IEEE 802.22 will switch to a new channel if a primary user is detected on that channel, based on the threshold detection. If no channels are available IEEE 802.22 will cease transmission.

8.3.7.1 Spectrum Manager and Spectrum Automation

The SM is implemented in the BS and is responsible for deciding which channel to use based on information from the SSF. The SM specifies the set of channel lists, that is, the backup, candidate, and protected channel lists. This is done by communicating with GEO and SSF functions. In IEEE 802.22 a channel will originally get status as backup channel when sensed as unused every six seconds over a period of 30 seconds. An algorithm can be applied to optimize which of the backup channels will be selected as the operating one. The first available channel in the list will be selected as the operating one, where the list is in increasing order.

In the CPE, SA is the cognitive function which actually is a lightweight version of the SM in the BS. The SA is controlled by the BS and is mostly responsible for reporting information to the SM. In rare cases, SA itself is responsible for sensing at initial CPE power on, when it loses contact with the BS or during an idle time when there are no tasks pending.

8.3.7.2 Two-Stage Spectrum Sensing Implementation

The SSF can first be classified into in-band sensing, which senses the operating channel, and out-of-band sensing, which senses activity on other channels that potentially can be used by the IEEE 802.22 system. For in-band sensing the two-stage spectrum sensing approach, as specified by the standard, is applied and illustrated in Figure 8.2, where (i) coarse sensing uses simple energy detection for frequent and short sensing periods t_c. If coarse sensing detects an incumbent signal it can switch to (ii) fine sensing that uses more detailed detection for a longer period t_s. If an incumbent signal is detected by fine sensing, then the operating channel is switched to one of the backup channels. Thus the two-stage spectrum sensing strategy is event-based where a detection by coarse sensing triggers a fine sensing period. However, if coarse sensing does not detect any incumbents, an optional time-based

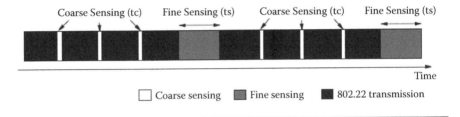

Figure 8.2 **Illustration of IEEE 802.22 two-stage spectrum sensing mechanism [38, Sec. 6.23.3.1].**

approach has been implemented where fine sensing periods can be started at specific times. Coarse and fine sensing are often referred to as intraframe and interframe sensing, respectively.

Probability of detection p_d and probability of false alarm p_f are implemented separately for coarse and fine sensing. In the simulation we were able to manipulate p_d and p_f in order to simulate the performance with different sensing constraints. The effect of p_d and p_f on the IEEE 802.22 will mostly be considered in unreliable coarse sensing simulation scenarios. For fine sensing that uses more advanced sensing techniques (often over longer periods), the sensor is considered to be more reliable, and p_d and p_f are therefore set to one and zero, respectively. Further, cooperative sensing with the "OR" rule is used for all sensing stages so that, if one node detects a primary user, the BS decides that it is actually present. Therefore, probabilities of the total system false alarm and detection are calculated as $p_f = 1 - (1 - p_{f,i})^N$ and $p_d = 1 - (1 - p_{d,i})^N$, respectively, where N is the number of cooperative sensing nodes and $p_{f,i}$ and $p_{d,i}$ are the probability of false alarm and detection for each single node, respectively.

For the cooperative sensing case with a total of $N = 5$ users (i.e., one BS and four CPEs), we consider $p_d = 0.99$ for the whole system. To find p_d for each user we use the "OR" rule to calculate individual probability of detection as $p_{d,i} = 1 - \sqrt[N]{1 - p_d} = 0.6019$. From [49, Eq. (1)], assuming intraframe sensing time of $t_s = 0.5$ ms and SNR=-10 dB, we get $p_{f,I} = 0.0024$.

Coarse sensing is carried out during allocated time periods at the end of the uplink OFDMA subframe, typically for 1 ms. Sensing occurs at every second OFDMA frame. Quiet periods for fine sensing are scheduled by the MAC layer. A sensing window specification of IEEE 802.22 specifies the sensing duration and the intervals between consecutive sensing intervals. In the simulations, this sensing window specification will remain fixed during every simulation. The fine sensing period is typically set to 30 ms, spanning three OFDMA frames. A predefined interval might be set. It might also be infinite if the event-based strategy is used.

Out-of-band sensing is performed during quiet periods allocated for the fine sensing periods, and all relevant channels are sensed during one fine sensing period. Note that for the self coexistence protocol, needed when multiple BSs are used, predefined intervals between fine sensing periods should be scheduled. This would, however, have a minor impact on the system throughput.

8.3.8 Error Protection

In the analysis, we assume that SUs can transmit data on the misdetected subchannels with the same performance as on the idle subchannels, and thus all subchannels detected as idle can be utilized for data transmission in our model. When subchannels are detected as occupied by any of the PUs, they are automatically excluded from OSA operation until PUs vacate them, just as assumed in, e.g., [17]. In analysis we assume a perfect channel, thus no errors are considered. However, in the simulations, we consider channel errors. For error protection IEEE 802.22 uses automatic repeat request (ARQ), which is supported by the ns-2 simulator. For error correction coding IEEE 802.22 uses convolutional turbo codes (CTC) and block turbo code (BTC), although error correction coding will not be used in the ns-2 simulations. For the subcarrier allocation strategy, IEEE 802.16 supports partially used subcarrier (PUSC) and fully used subcarrier (FUSC) [39], whereas IEEE 802.22 uses PUSC, which also is implemented in the ns-2 simulator. This, however, is irrelevant for the analysis, because only one BS is considered, and subcarrier randomization is needed to minimize probability of same subcarrier selection by different IEEE 802.22 systems.

8.4 Analytical Evaluation of OS-OFDMA System

Given the system model, we can proceed with the derivation of the metrics of interest (i.e., average network throughput and blocking probabilities). Note, again, that the model presented in this section is taken from [10].

8.4.1 Throughput Analysis

Generally, in OFDMA-based wireless networks, throughput depends on how many subchannels are utilized by CBR and VBR connections [22, Sec. III-B]. Let $\Pr_1(m_c, m_v)$ be the probability that m_c and m_v number of subchannels are utilized by CBR and VBR connections, respectively. Then the total system throughput H can be calculated as

$$H = \eta \sum_{m_c=0}^{M} \sum_{m_v=0}^{M} C(m_c + m_v) \Pr_1(m_c, m_v) \tag{8.2}$$

where $\eta = (t_f - t_s)/t_f$ is the ratio of data transmission time to total frame length. To compute metric H we need to decompose $\Pr_1(\cdot)$ into multiple probabilities that describe relations between different network functionalities and users.

Values of m_c and m_v are easily determined if we know the number of CBR connections connected with the base station, U_c, the number of connected VBR connections connected with the base station, U_v, and the number of subchannels detected as idle, M_a. Thus, to compute $\Pr_1(\cdot)$, we need to decompose it as

$$\Pr_1(\cdot) \triangleq \sum_{U_c=0}^{U_{c,\max}} \sum_{U_v=0}^{U_{v,\max}} \sum_{M_a=0}^{M} \Pr_2(m_c, m_v | U_c, U_v, M_a) \Pr_3(U_c, U_v, M_a) \qquad (8.3)$$

To derive $\Pr_2(\cdot)$ used in (8.3) we observe that the total number of subchannels used by all CBR users, $U_c\, l_c$, cannot be greater than the number of the available subchannels, M_a, because if there is no subchannel available the CBR connection will be blocked. Thus we only consider the case when $U_c\, l_c \le M_a$. Because a CBR connection has higher priority than a VBR connection, all CBR connections can transmit data through overall $m_c = U_c\, l_c$ subchannels. Then, remaining subchannels are used by VBR connections (i.e., $m_v = M_a - U_c\, l_c$). On the other hand, if there are no VBR connections in the system, $m_v = 0$. Therefore $\Pr_2(\cdot) = 1$ when $m_c = U_c\, l_c$, $m_v = M_a - U_c\, l_c$ for $U_c\, l_c < M_a$, $Uv > 0$, or $m_c = U_c\, l_c$, $m_v = 0$ for $U_c\, l_c < M_a$, $Uv = 0$, and zero otherwise.

To calculate $\Pr_3(\cdot)$ in (8.3), we model SU traffic as Markov chain with the state $\{U_c,\ U_v,\ M_a\}$. The state transition probability is denoted as $\Pr_4(U_c^{(t)}, U_v^{(t)}, M_a^{(t)} | U_c^{(t-1)}, U_v^{(t-1)}, M_a^{(t-1)})$, describing the state transition from the state $\{U_c^{(t-1)}, U_v^{(t-1)}, M_a^{(t-1)}\}$ at time $t-1$ to state $\{U_c^{(t)}, U_v^{(t)}, M_a^{(t)}\}$ at time t. Based on the conditional probability property, we decompose probability $\Pr_4(\cdot)$ as

$$\Pr_4(\cdot) \triangleq \Pr_5(U_c^{(t)}, U_v^{(t)} | U_c^{(t-1)}, U_v^{(t-1)}, M_a^{(t)}, M_a^{(t-1)}) \Pr_6(M_a^{(t)} | M_a^{(t-1)}) \qquad (8.4)$$

Probability $\Pr_6(\cdot)$ allows us to compute the certain state evolution of M_a. Before defining $\Pr_6(\cdot)$ we focus on describing $\Pr_5(\cdot)$. We decompose it as

$$\Pr_5(\cdot) \triangleq \Pr_7(U_c^{(t)} | U_c^{(t-1)}, M_a^{(t)}, M_a^{(t-1)}) \Pr_8(U_v^{(t)} | U_v^{(t-1)}, U_c^{(t)}, U_c^{(t-1)}, M_a^{(t)}, M_a^{(t-1)}) \quad (8.5)$$

Before deriving $\Pr_7(\cdot)$ and $\Pr_8(\cdot)$, we need to introduce general equations for connection generation and connection release probability.

If the inter-arrival time has a negative exponential distribution with average arrival rate λ_x, the number of connections generated in a frame of length t_f has a Poisson distribution. However, because we limit the maximum number of users to $U_{x,\max}$, the number of users including newly generated connections, U, cannot exceed $U_{x,\max}$. Considering this limit, we derive the probability of k new connections being generated in a frame, $G_x(k|U)$, as

$$G_x(\cdot) = \begin{cases} \dfrac{(\lambda_x t_f)^k e^{-\lambda_x t_f}}{k!}, & k \geq 0,\ U < U_{x,\max} \\[2ex] \displaystyle\sum_{i=k}^{\infty} \dfrac{(\lambda_x t_f)^i e^{-\lambda_x t_f}}{i!}, & k \geq 0,\ U = U_{x,\max} \\[2ex] 0, & k < 0 \end{cases} \tag{8.6}$$

Note that the subscript $x = \{w, n, c, v\}$ indicates the class of users, i.e., w for WPU, n for NPU, c for CBR, and v for VBR.

Further, denoting the departure rate of each connection as μ, in general, the probability that j connections are released from m_x active connections during a frame of length t_f, $T_x(\ |\ m_x, \mu, t_f)$, can be calculated recursively as a product between the probability that one connection is released from m_x active connections at time t_1, and the probability that $j{-}1$ connections are released from $m_x{-}1$ active connections during remaining time $t_f{-}t_1$ as

$$T_x(\cdot) = \begin{cases} \displaystyle\int_0^{t_f} m_x \mu e^{-m_x \mu t_1} T_x(j-1 \big| m_x - 1, \mu, t_f - t_1) dt_1, & j > 0 \\[2ex] e^{-m_x \mu t_f}, & j = 0 \end{cases} \tag{8.7}$$

which after some manipulation reduces to

$$T_x(\cdot) = \binom{m_x}{j} e^{-m_x \mu t_f} (e^{\mu t_f} - 1)^j \tag{8.8}$$

From now on, we consider the connection release probability for $U_x^{(t-1)}$ users within the duration of t_f, and thus we abbreviate $T_x(j|\mu) \triangleq T_x(j|U_x^{(t-1)}, \mu, t_f)$.

To derive the state transition probabilities $\Pr_7(\cdot)$ and $\Pr_8(\cdot)$ in (8.5), we introduce supporting variables i and j denoting that i connections are newly generated and j connections are released. We now need to calculate all possible sets for i and j and apply them later to (8.6) and (8.7) to calculate $\Pr_7(\cdot)$ and $\Pr_8(\cdot)$.

Before finding possible sets for i and j, we consider valid conditions for $U_c^{(t)}, U_c^{(t-1)}, M_a^{(t)}$, and $M_a^{(t-1)}$ to derive $\Pr_7(\cdot)$. We denote the number of users being able to utilize all available subchannels as $U_a^{(t)} = \langle M_a^{(t)} / l_c \rangle$ at time t and $U_a^{(t-1)} = \langle M_a^{(t-1)} / l_c \rangle$ at time $t{-}1$. Then, because $U_a^{(t-1)}$, $U_a^{(t)}$ is the maximum number of users, the number of active CBR users cannot be greater than $U_a^{(t-1)}$, and $U_a^{(t)}$, respectively, i.e., $U_c^{(t)} \leq U_a^{(t)}$ and $U_c^{(t-1)} \leq U_a^{(t-1)}$. Now, because the possible sets of i and j are different for the cases $U_c^{(t)} < U_a^{(t)}$ and $U_c^{(t)} = U_a^{(t)}$, we consider them separately.

The first case $U_c^{(t)} < U_a^{(t)}$ represents the situation when the number of subchannels detected as idle is more than the number of all the subchannels that are going to be utilized by CBR connections before spectrum sensing. In other words no CBR connection is blocked due to the PU appearance. Because the number of CBR connections is $U_c^{(t-1)}$ at time $t-1$ and $U_c^{(t)}$ at time t, the change of the number of CBR connections must be the same as $U_c^{(t)} - U_c^{(t-1)}$, i.e., $i - j = U_c^{(t)} - U_c^{(t-1)}$. In addition, because there are $U_c^{(t-1)}$ active connections at time $t-1$, more than $U_c^{(t-1)}$ connections cannot be released, i.e., $j \leq U_c^{(t-1)}$. Therefore $\{i, j\} \in K_{ca}\{i, j | i - j = U_c^{(t)} - U_c^{(t-1)}, j \leq U_c^{(t-1)}\}$.

Next, we consider the other case, $U_c^{(t)} = U_a^{(t)}$, that CBR connections may be blocked due to the PU appearance. In this case, before spectrum sensing, the total number of connections including newly generated connections is $U_c^{(t-1)} + i - j$. However, after spectrum sensing, the connections that utilize subchannels detected as busy are blocked, and the remaining connections $U_c^{(t)}$ utilize all available subchannels $M_a^{(t)}$. Thus, the number of CBR connections before spectrum sensing, $U_c^{(t-1)} + i - j$, can be greater than or equal to the number of CBR connections after spectrum sensing, $U_c^{(t)}$, but should be less than or equal to $U_{c,\max}$, i.e., $U_c^{(t)} \leq U_c^{(t-1)} + i - j \leq U_{c,\max}$. Therefore $\{i, j\} \in \Gamma_{cb}\{i, j | U_c^{(t)} - U_c^{(t-1)} \leq i - j \leq U_{c,\max} - U_c^{(t-1)}, j \leq U_c^{(t-1)}\}$. Thus, considering all possible cases,

$$\Pr_7(\cdot) \triangleq \begin{cases} \sum_{\{i,j\} \in \Gamma_{ca}} T_c(j|\mu_c) G_c(i|U_c^{(t)}), & U_c^{(t-1)} \leq U_a^{(t-1)}, U_c^{(t)} < U_a^{(t)} \\ \sum_{\{i,j\} \in \Gamma_{cb}} T_c(j|\mu_c) G_c(i|U_c^{(t-1)} + i - j), & U_c^{(t-1)} \leq U_a^{(t-1)}, U_c^{(t)} = U_a^{(t)} \end{cases} \tag{8.9}$$

For VBR traffic, assuming all VBR connections share the same portion of the idle bandwidth, we calculate the number of subchannels for one VBR connection, l_v, as

$$l_v = \begin{cases} \dfrac{M_a^{(t-1)} - U_c^{(t-1)}}{U_v^{(t-1)}}, & U_v^{(t-1)} > 0 \\ 0, & U_v^{(t-1)} = 0 \end{cases} \tag{8.10}$$

For VBR connections, we do not need to consider the case that the VBR connection is blocked by the PU, because VBR connections are buffered instead of blocked when there is no available subchannel. Thus, with the l_v subchannels, $\Pr_8(\cdot)$ is defined as

$$\text{Pr}_8(\cdot) \triangleq \sum_{\{i,j\}\in\Gamma_v} G_v(i|U_v^{(t)})T_v(j|l_v\mu_v) \tag{8.11}$$

where $\Gamma_v \triangleq \{i,j|i-j=U_v^{(t)}-U_v^{(t-1)}, j \le U_v^{(t-1)}\}$. Finally, substituting (8.9) and (8.11) into (8.5), we calculate $\text{Pr}_5(\cdot)$.

We can now derive the state transition probability of the subchannels detected as idle, $\text{Pr}_6(\cdot)$ in (8.4). This conditional probability can be expressed as

$$\text{Pr}_6(\cdot) \triangleq \frac{\text{Pr}_9(M_a^{(t)}, M_a^{(t-1)})}{\text{Pr}_{10}(M_a^{(t-1)})} \tag{8.12}$$

We now need to define $\text{Pr}_9(\cdot)$ and $\text{Pr}_{10}(\cdot)$ from (8.12). The number of subchannels that are detected as idle depends on the PU traffic and the spectrum sensing. Thus, by denoting $U_w^{(t)}$ and $U_n^{(t)}$ as the number of WPUs and NPUs at time t, respectively, and $U_w^{(t-1)}$ and $U_n^{(t-1)}$ at time $t-1$, we have

$$\text{Pr}_9(\cdot) \triangleq$$

$$\sum_{\Theta} \text{Pr}_{11}(M_a^{(t)}, M_a^{(t-1)}|U_w^{(t)}, U_w^{(t-1)}, U_n^{(t)}, U_n^{(t-1)})\text{Pr}_{12}(U_w^{(t)}, U_n^{(t)}, U_w^{(t-1)}, U_n^{(t-1)}) \tag{8.13}$$

and Θ is the set of all possible $U_w^{(t)}$, $U_n^{(t)}$, $U_w^{(t-1)}$, $U_n^{(t-1)}$.

Probability $\text{Pr}_{11}(\cdot)$ in (8.13) can be decomposed as

$$\text{Pr}_{11}(\cdot) \triangleq \text{Pr}_{13}(M_a^{(t)}|U_w^{(t)}, U_n^{(t)})\text{Pr}_{13}(M_a^{(t-1)}|U_w^{(t-1)}, U_n^{(t-1)}) \tag{8.14}$$

In expression $\text{Pr}_{13}(\cdot)$ in (8.14), M_a subchannels include M_m subchannels that are occupied by PUs but misdetected and M_0 subchannels correctly detected as idle. Therefore $M_a = M_m + M_0$. We will define a supporting function to compute $\text{Pr}_{13}(\cdot)$. Using the detection probability p_d and the false alarm probability p_f, we can derive the probability that M_m busy subchannels are misdetected and M_0 idle subchannels are correctly detected for given U_w WPUs and U_n NPUs as

$$\text{Pr}_{14}(M_m, M_0|U_w, U_n) \triangleq$$

$$\binom{M_p}{M_m}(1-p_d)^{M_m} p_d^{M_p-M_m}\binom{M-M_p}{M_0}(1-p_f)^{M_0} p_f^{M-M_p-M_0} \tag{8.15}$$

where $M_p = U_w l_w + U_n l_n$ is the number of subchannels actually occupied by PUs. Therefore, $\text{Pr}_{13}(\cdot)$ in (8.14) can be derived as

$$\text{Pr}_{13}(\cdot) \triangleq \sum_{m=0}^{M_a} \text{Pr}_{14}(m, M_a - m | U_w, U_n) \tag{8.16}$$

The probability $\text{Pr}_{12}(\cdot)$ in (8.13) can be then decomposed into conditional probabilities as follows:

$$\text{Pr}_{12}(\cdot) \triangleq \text{Pr}_{15}(U_w^{(t)}, U_n^{(t)} | U_w^{(t-1)}, U_n^{(t-1)}) \text{Pr}_{16}(U_w^{(t-1)}, U_n^{(t-1)}) \tag{8.17}$$

Probability $\text{Pr}_{16}(\cdot)$ in (8.17) is computed by constructing a Markov chain using the state transition probability $\text{Pr}_{15}(\cdot)\text{Pr}_{17}(U_w^{(t)} | U_w^{(t-1)})\text{Pr}_{18}(U_n^{(t)} | U_n^{(t-1)}, U_w^{(t)}, U_w^{(t-1)})$ based on the assumption that the WPU has a higher priority of channel utilization than the NPU. Because we assume that the inter-arrival time and departure time follows the negative exponential distribution, we can use equation (8.6) and (8.7) in a similar way as in the derivation of (8.9) and (8.11). By denoting the available subchannels for NPU as $U_{e2} = (M - U_w^{(t)} l_w)/l_n$ at time t and $U_{e1} = (M - U_w^{(t-1)} l_w)/l_n$ at time $t-1$, we can derive $\text{Pr}_{17}(\cdot)$ and $\text{Pr}_{18}(\cdot)$ as

$$\text{Pr}_{17}(\cdot) \triangleq \sum_{\{i,j\} \in K_w} G_w(i | U_w^{(t)}) T_w(j | \mu_w) \tag{8.18}$$

$$\text{Pr}_{18}(\cdot) \triangleq \begin{cases} \displaystyle\sum_{\{i,j\} \in K_{na}} T_n(j | \mu_n) G_n(i | U_n^{(t)}), & U_n^{(t-1)} \leq U_{e1}, U_n^{(t)} < U_{e2} \\ \displaystyle\sum_{\{i,j\} \in K_{nb}} T_n(j | \mu_n) G_n(i | U_n^{(t-1)} + i - j), & U_n^{(t-1)} \leq U_{e1}, U_n^{(t)} = U_{e2} \end{cases} \tag{8.19}$$

where $K_w \triangleq \{i, j | i - j = U_w^{(t)} - U_w^{(t-1)}, j \leq U_w^{(t-1)}\}$, $K_{na} \triangleq \{i, j | i - j = U_n^{(t)} - U_n^{(t-1)}, j \leq U_n^{(t-1)}\}$, and $K_{nb} = \{i, j | U_n^{(t)} - U_n^{(t-1)} \leq i - j \leq U_{n,\max} - U_n^{(t-1)}, j \leq U_n^{(t-1)}\}$. Note that $\text{Pr}_{10}(M_a^{(t-1)})$ in (8.12) is simply calculated by summing up the product of $\text{Pr}_{13}(\cdot)$ and $\text{Pr}_{16}(\cdot)$ for all U_w and U_n.

8.4.2 Connection Blocking Rate Analysis for CBR Traffic

Due to the assumption that CBR traffic cannot be buffered once all subchannels are occupied by PU, we define CBR connection blocking rate as the average number of CBR connections blocked by the PU appearance per second.

By denoting $\text{Pr}_B(k)$ as the probability that k CBR connections are blocked in a frame, we have

$$R_B = \sum_{k=0}^{U_{c,\max}} k \Pr_B(k) / t_f \tag{8.20}$$

Because the number of blocked connections k depends on how many CBR users are in the system and how many subchannels are available, we decompose $\Pr_B(k)$ in (8.20) as

$$\Pr_B(k) \triangleq$$
$$\sum_W \Pr_b(k, U_c^{(t)} | U_c^{(t-1)}, M_a^{(t)}, M_a^{(t-1)}) \Pr_6(M_a^{(t)} | M_a^{(t-1)}) \Pr_c(U_c^{(t-1)}, M_a^{(t-1)}) \tag{8.21}$$

where W is the set of all possible values of $U_c^{(t)}$, $U_c^{(t-1)}$, $M_a^{(t)}$, and $M_a^{(t-1)}$. $\Pr_c(\cdot)$ can be calculated by summing up $\Pr_3(\cdot)$ for all $U_v^{(t-1)}$.

In (8.20) $\Pr_B(0)$ does not affect R_B calculation because it is multiplied by $k = 0$, and thus we focus on the case $k > 0$. We define $\Pr_{b+}(k, U_c^{(t)} | U_c^{(t-1)}, M_a^{(t)}, M_a^{(t-1)})$ as $\Pr_b(k, U_c^{(t)} | U_c^{(t-1)}, M_a^{(t)}, M_a^{(t-1)})$ for $k > 0$. This conditional probability is derived in a similar way as for $U_c^{(t)} = U_a^{(t)}$ in (8.9). Recall that $U_a^{(t)} = M_a^{(t)} / l_c$ and $U_a^{(t-1)} = M_a^{(t-1)} / l_c$. For supporting variables i and j, denoting that i connections are newly generated and j connections are released, the instant number of CBR connections before spectrum sensing is $U_c^{(t-1)} + i - j$, and the number of actual CBR connections after spectrum sensing is $U_c^{(t)}$. Thus, $i - j = k + U_c^{(t)} - U_c^{(t-1)}$. Therefore, by defining $K_b(k)\{i, j | i - j = k + U_c^{(t)} - U_c^{(t-1)}, j \leq U_c^{(t-1)}\}$, we can derive the conditional probability that k connections are blocked by the PU appearance as

$$\Pr_{b+}(\cdot) = \sum_{\{i,j\} \in K_b(k)} G_c(i | U_c^{(t-1)} + i - j) T_c(j | \mu_c)$$

for $U_c^{(t-1)} \leq U_a^{(t-1)}$, $U_c^{(t)} = U_a^{(t)}$ and zero otherwise.

8.5 Results on OS-OFDMA System

8.5.1 Numerical Results: Impact of Traffic Characteristics of OS-OFDMA System

To present numerical results, we set up traffic parameters taking into consideration an IEEE 802.22 network case study, where many active licensed incumbents operate over a single empty TV band (following the system model presented in Section 8.3). Since the traffic pattern of PUs for such a case is not known well,

Table 8.2 Summary of Parameters Used in the Numerical Evaluation

Parameter	Value	Parameter	Value	Parameter	Value
X	1	$1/\lambda_w$	144 s	$U_{w,max}$	1
Y	8	$1/\lambda_n$	12 s	$U_{n,max}$	4
M	8	$1/\lambda_c$	1 s	$U_{c,max}$	4
C	460.8 kbps	$1/\lambda_v$	24 s	$U_{v,max}$	4
p_f	0.05	$1/\mu_w$	48 s	l_w	8
p_d	0.99	$1/\mu_n$	12 s	l_n	2
t_s	2 ms	$1/\mu_c$	1 s	l_c	2
t_f	200 ms	$1/\mu_v$	24 s		

for the WPU we keep the wireless assist video devices [40] in mind, which can be assumed to broadcast on average 4 hours of signal transmission for every 12 hours in this scenario. For the NPU, we consider environments with numerous microphones and assume that they appear every hour and utilize channels for one hour on average. For CBR, considering voice or video transmission, we assume that on average 5-minute-long CBR connection is generated for every 5 minutes. For VBR, we assume that two-hour data traffic is generated for every two hours on average. Based on the time values described above, we set the inter-arrival time and the departure time for each user class, and then the relative ratios between inter-arrival times and departure times of the different user classes are determined. For simulation efficiency, because we operate in large parameter ranges, we scale them down by setting the CBR connection arrival rate to 1 second while preserving the ratios between all traffic parameters. We contrast the bandwidths of WPU and NPU by setting $l_w = 8$ and $l_n = 2$. We also assume that the CBR connection utilizes the same bandwidth as the NPU, i.e., $l_c = 2$. We choose the other parameters, as summarized in Table 8.2 to investigate the impact of the traffic of each class on the throughput and the CBR connection blocking rate.

The results of the investigation are presented in Figure 8.3 and Figure 8.4. The individual throughputs of CBR and VBR are computed by averaging either, respectively, m_c only or m_v only from (8.2). Note that each analytical result was verified by simulation using method of batch means with 90% confidence level. Warm-up time consisted of 100 iterations, while each batch size was equal to 10000 iterations with 100 batches per simulation. Simulations were implemented in MATLAB® and reflected exactly considered system models. We conclude that we obtained a very good match between the simulations and the analysis.

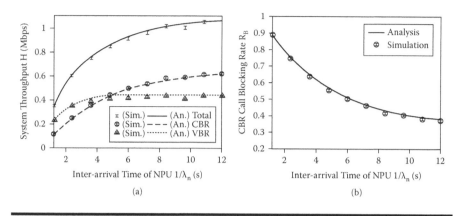

Figure 8.3 **Performance of OS-OFDMA system as a function of the inter-arrival time of the NPU $1/\lambda_n$ for (a) system throughput, and (b) CBR connection blocking rate.**

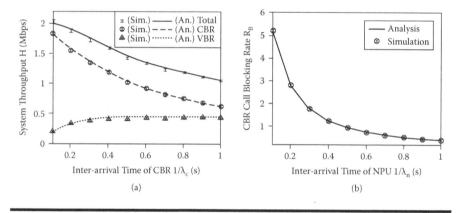

Figure 8.4 **Performance of OS-OFDMA system as a function of the inter-arrival time of the CBR connections $1/\lambda_c$ for (a) system throughput, and (b) CBR connection blocking rate.**

8.5.1.1 Impact of Narrowband PU Traffic on OS-OFDMA System

In order to see the effect of the NPU traffic on the performance of the SU, we sweep the average inter-arrival time of the NPU $1/\lambda_n$ from 0.1 to 1 of the reference value in Table 8.2, i.e., 1.2 seconds $\leq 1/\lambda_n \leq 12$ seconds.

The results are presented in Figure 8.3. As the inter-arrival time of NPU increases, the throughput of SU also increases, while the CBR connection blocking rate decreases. This is because the subchannels can remain idle for long periods. Also we observe in Figure 8.3(a) that for the small inter-arrival time, the VBR throughput is higher than the CBR, while for the large inter-arrival time it is the opposite. This is because, when there is no subchannel available, the VBR connections are buffered but the CBR connections are dropped. If a PU disappears so that more idle subchannels are available, the VBR class can utilize the idle subchannels immediately through a connection buffering, but the CBR class may not utilize them until the next CBR connection arrives.

8.5.1.2 Behavior of CBR Traffic in OS-OFDMA System

In the next experiment we observe the performance of the OS-OFDMA system as a function of SU traffic. We set inter-arrival time of the VBR connection, i.e., $1/\lambda_n$ = 2.4 s, and sweep the inter-arrival time of the CBR connection from 0.1 to 1 of the reference value in Table 8.2, i.e., 0.1 s ≤ $1/\lambda_n$ ≤1 s.

The results are given in Figure 8.4. We can observe that the individual throughputs of the CBR and the VBR change significantly. As the inter-arrival time of the CBR connection increases, the throughput of the VBR increases because the VBR connections have more chances to utilize idle subchannels.

An interesting observation is that, as the inter-arrival time of the CBR connection decreases, the throughput and the CBR connection blocking rate increase together. This result means that, for example, if lots of CBR connections concentrate on a small portion of the bandwidth, more idle subchannels can be utilized for the CBR connections; however, at the same time many CBR users can experience frequent connection drops caused by the PU appearance. In other words, OS-OFDMA system designed only based on throughput maximization may not guarantee the high quality of service because connections may be dropped often. One of the ideas to resolve the problem is to assign new connections uniformly over available channels, or to assign new connection to a free channel least utilized by PU.

8.5.2 Simulation Results: Efficiency of PU Detection Process

This section presents simulation results evaluating the effects of physical layer setup, like modulation type and detection efficiency, on OS-OFDMA performance. Due to the vast number of parameters to consider in this chapter we present results for two case studies only. Before we proceed with describing the results, we note that each simulation run considers network operating for 500 seconds. This requires about 1 hour of simulation time using 64-bit machine with 4 GB RAM. Each simulation scenario is run at least 10 times and averaged in order to get reliable results. In total, each point in the plots presented in the subsequent sections takes about 10 hours to generate.

8.5.2.1 OS-OFDMA Throughput with Increasing Distance between BS and CPE

We study the IEEE 802.22 performance when the transmission distance and, in turn, propagation effects between BS and CPE vary. As the distance between the

BS and CPE, d_{cpe}, increases the propagation delay will increase and the signal quality will generally decrease. The power delay spread will increase due to multipath effects, which as a result will decrease the signal quality. Interference from PUs will also impact the signal quality of IEEE 802.22 system. In total, SINR will increase as the distance between BS and CPE increases. Therefore, it is interesting to analyze the reduction in throughput due to PU appearance, sensing strategies, and sensing quality together, in addition to assessing the delay characteristics.

In Figure 8.5(a) the throughput for the total IEEE 802.22 system is measured as a function of increasing distance between the CPEs and BS, while Figure 8.5(b) shows the corresponding SIR for the considered scenario. The distance between the CPE and PU is fixed to d_{wm} = 250 meters; therefore when the CPE moves the PU also moves. The cooperative case described above is considered with $p_{d,I}$ = 0.6019 and $p_{f,I}$ = 0.0024. Three modulation and coding rates are used, i.e., 16-QAM 1/2,

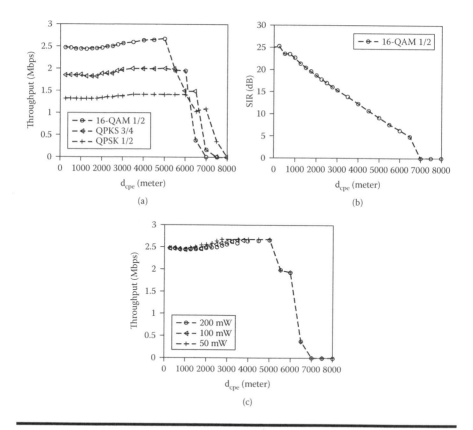

Figure 8.5 Simulation results: (a) Aggregate IEEE 802.22 throughput as a function of distance between CPE and BS (d_{cpe}), $1/\mu_w$ = 1 and $1/\lambda_w$ = 1.6, (b) corresponding SIR, and (c) throughput for varying PU power levels.

QPSK 3/4, and QPSK 1/4. PU inter-arrival and departure times are fixed at $1/\lambda_w$ = 1.6 and $1/\mu_w$ = 1 seconds, respectively.

It can be observed that between d_{cpe} = 250 and 1250 m all the IEEE 802.22 nodes are able to detect all the PUs. Therefore the decrease in throughput from d_{cpe} = 250 to 1250 m, especially visible for 16-QAM 1/2, is due to the reduced received signal quality at the CPE.

The plot for 16-QAM 1/2 shows that the throughput gradually increases from a certain point on the plot d_{cpe} = 250 up to d_{cpe} = 5000 m, for QPSK 3/4 to d_{cpe} = 5500 m and for QPSK 3/4 to d_{cpe} = 6000 m. The reason for this constant increase from a certain point on the plot is that, when d_{cpe} increases, only the CPEs that are closest to the actual PU will be able to detect the PU. Because cooperative sensing is used with a relatively low probability of detection, more spectrum opportunities due to missed detections cause more throughput at the greater distances. Also, it is observed that at greater distances throughput drops considerably and eventually to zero, which is due to the propagation effect and reduction in signal quality at higher distances (note that no adaptive modulation and coding is used). For the lower modulation and coding rate profiles a longer coverage range can be observed, which is due to the more robust modulation and coding rates that allows for lower SIR values.

As the distance between the BS and CPE increases, the CPE furthest away from the PU stops detecting the PU such that the amount of detecting IEEE 802.22 nodes that contributes to the cooperative detection with the OR rule decreases to N = 4 and p_d = 0.975. At distances above that, first only the BS and the closest CPE are able to detect the PU, i.e., N = 2 and p_d = 0.84; finally only the closest CPE is able to detect the PU, i.e., N = 1 and $p_d = p_{d,1}$ = 0.6. Therefore a stair-like pattern is observed in the plot.

It can be observed in Figure 8.5(b) that average SIR for the CPEs constantly decreases as the distance between BS and CPE increases. In Figure 8.5(c), the throughput for IEEE 802.22 is plotted as a function of d_{cpe} for three different PU power levels; 50, 100, and 200 mW, where the latter is the same as in Figure 8.5(a). Modulation and coding rate is 16-QAM 1/2. It can be observed that throughput increases at shorter d_{cpe} and peak throughput is achieved at shorter d_{cpe} for the lower PU power levels. This can be explained by the shorter coverage for the lower PU power levels. Also, we observe a slightly higher peak throughput when PU power is set to 50 mW compared to 100 mW and 200 mW.

8.5.2.2 OS-OFDMA Throughput with Increasing Distance between CPE and PU

Second, we study the impact of propagation delay between the IEEE 802.22 nodes and the PUs. As the PU moves further away from the IEEE 802.22 nodes, fewer nodes are able to detect the presence of the PU. Therefore the positive effect where the probability of detection increases due to multiple sensors able to detect the PU in the cooperative sensing scheme will decrease. In Figure 8.6(a) the throughput for

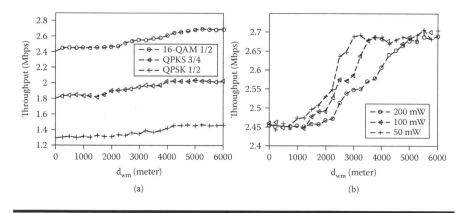

Figure 8.6 Simulation results: (a) Aggregate IEEE 802.22 throughput as a function of distance between PU and CPE (d_{wm}), $1/\mu_w = 1$ and $1/\lambda_w = 1.6$, and (b) throughput for varying PU power levels.

the IEEE 802.22 system is measured for increased distances between the PU and CPE (d_{wm}). The CPEs are always located at $d_{cpe} = 1000$ m from the BS.

We observe that as the PU moves further away from the CPE the throughput of the IEEE 802.22 system increases because fewer CPEs are able to detect the PU. Also, at higher distances even the closest CPE cannot detect the PU. For 16-QAM 1/2, throughput reaches a maximum at 2.7 Mbps when $d_{wn} \leq 5250$ m. Similarly we observe a maximum at 2 Mpbs for QPSK 3/4 and 1.4 Mpbs for QPSK 1/2. An observation is that the lower modulation and coding rate profiles seem to congest at shorter distances than for 16-QAM 1/2, which can be explained by the fact that these profiles are more robust at lower SIR values. A stair-like pattern can also be observed in this plot for all the modulation and coding rate profiles as the number of CPEs able to detect the PU decreases as d_{wm} increases.

In Figure 8.6(b), the throughput for IEEE 802.22 is plotted as a function of d_{wm} for three different PU power levels; 50, 100, and 200 mW, where latter is the same as in Figure 8.6(a). Modulation and coding rate is 16-QAM 1/2. A similar pattern as discussed above for the PU power 200 mW case is observed for the lower PU power levels, but because the PU has shorter range, the throughput increases at shorter d_{wm} and throughput also peaks at shorter d_{wm}.

8.6 Conclusions

In this chapter we have presented results of analytical investigation, as well as simulation results, for OS-OFDMA network performance, with a focus on IEEE 802.22. Specifically, we have investigated average throughput obtained by OS-OFDMA network as a function of primary user activity level (arrival and departure rate) and the effect of physical separation between primary user (wireless microphone) and secondary user (OS-OFDMA network). Both analytical and simulation models considered detailed implementation of OS-OFDMA network. In particular, different priority levels for various primary user types and secondary user classes were considered in the analytical model. Simulations were performed with the aid of ns-2 simulator, considering propagation model and physical layer implementation of OFDMA in OS-OFDMA context, following the IEEE 802.22 draft standard.

We summarize some of the important conclusions drawn from the investigation. First, we conclude that adaptive deactivation of subchannels in wide frequency bands is a promising access technique for OS-OFDMA networks, when highly active primary users are present. Second, for low activities of narrowband primary users VBR traffic of OS-OFDMA network obtains higher throughput than CBR because more spectral opportunities are present for VBR than for CBR. And finally, the effect of spatial separation between wireless microphones and CPE of IEEE 802.22 on network throughput is strictly not linear. That is, as the distance decreases the throughput of IEEE 802.22 starts to decrease, but after some point it starts to slowly rise again, while later it drops sharply to almost zero due to the broken connectivity between CPE and BS.

References

1. Q. Zhao and B. M. Sadler, "A Survey of Dynamic Spectrum Access: Signal Processing, Networking, and Regulatory Policy," *IEEE Signal Processing Mag.*, vol. 24, no. 3, pp. 79–89, May 2007.
2. P. Pawełczak, S. Pollin, H.-S. W. So, A. Bahai, R. V. Prasad, and R. Hekmat, "Quality of Service of Opportunistic Spectrum Access: a Medium Access Control Approach," *IEEE Wireless Commun. Mag.*, vol. 15, no. 5, pp. 20–29, Oct. 2008.
3. G. Staple and K. Werbach, "The End of Spectrum Scarcity," *IEEE Spectr.*, vol. 41, no. 3, pp. 48–52, Mar. 2004.
4. L. Nuyami, *WiMAX: Technology for Broadband Wireless Access.* Hoboken, NJ: Wiley, 2007.
5. IEEE Standards Association, *Draft Standard for Local and Metropolitan Area Networks: Standard Air Interface for Mobile Broadband Wireless Access Systems Supporting Vehicular Mobility: Physical and Media Access Control Layer Specification*, IEEE Std. P802.20 Draft 3.0, Mar. 2007.
6. D. Astély, E. Dahlman, A. Furuskär, Y. Jading, M. Lindström, and S. Parkvall, "LTE: The Evolution of Mobile Broadband," *IEEE Commun. Mag.*, vol. 47, no. 4, pp. 44–51, Apr. 2009.
7. IEEE Standards Association, *Draft Standard for Wireless Regional Area Networks Part 22: Cognitive Wireless RAN Medium Access Control (MAC) and Physical Layer (PHY) Specifications: Policies and Procedures for Operation in the TV Bands*, IEEE Std. P802.22 Draft 3.0, Apr. 2010.
8. C. R. Stevenson, G. Chouinard, Z. Lei, W. Hu, S. J. Shellhammer, and W. Caldwell, "IEEE 802.22: The First Cognitive Radio Wireless Regional Area Network Standard," *IEEE Commun. Mag.*, vol. 47, no. 1, pp. 130–138, Jan. 2009.

9. C. Cordeiro, D. Cavalcanti, and S. Nandagopalan, "Cognitive Radio for Broadband Wireless Access in TV Bands: The IEEE 802.22 Standard," in *Cognitive Radio Communications and Networks: Principles and Practice*, A. M.Wyglinski, M. Nekovee, and Y. T. Hou, Eds. Amsterdam, The Netherlands: Elsevier, Inc., 2009.

10. J. Park, P. Pawełczak, P. Grønsund, and D. Čabrić, "Performance of Opportunistic Spectrum OFDMA Network with Users of Different Priorities and Traffic Characteristics," in *Proc. IEEE GLOBECOM,* Miami, FL, USA, Dec. 6–10, 2010.

11. J. Park, P. Pawełczak, P. Grønsund, and D. Čabrić. Subchannel Notching and Channel Bonding: Analysis Framework for Opportunistic Spectrum OFDMA and Its Application to IEEE 802.22 Standard. [Online]. Available: http://arxiv.org/abs/1007.5080.

12. The Network Simulator NS-2. [Online]. Available: http://www.isi.edu/nsnam/ns/.

13. T. A. Weiss and F. K. Jondral, "Spectrum Pooling: an Innovative Strategy for the Enhancement of Spectrum Efficiency," *IEEE Commun. Mag.,* vol. 42, no. 3, pp. S8–S14, Mar. 2004.

14. D. T. Ngo, C. Tellambura, and H. H. Nguyen, "Resource Allocation for OFDMA-Based Cognitive Radio Multicast Networks with Primary User Activity Consideration," *IEEE Trans. Veh. Technol.,* vol. 59, no. 4, pp. 1668–1679, May 2010.

15. A. E. Leu, B. L. Mark, and M. A. McHenry, "A Framework for Cognitive Wimax Frequency Agility," *Proc. IEEE,* vol. 97, no. 4, pp. 755–773, Apr. 2009.

16. L. Song, C. Feng, Z. Zeng, and X. Zhang, "Research on WRAN System Level Simulation Platform Design," in *Proc. ICST Chinacom,* Hangzhou, China, Aug. 25–27, 2008.

17. Y. Zhang and C. Leung, "Resource Allocation for Non-Real-Time Services in OFDM-Based Cognitive Radio Systems," *IEEE Commun. Lett.,* vol. 13, no. 1, pp. 16–18, Jan. 2009.

18. A. Durresi and V. Paruchuri, "Broadcast Protocol for Energy-Constrained Networks," *IEEE Trans. Broadcast.,* vol. 53, no. 1, pp. 112–119, Mar. 2007.

19. Y.-C. Liang, A. T. Hoang, and H.-H. Chen, "Cognitive Radio on TV Bands: A New Approach to Prove Wireless Connectivity for Rural Areas," *IEEE Wireless Commun. Mag.,* vol. 15, no. 3, pp. 16–22, June 2008.

20. J. Wang, M. S. Song, S. Santhiveeran, K. Lim, G. Ko, K. Kim, and S. H. Hwang, "First Cognitive Radio Networking Standard for Personal/Portable Devices in TV White Spaces," in *Proc. IEEE DySPAN,* Singapore, Apr. 6–9, 2010.

21. IEEE Standards Association, *Draft Standard for Information Technology; Telecommunications and Information Exchange Between Systems; Local and Metropolitan Area Networks; Specific Requirements; Part 11: Wireless LAN Medium Access Control (MAC) and Physical Layer (PHY) Specifications; Amendment 5: TV White Spaces Operation,* IEEE Std. P802.11af Draft 0.02, May 2010.

22. S.-E. Elayoubi and B. Fouresité, "Performance Evaluation of Admission Control and Adaptive Modulation in OFDMA WiMax Systems," *IEEE/ACM Trans. Networking,* vol. 16, no. 5, pp. 1200–1211, Oct. 2008.

23. H.-S. Chen, W. Gao, and D. G. Daut, "Spectrum Sensing Using Cyclostationary Properties and Application to IEEE 802.22 WRAN," in *Proc. IEEE GLOBECOM,* Washington, DC, USA, Nov. 26–30, 2007.

24. H.-S. Chen, W. Gao, and D. G. Daut, "Signature Based Spectrum Sensing Algorithms for IEEE 802.22 WRAN," in *Proc. IEEE ICC,* Glasgow, Scotland, UK, EU, June 24–28, 2007.

25. J. Park, T. Song, J. Hur, S. M. Lee, J. Choi, K. Kim, K. Lim, C.-H. Lee, H. Kim, and J. Laskar, "A Fully Integrated UHF-Band CMOS Receiver with Multi-Resolution Spectrum Sensing (MRSS) Functionality for IEEE 802.22 Cognitive Radio Applications," *IEEE J. Solid-State Circuits,* vol. 44, no. 1, pp. 258–268, Jan. 2009.

26. H. Kim, J. Kim, S. Yang, M. Hong, and Y. Shin, "An Effective MIMO-OFDM System for IEEE 802.22 WRAN Channels," *IEEE Trans. Circuits Syst. II,* vol. 55, no. 8, pp. 821–825, Aug. 2008.

27. D. Niyato, E. Hossain, and Z. Han, "Dynamic Spectrum Access in IEEE 802.22-Based Cognitive Wireless Networks: A Game Theoretic Model for Competitive Spectrum Bidding and Pricing," *IEEE Wireless Commun. Mag.,* vol. 16, no. 2, pp. 16–23, Apr. 2009.

28. R. Al-Zubi, M. Z. Siam, and M. Krunz, "Coexistence Problem in IEEE 802.22 Wireless Regional Area Networks," in *Proc. IEEE GLOBECOM,* Honolulu, HI, USA, Nov. 30–Dec. 4, 2009.

29. D. Huang, C. Miao, Y. Miao, and Z. Shen, "A Game Theory Approach for Self-Coexistence Analysis Among IEEE 802.22 Networks," in *Proc. IEEE ICICS,* Macau, China, Dec. 7–10, 2009.

30. S. Sengupta, R. Chandramouli, S. Brahma, and M. Chatterjee, "A Game Theoretic Framework for Distributed Self-Coexistence Among IEEE 802.22 Networks," in *Proc. IEEE GLOBECOM,* New Orleans, LA, USA, Nov. 30–Dec. 4, 2008.

31. H. Kim and K. G. Shin, "Asymmetry-Aware Real-Time Distributed Joint Resource Allocation in IEEE 802.22 WRANs," in *Proc. IEEE INFOCOM,* San Diego, CA, USA, Mar. 15–19, 2010.

32. C.-H. Ko and H.-Y. Wei, "Game Theoretical Resource Allocation for Inter-BS Coexistence in IEEE 802.22," *IEEE Trans. Veh. Technol.,* vol. 59, no. 4, pp. 1729–1744, May 2010.

33. W. Hu, D. Willkomm, M. Abusubiah, J. Gross, G. Vlantis, M. Gerla, and A.Wolisz, "Dynamic Frequency Hopping Communities for Efficient IEEE 802.22 Operation," *IEEE Commun. Mag.,* vol. 45, no. 5, pp. 80–86, May 2007.

34. S. Segupta, M. Chatterje, and R. Chandramouli, "A Coordinated Distributed Scheme for Cognitive Radio Based IEEE 802.22 Wireless Mesh Networks," in *Proc. IEEE CogNet (IEEE ICC 2008 Workshop),* Beijing, China, May 19–23, 2008.

35. IEEE Standards Association, *Standard for Local and Metropolitan Area Networks Part 16: Air Interface for Fixed and Mobile Broadband Wireless Access Systems Amendment 2: Physical and Medium Access Control Layers for Combined Fixed and Mobile Operation in Licensed Bands and Corrigendum 1,* IEEE Std. 802.16e, Apr. 2006.

36. N. Golmie, R. Rouil, D. Doria, X. Guo, R. Iyengar, S. Kalyanaraman, S. Krishnaiyer, S. Mishra, B. Sikdar, R. Jain, R. Patneyand, C. So-In, and S. Parekh, WiMAX Forum System Level Simulator NS-2 MAC+PHY Add-On for WiMAX (IEEE 802.16), 2009.

37. WiMAX Forum. [Online]. Available: http://www.wimaxforum.org/.

38. IEEE Standards Association, *Draft Standard for Wireless Regional Area Networks Part 22: Cognitive Wireless RAN Medium Access Control (MAC) and Physical Layer (PHY) Specifications: Policies and Procedures for Operation in the TV Bands,* IEEE Std. P802.22 Draft 3.0, Apr. 2010.

39. WiMAX Forum, "Mobile WiMAX—Part 1: A Technical Overview and Performance Evaluation," Aug. 2006. [Online]. Available: http://www.wimaxforum.org/resources/documents/marketing/whitepapers.

40. E. Reihl, "Wireless Microphone Characteristics," IEEE, Tech. Rep. 802.22-06/0070r0, May 2006.

41. D. Lam, D. C. Cox, and J. Widom, "Teletraffic Modeling for Personal Communication Services," *IEEE Commun. Mag.,* vol. 35, no. 2, pp. 79–87, Feb. 1997.

42. P. Grønsund, H. N. Pham, and P. E. Engelstad, "Towards Dynamic Spectrum Access in Primary OFDMA Systems," in *Proc. IEEE PIMRC,* Tokyo, Japan, Sept. 13–16, 2009.

43. J. Park, P. Pawełczak, and D. Čabrić. "Performance of Joint Spectrum Sensing and MAC Algorithms for Multichannel Opportunistic Spectrum Access Ad Hoc Networks," *IEEE Transactions on Mobile Computing,* vol. 10, no. 7, pp. 1011–1027.

44. J. Y. Won, S. B. Shim, Y. H. Kim, S. H. Hwang, and M. S. Song, "An Adaptive OFDMA Platform for IEEE 802.22 Based on Cognitive Radio," in *Proc. APCC,* Busan, Republic of Korea, Aug. 31–Sep. 1, 2006.

45. E. Demosso, *COST 231. Digital Mobile Radio Towards Future Generation Systems. Final Report.* European Commission, EUR 18957. Brussels, Belgium, 1999. [Online]. Available: http://www.lx.it.pt/cost231.

46. R. H. Clarke, "A Statistical Theory of Mobile-Radio Reception," *Bell System Technical Journal,* vol. 47, pp. 957–1000, 1968.

47. "Guidelines for Evaluations of Radio Transmission Technologies for IMT-2000," ITU ITU-R, Tech. Rep. M.1225, 1997.

48. "Effective SIR Computation for OFDM System-Level Simulations," 3GPP-TSG-RAN-1, Tech. Rep. RI-03-1370, 2003.

49. P. Pawełczak, S. Pollin, H.-S. W. So, A. Bahai, R. V. Prasad, and R. Hekmat, "Performance Analysis of Multichannel Medium Access Control Algorithms for Opportunistic Spectrum Access," *IEEE Trans. Veh. Technol.,* vol. 58, no. 6, pp. 3014–3031, July 2009.

Chapter 9

Inter-Network Spectrum Sharing and Communications in Cognitive Radio Networks Using On-Demand Spectrum Contention and Beacon Period Framing

Wendong Hu, Gregory J. Pottie, and Mario Gerla

Contents

9.1 Introduction

Cognitive Radio (CR) [1] has been considered as an enabling technology of dynamic spectrum access (DSA) networks that allows unlicensed radio transmitters to operate in the licensed bands at locations where that spectrum is temporarily not in use. Based on cognitive radio technology, the Institute of Electrical and Electronics Engineers (IEEE), following a Federal Communication Commission (FCC) Notice of Proposed Rulemaking in 2004 [2], has fostered 802.22 [3,4]. It is an emerging standard for Wireless Regional Area Networks (WRAN) aiming to provide alternative broadband wireless access in, among other places, rural areas. CR operates on a license-exempt and noninterference basis in the TV band (between 47 and 910 MHz) without creating harmful interference to the licensed services, which include, among others, digital TV (DTV) and low-power licensed devices such as wireless microphones (licensed under Part 74 of FCC rules). Most recently, the IEEE 802.11af task group [5] was formed in January 2010, aiming to

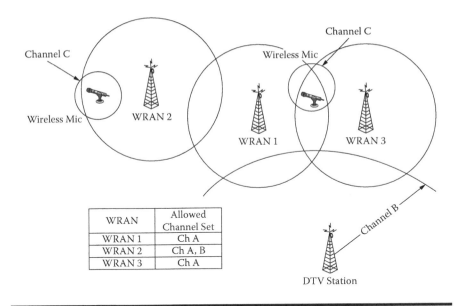

WRAN	Allowed Channel Set
WRAN 1	Ch A
WRAN 2	Ch A, B
WRAN 3	Ch A

Figure 9.1   A typical deployment scenario where multiple IEEE 802.22 WRAN cells are coexisting with digital TV and wireless microphone services.

adapt the 802.11 technologies to TV band operation, while IEEE 802.19.1 task group [6] has started developing coexistence mechanisms since December 2009 for the dissimilar or independently operated IEEE 802 family of networks and devices in the TV band.

In a typical deployment scenario of DSA networks, multiple network cells, each of which may consist of a base station (BS) or access point (AP), and the associated customer premise equipment (CPE) with a potentially large communication range of up to 100 km,* may operate in the same vicinity while coexisting with DTV and licensed wireless microphones. In order to effectively avoid harmful interference to these licensed incumbents, the set of channels on which the network cells are allowed to operate could be quite limited. For example as shown in Figure 9.1, residing within the protection contours of DTV and wireless microphones, both IEEE 802.22 network cells, WRAN1 and WRAN3, are only allowed to operate on channel A, while network cell WRAN2 may occupy either channel A or B, assuming that in total only 3 channels (channels A, B, and C) are available. If, for instance, WRAN1 and WRAN3 (or WRAN1 and WRAN2) attempt to perform data transmissions on channel A simultaneously, mutual interference between these collocated network cells could degrade the system performance significantly.

---

* Coverage range as specified in IEEE802.22 is 33 km at 4 watts CPE EIRP (equivalent isotropically radiated power) under the FCC rules in the United States. The coverage range may be extended to 100 km outside of the United States, where much higher transmit power limits are allowed.

Although avoiding harmful interference to licensed incumbents is the prime concern in the system design, another key design challenge to cognitive radio-based DSA network systems, with the scenario illustrated above in mind, is how to dynamically share the scarce spectrum among the collocated network cells so that performance degradation, due to mutual co-channel interference, is effectively mitigated. Moreover, it is important that the inter-network (i.e., among nearby network cells) spectrum sharing scheme be developed to maintain efficient spectrum usage, to accommodate a large scale of networks with various coexistence scenarios, and to provide fairness in spectrum access among the coexisting network cells.

To that end, we describe in this chapter a distributed, message-based, cooperative, and real-time spectrum sharing concept called On-Demand Spectrum Contention (ODSC) [4,7] that has been adopted by the IEEE 802.22 draft standard [3]. ODSC protocols are designed for a fair, low-overhead, quality-of-service (QoS)-aware spectrum sharing among the coexisting network cells, and can be adapted to provide coarse-grain (e.g., channel-based) and fine-grain (e.g., frame-based) sharing mechanisms.

The basic mechanism of ODSC is as follows: on a demand-driven basis, base stations (or APs) of the coexisting network cells contend for the shared spectrum resource (e.g., a TV channel to operate for a particular period of time) by comparing the spectrum access priority numbers (SAPN) or alternatively the spectrum contention numbers (SCN) that are randomly generated by these contending network cells. The spectrum access priority numbers used to resolve the contention for the shared spectrum resource are carried by MAC (medium access control) messages exchanged among the contending network cells through an independently accessible inter-network communication channel. The contention decisions are made by the coexisting network cells in a distributed way. Only the winner cell, which possesses a higher spectrum access priority compared to those of the other contending cells (the losers), can occupy the shared spectrum resource.

The effectiveness of the ODSC protocols relies on the availability of an efficient and reliable inter-network communication channel for the interactive MAC message exchanges among network cells. In fact, in addition to supporting cooperative spectrum sharing protocols such as ODSC, a reliable inter-network communication channel is also indispensable to other inter-network coordination functions for cognitive radio-based networks, such as IEEE 802.22 (examples of such coordination functions are inter-network synchronization of quiet periods for spectrum sensing [4], and coordinated Dynamic Frequency Hopping [8]). In the second part of this chapter, we introduce a beacon-based inter-network communication protocol called Beacon Period Framing (BPF) Protocol that realizes a reliable, efficient, and scalable inter-network communication channel reusing the radio frequency channels occupied by the network cells (i.e., the in-band radio frequency channel). Although only over-the-air solutions using the in-band RF channel are described in this chapter, inter-network communication channel can also be realized using the IP (Internet protocol)-based backhaul approach or over-the-air solutions with dedicated radio frequency channels. We will further discuss and compare these different approaches in Section 9.5.

The remainder of this chapter is organized as follows. In Section 9.2 we describe the details of the ODSC protocols, providing both the general mechanisms of operations and the inter-network message flows, and in particular, the functional procedures for a frame-based (fine-grained) ODSC operation. Section 9.3 presents the concepts of BPF Protocol. Performance analyses and discussion are given in Section 9.4. A number of interesting system issues are further discussed by comparing concepts and techniques proposed in this chapter with other alternative approaches for dynamic spectrum sharing and inter-network communications in Section 9.5, which also concludes the chapter.

9.2 On-Demand Spectrum Contention Protocols

9.2.1 Overview

ODSC is a family of DSA coexistence protocol that employs interactive MAC control messaging on an independently accessible inter-network communication channel to provide efficient, scalable, and fair inter-network spectrum sharing among the coexisting network cells. To achieve these design goals, ODSC allows the coexisting network cells to compete for the shared spectrum resource by exchanging and comparing randomly generated contention access priority numbers carried in the over-the-air MAC control messages. This spectrum contention process is iteratively driven by spectrum contention demands (i.e., intra-cell demands for additional spectrum resources to support increasing data services, and inter-cell demands requesting spectrum resource acquisitions). The contention decisions are made by the coexisting network cells in a distributed way, which allows an arbitrary number of cells to contend for the shared spectrum resources in their proximities without relying on a central arbiter. Instead of behaving selfishly, the competing network cells cooperate with one another to achieve the goals of fair spectrum sharing and efficient spectrum utilization.

Unlike the traditional contention-based medium access schemes such as Aloha [9] and Carrier Sensing Multiple Access (CSMA) [10,11], the ODSC protocol, as highlighted above, takes distinguishable approaches for resolving channel (spectrum resource) access contention and mitigating access collisions, aiming to improve the spectrum access efficiency. Due to the lack of exact knowledge of when the neighboring systems will transmit, a wireless system using Aloha or CSMA protocols generally resolves the access contention by deferring data transmission for a random period, and in the case of a collision, reinitiates the contention process by setting a potentially much larger random period for the transmission deferral (e.g., using the exponential random back-off mechanism) in order to reduce the chance of reoccurring collisions.

Taking better approaches, the ODSC protocols allow a wireless system (IEEE 802.22 WRAN, for example) to compete for the spectrum resource access without resorting to any transmission deferral which clearly sacrifices the spectrum efficiency, thanks to the contention resolution process carried out in parallel with the ongoing data services. Moreover, as the spectrum contention is resolved by comparing the spectrum access priority numbers that are randomly selected from a very large pool of values, the likelihood of an access collision among coexisting wireless systems using ODSC is effectively minimized.

In the subsequent sections, we will describe the details of the general principles of ODSC protocols and particularly the functional procedures for frame-based ODSC operations.

9.2.2 On-Demand Spectrum Contention Principles

9.2.2.1 Model of DSA Coexistence Scenarios

We model DSA coexistence scenarios using the concept of the contention graph. It is defined as a unidirectional graph $G = (V, E)$, where vertices V denote a set of network cells and an edge connecting two vertices belongs to E if and only if the mutual interference (i.e., the contention condition) exists between the network cells represented by the vertices when these cells are operating on the same spectrum resource (e.g., a TV channel) simultaneously. We also refer to the vertices connected by an edge in the contention graph as "one-hop" neighbors.

Without loss of generality, a simple contention graph that models a coexistence scenario containing three network cells is used to illustrate the basic principles of the on-demand spectrum contention protocols. As shown in Figure 9.2(a), the edges in this simple graph, (Network 1, Network 2) and (Network 1, Network 3), indicate the mutual interference condition exists for these network pairs if a common spectrum resource is accessed at the same time by their individual network cells.

Based on the contention graph model, we further introduce the concepts of "Contention Source" and "Contention Destination" that are used extensively in the ODSC operations to classify the coexisting network cells into two categories according to their roles and functionalities in the spectrum resource sharing process:

- Contention Source (SRC, or ODSC-SRC): A network cell that is requesting additional spectrum resources and is initiating an interactive contention process with the target Contention Destination (defined below) that occupies the spectrum resource of interest
- Contention Destination (DST, ODSC-DST): A network cell that is the target of a received contention request initiated by a Contention Source (SRC) asking for a spectrum resource, and is the occupier of the spectrum resource being demanded to be shared with a requesting Contention Source (SRC)

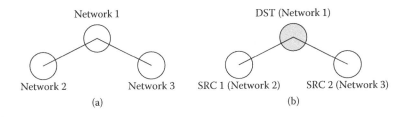

Figure 9.2 A simple model of coexistence scenario using contention graph.

Take our simple model of coexistence scenario as shown in Figure 9.2(b) as an example, and assume that a spectrum resource (e.g., TV Channel A) is being occupied by Network 1. With such an assumption, Network 1 is the only occupier of Channel A due to its mutual interference conditions with Network 2 and Network 3. Network 2 and Network 3 can both behave as Contention Sources (i.e., SRC 1 and SRC 2) through independently initiating a contention process targeted to Network 1 for requesting to acquire the spectrum resource (e.g., Channel A) that is currently operated by Network 1. After receiving the contention request, Network 1 will take the role of a contention destination (DST).

It is important to note that the concepts of "Contention Source" and "Contention Destination" are only associated with

1. An active spectrum contention process (a process in which the contention has not been resolved) for sharing a specific spectrum resource
2. A specific set of network cells that are involved in such a process

In other words, a network cell behaving as a DST in an active contention process can also be simultaneously functioning as an SRC (or another DST) contending, for example, for a different spectrum resource with a different set of network cells.

9.2.2.2 General ODSC Message Flow and Mechanisms

In this section, we describe the general ODSC message flow and mechanisms of the generic ODSC protocol in detail. Without loss of generality, we refer to the shared spectrum resource as a TV channel in this section to flexibly represent a spectrum resource, which can be scalable in terms of the granularity of its usage.

Before initiating MAC layer control messaging of the ODSC protocol, a network cell that is demanding additional spectrum resources first evaluates and selects a TV channel on which no incumbent is detected. The cell then verifies if the selected channel can be shared, employing techniques such as transmit power control, with all other co-channel communication systems without causing any mutually harmful interference. If it is feasible, the network cell schedules its data transmissions on the selected channels with appropriate operational (e.g., transmit

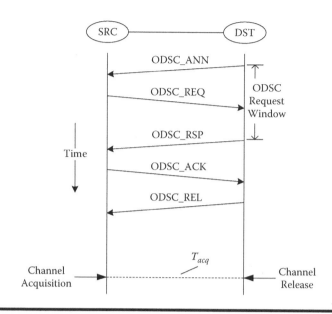

Figure 9.3   Basic ODSC message flow.

power control) settings. Otherwise, ODSC messaging takes place, allowing cooperative spectrum contention among the network cells to share the selected TV channel in a time-sharing manner.

Figure 9.3 depicts the basic MAC messaging flow of the generic ODSC protocol between two network cells that are within interference range of each other (i.e., within a "one-hop" distance) modeled by the contention graph. We assume that the MAC messages are delivered by robustly designed coexistence beacons such that the MAC messages can be reliably received by all coexisting cells within a "one-hop" distance.

9.2.2.2.1 Resource Availability Announcement

First, a spectrum-demanding network cell, a potential ODSC contention source (SRC), captures the ODSC announcement messages (ODSC_ANN) regularly broadcast by a spectrum occupier network cell, a potential ODSC contention destination (DST).

9.2.2.2.2 Contention Request

If a spectrum-demanding network cell receives ODSC_ANN messages from multiple spectrum resource occupiers indicating the availability of the same TV channel of interest, it randomly selects one of the spectrum resource occupiers (see Figure 9.4). The spectrum-demanding network cell, starting to function as a contention source (SRC), then sends an ODSC request message (ODSC_REQ),

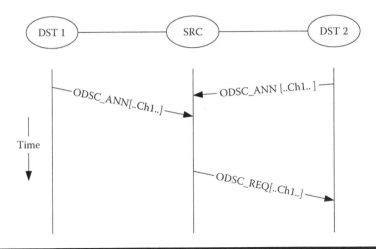

Figure 9.4 Selection of contention destination.

including a spectrum access priority number (SAPN), which is for example a fixed-point number uniformly selected from $[0, 2^{16}-1]$, to the selected spectrum occupier network cell, which will take a role of DST.

ODSC_REQ message is designed as a "Request to Send" message.

Each DST maintains an ODSC request window (see Figure 9.5) so as to allow multiple SRCs to submit ODSC_REQ messages at different time instances without losing the fair chance to participate in a contention process.

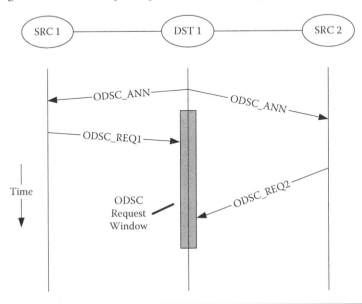

Figure 9.5 ODSC request window.

9.2.2.2.3 Contention Resolution and Response

At the end of an ODSC request window, if any ODSC_REQ is received, the DST then randomly generates its own SAPN and compares it with the smallest SAPN carried in the received ODSC_REQ messages.

If the DST's SAPN is smaller (i.e., possesses higher priority), DST sends each SRC an ODSC_RSP message indicating a contention failure. Otherwise, the SRC with the smallest SAPN will receive an ODSC_RSP message with an indication of contention success, and all the other SRCs will be informed of a contention failure.

9.2.2.2.4 Contention Acknowledgment

Upon receiving a success notice, the winner SRC, as depicted in Figure 9.6, broadcasts an ODSC acknowledgment message (ODSC_ACK) indicating the time, T_{acq} (e.g., T0 in Figure 9.6), at which it intends to acquire the TV channel from the selected DST.

It is important to note that the ODSC_ACK message is designed as a "Clear to Send" message, which is used to notify all the neighbor network cells of an intention of acquiring a TV channel at a particular time.

9.2.2.2.5 Announcement of Spectrum Resource Release

All DSTs that are on the same TV channel as the one being contended for and are within a one-hop distance of the winner SRC respond to the ODSC_ACK message by scheduling a channel release to occur at T_{acq} and broadcast an ODSC_REL message to the neighborhood (see Figure 9.6). The ODSC_REL message contains information about the channel to release, the channel release time (set to T_{acq}), and

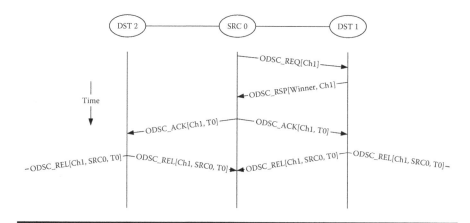

Figure 9.6 Contention acknowledgment and announcement of spectrum resource release.

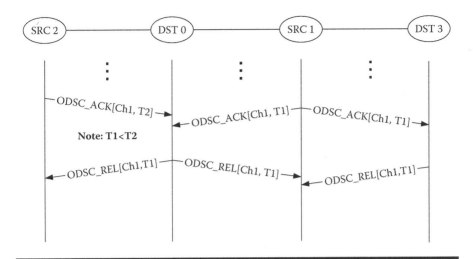

Figure 9.7 Overriding of channel release time.

the ID of the winner SRC that will acquire the channel. As indicated in Section 9.2.2.2.6, the ODSC_REL message also serves to announce the availability of the TV channel and therefore can be used to enhance the TV channel reuse.

If ODSC_ACK messages are received from multiple SRCs before the TV channel is released, a DST selects the earliest T_{acq} specified in the received ODSC_ACK message as the channel release time. This avoids collisions between the neighboring DST and SRC when their channel switching times do not agree (see Figure 9.7).

9.2.2.2.6 Scheduling of Spectrum Resource Acquisition

As shown in Figure 9.8, all SRCs that capture the ODSC_REL message from a "one-hop" DST will also schedule channel acquisitions at T_{acq} as long as it is determined from the ODSC_REL message that the "one-hop" DST is releasing the TV channel to either itself or to a winner SRC that is more than one hop away. On the other hand, if multiple ODSC_REL with different T_{acq} are received before the channel switching, the later T_{acq} is taken for channel acquisition.

9.2.2.3 *Contention Resolution*

As previously described, the network cell in control of a spectrum resource, the DST, determines the winner of contention over that resource. In doing so, the DST gathers spectrum access priority numbers (or alternatively, the spectrum contention numbers) from each participating network cell and resolves the contending requests based on a spectrum contention algorithm, as shown in Figure 9.9.

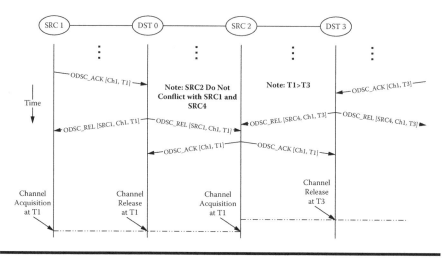

Figure 9.8 Scheduling of spectrum resource acquisition.

9.2.2.3.1 Contention Resolution Algorithm

The Spectrum Contention Resolution algorithm takes as inputs an array of N network IDs of the contending network cells (*Network*), an array of N spectrum contention numbers of the contending network cells (*SCN*), the spectrum resource (e.g., a TV channel) being contended for (*Channel*), and the parameter (N), and it computes the ID of the "Winner" network, $SCN_{winner}(Channel)$ and its corresponding "Winner" spectrum contention number for accessing the spectrum resource, $Network_{winner}(Channel)$.

Step 1 of the algorithm identifies the $SCN_{winner}(Channel)$, which is the smallest contention number in *SCN*. In Steps 2 to 4, the algorithm searches for the $SCN_{winner}(Channel)$ who owns the $SCN_{winner}(Channel)$, and generates the outputs as soon as the search succeeds.

Where in the algorithm (Figure 9.9),

N: Total number of contending network cells

Network: The array of IDs of the contending network cells, *Network*[i] for $i \leftarrow$ 1 to N

SCN: The array of spectrum contention numbers of the contending network cells, in which *SCN*[i] is the spectrum contention number of *Network*[i] for $i \leftarrow$ 1 to N

Channel: The spectrum resource (e.g., a TV channel or a data frame) being contended for

$SCN_{winner}(Channel)$: The winner spectrum contention number for accessing the *Channel*

$Network_{winner}(Channel)$: The ID of the winner Network cell to access the *Channel*

Spectrum Contention Resolution (*N, Network, SCN, Channel*)

1 $SCN_{winner}(Channel) \leftarrow \min_{i=0}^{N} \{SCN[i]\}$

2 **for** $i \leftarrow 1$ **to** N

   3 **if** $SCN[i] = SCN_{winner}$

      4 **return** $Network_{winner}(Channel) \leftarrow Network[i]$ and $SCN_{winner}(Channel)$

Figure 9.9 On-demand spectrum contention algorithm

9.2.2.3.2 Spectrum Contention Number

Each contending network cell, SRC or DST, produces a spectrum contention number. The spectrum contention number, SCN_i, for network cell i, $Network_i$, is generated randomly as:

$$SCN_i = \textbf{RANDOM}\ [0, 2^X - 1]$$

for a fixed-point representation using data size of x bits.

9.2.2.4 Considerations

9.2.2.4.1 Coordination of Distributed Networks

In a large-scale network, it is likely that multiple DSTs and multiple SRCs coexist. As the contention processes are fully random and independent, different SRCs could select their own DSTs to contend for the same spectrum resource and the contentions outcomes (i.e., winners of the contention and channel acquisition/release times) could be in conflict. The ODSC message flow described above is designed to coordinate the discrepancies between the conflicting contention decisions in order to ensure the stability of the coexistence behaviors and avoid loss of spectrum efficiency across the networks.

9.2.2.4.2 Granularity of Spectrum Sharing

At any one time only one network cell can utilize the shared channel in a close proximity. While the network cell occupying the channel sends and receives data over a particular period of time, other neighboring network cells remain idle. This is true even when the network cell occupying the channel may not be fully utilizing the bandwidth of the channel over its allocated period of time. Additionally a network cell demanding a spectrum resource would have to remain idle for a relatively long duration (in the order of plurality of data frames) until the channel to share is released by the occupying network cell. As a consequence, such a potentially long turnaround time of channel acquisition may negatively impact the quality of service (QoS) of time-sensitive applications due to the long service interruptions. Spectrum sharing on a finer granularity than operating on a channel for relatively long duration (such as on a frame-by-frame basis) is advantageous to enhance both the utilization of the operating spectrum and the QoS of the application.

What is needed is a set of general mechanisms for an arbitrary number of distributed network devices to share limited spectrum resources. Although the above description of the generic ODSC protocol outlines how a protocol is employed for resolving problems of radio resource sharing where the basic unit of the spectrum resource is a radio frequency channel (i.e., operating on an RF channel for an arbitrarily long period of time before releasing it), the same principle of ODSC applicability, without loss of generality, is desirable for other, more fine-grained, apportionment of the shared spectrum. It is desirable to apportion the shared radio spectrum so that any effective combination of radio spectrum resource in both the time and frequency domain, such as a frame on a frequency channel or multiple frames on multiple frequency channels, can be effectively shared.

9.2.2.4.3 Frame-Based ODSC Operation—A First Glance

Figure 9.10(a) and (b) show a simple example of inter-network coexistence operation using the ODSC protocol, in which 3 coexisting network cells share a TV channel at a granularity of one data frame. Instead of being occupied by a network cell exclusively (within the range of mutual interference) for a relatively long duration before being released to another network cell, a TV channel is allowed to be shared by multiple contending network cells simultaneously (in a fine-grained interleaving manner) within a particular time duration. Spectrum resource sharing mechanisms aim to reduce the turnaround time of channel acquisition. Therefore QoS of time-sensitive applications can be better supported. Note that ODSC messages exchanged among the contending network cells are transmitted and received through an independently accessible inter-network communication channel, which will be described in Section 9.3.

The frame-based ODSC operation as illustrated in this example will be described in detail in Section 9.2.3 and is designed based on the following basic concepts:

- Super-frame structure—a collection of multiple data frames (16 data frames per super-frame as used in this example).
- Synchronized super-frames and data frames across all coexisting network cells.
- Super-frame based synchronized data frame allocation—Data frame allocation is contended and updated on a super-frame by super-frame basis.

As can be observed in Figure 9.10(b), during the time period of super-frame N, potentially multiple contention processes are going on among the coexisting

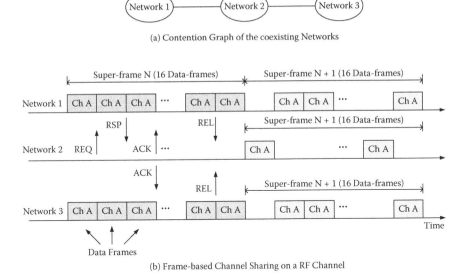

(a) Contention Graph of the coexisting Networks

(b) Frame-based Channel Sharing on a RF Channel

Figure 9.10   An example of Frame-Based ODSC operation.

network cells for some or all of the data frames in a super-frame, without affecting data frame assignments of the current super-frame (super-frame N). At the end of super-frame N, the results of all resolved contention processes are synchronized among all contending network cells (e.g., between Network 1 and Network 2, and between Network 2 and Network 3), and are providing new and synchronized data-frame allocations to be used for the next super-frame. In super-frame N+1, the new data frame allocations resolved and synchronized in the previous super-frame are updated and used by the corresponding network cells, respectively.

9.2.3 Frame-Based ODSC Protocol

Following the brief introduction in Section 9.2.2.4.3, this section describes in detail the mechanisms and functional procedures of the ODSC protocol for spectrum resource sharing on a frame-by-frame basis, referred to as frame-based, on-demand spectrum contention. Note that the Frame-based ODSC protocol follows the general ODSC principles described in Section 9.2.2.

9.2.3.1 Fundamentals

9.2.3.1.1 Synchronized Super-Frame

The frame-based ODSC protocol described in this section is based on the concept of synchronized super-frames. Each of the synchronized super-frames, as shown in

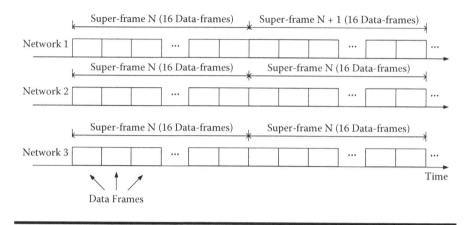

Figure 9.11 Synchronized super-frames.

Figure 9.11, is a collection of multiple data frames (16 data frames per super-frame as defined in IEEE 802.22) and is synchronized in time across all DSA coexisting network cells on all TV channels. Note that a data frame in Figure 9.11 may represent a network operation providing data services on any one of the available TV channels for a duration of a data frame or just an idle frame.

9.2.3.1.2 Top-Level Procedure of Inter-Network Coexistence

Before looking into the detailed mechanisms and procedure of the frame-based spectrum contention process for sharing a common TV channel, it is important to first understand the general inter-network coexistence procedure.

Figure 9.12 is a top-level flowchart detailing the inter-network coexistence procedure, which is described step by step as follows.

1. The network cell is powered on and starts its operation.
2. The network cell performs network discovery, which includes identifying
   a. TV channel occupancies of the neighboring network cells
   b. Pattern for coexistence communication (see Section 9.3) among the neighboring network cells
   c. Frame allocation patterns of the neighboring network cells on specific TV channels within a synchronized super-frame (this information can be obtained from the received ODSC announcement packets)
3. The network cell undergoes an etiquette-based channel acquisition [4] which is to identify and acquire an available TV channel without detrimentally affecting incumbent device operation and neighboring network cells.
4. If the network cell successfully acquires a TV channel, it goes to normal service operations on the acquired TV channel (Step 5 below). If, however, the network cell fails to acquire any TV channel, it selects a TV channel

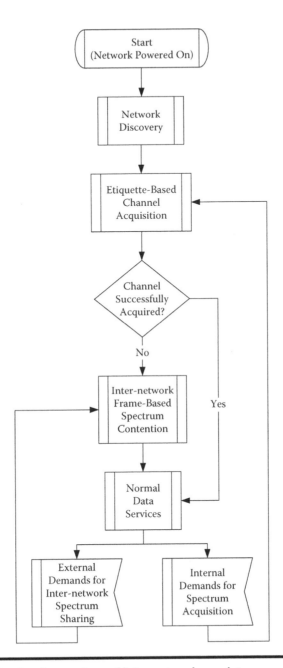

Figure 9.12   Top-level procedure of inter-network coexistence.

occupied by other network cells and performs the Inter-Network Frame-Based Spectrum Contention operations on the selected TV channel (Step 6 below).

5. The network cell performs normal service operations.

During the normal services operations, the network may receive internal demands (received from the inside of the network cell, e.g., from the CPEs) for additional spectrum resources (data frame transmission opportunities). When this occurs, the network cell reinitiates the spectrum acquisition process starting from Step 3 (etiquette-based channel acquisition).

On the other hand, the network cell may also receive external demands (received from other network cells) for sharing its occupied spectrum resources (data frame transmission opportunities) on the operating channel. When this occurs, the network cell performs the Inter-Network Frame-Based Spectrum Contention operations on its operating TV channel (Step 6).

6. The network cell performs the Inter-Network Frame-Based Spectrum Contention operations on the selected TV channel (as described in detail below), and then goes to the normal services operations (Step 5) after the frame contentions have been successfully resolved (that is, each contending network cell has obtained an appropriate number of the data frames in a super-frame).

9.2.3.2 Overall Procedure of Frame-Based ODSC Protocol

This section outlines the high-level procedure of Inter-Network Frame-Based Spectrum Contention operation, which is a key procedural step of Inter-Network Coexistence flow as indicated in Section 9.2.3.1.2 and Figure 9.12. Details of the protocol will be described in the subsequent sections. Figure 9.13 depicts the overall procedure for a network cell implementing the Frame-Based ODSC protocol.

Upon initiation of a contention process, the BS (or AP) of the network cell receiving the contention demand determines whether the initiated contention is based on an internal or external demand for accessing spectrum resources.

If the contention process is initiated by an external demand for acquiring access to the occupied frames, the network cell identifies the target data frames owned by itself and proceeds to the contention process as a contention destination (DST). The procedure of contention process as a contention destination will be described in Section 9.2.3.6.

On the other hand, if the contention process is initiated by an internal (intra-cell) demand for additional data frames, the network cell undergoes the following procedure steps:

1. Identify the available (non-occupied) data frames in the super-frame MAP. The super-frame MAP, shown in Figure 9.14, is a data structure maintained by the network cell and is providing a one-to-one mapping of the IDs of 16 data frames and IDs of their corresponding occupier network cells. For a

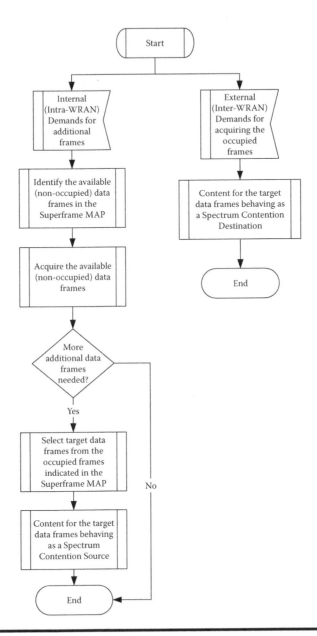

Figure 9.13 Overall procedure of frame-based spectrum contention protocol.

Data Frame ID	0	1	2	3	4	5	6	7	8	9	10	11	12	13	14	15
Network Cell ID																

Figure 9.14   Super-frame MAP structure.

data frame that is not occupied by any network cell, its network cell ID field is marked as "Non-occupied." Each network cell maintains a super-frame MAP that is synchronized across all network cells in its proximity and is constructed based on the spectrum usage information announced by the neighboring cells (e.g., ODSC_ANN, ODSC_REL messages).

2. Acquire the available (non-occupied) data frames (as described in Section 9.2.3.3 and 9.2.3.4).

   If no more additional data frames are needed, the network cell terminates the contention process (going to Step 5). However, if the network cell is not satisfied with the currently acquired data frames, it continues with the next step below.

3. Select target data frames from the occupied frames indicated in the super-frame MAP.

4. Content for the target data frames behaving as a Contention Source (SRC). The procedure for this step is described in Section 9.2.3.5.

5. Terminate the contention process.

9.2.3.3 Procedure of Available Frame Acquisition

The procedure to prepare acquiring one or multiple available frames is illustrated in Figure 9.15 and described below.

The network cell, maintaining a list of "Available Frames," first marks all of the frames on this list as "Winner Frames," which the network cell will attempt to acquire at the time triggered by the timeout of a specific timer.

Before the timer to perform frame acquisition times out, if the network cell received an ODSC_REL message indicating a frame, "Fx," is being released to another network cell, "Network N," it determines if "Network N" is one of its one-hop neighbors (i.e., within the mutual interference range of each other). There are two possible conditions at this point:

■ If the "Network N" is one of its one-hop neighbors, the network cell then checks if the frame, "Fx," is on the list of "Available Frames." If such condition is true, the network cell removes frame "Fx" from the list of "Available Frames"; otherwise, it leaves the list of "Available Frames" untouched.

■ If the "Network N" is not its one-hop neighbor, the network cell then checks if the frame "Fx" is on the list of "Available Frames." The network cell will add frame "Fx" to the list of "Available Frames" if this condition is not true; otherwise, it leaves the list of "Available Frames" untouched.

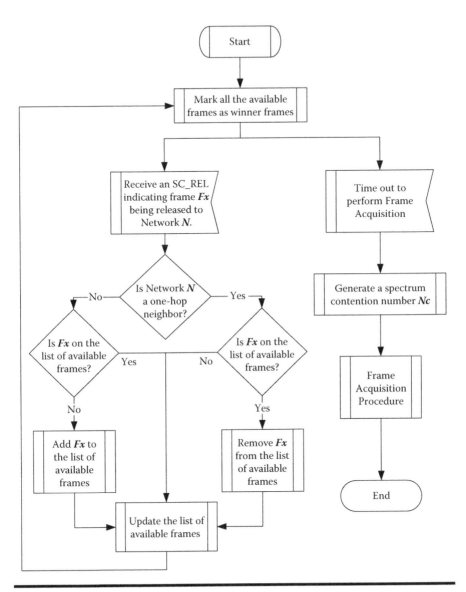

Figure 9.15 Available frame acquisition procedure.

At the time the frame acquisition timer times out, the network cell first generates a spectrum contention number "Nc," and then proceeds to the Frame Acquisition Procedure with the list of "Winner Frames" as will be described in Section 9.2.3.4. The spectrum contention number "Nc" will be used to resolve a potential conflict with another network cell that also attempts to acquire the same data frame simultaneously.

9.2.3.4 Procedure of Frame Acquisition

The procedure to occupy one or multiple frames is illustrated in Figure 9.16 and described below.

The network cell, maintaining a list of "Winner Frames," first broadcasts an ODSC_ACK message containing the contention number "Nc" (generated in a previous procedure) to announce the intention to occupy the winner frames starting from the subsequent super-frame (i.e., the super-frame after the current contention process is completed). Recall that the ODSC_ACK message is served as a "Clear to Send" message. The network cell will continue broadcasting the ODSC_ACK messages until a specific timer for frame occupancy timeouts, at which the network cell will attempt to occupy the "Winner Frames."

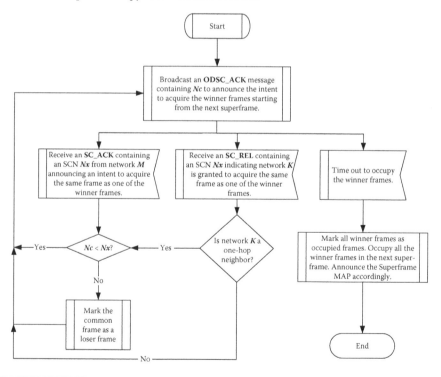

Figure 9.16 Frame acquisition procedure.

Before the time out of the timer for occupying the winner frames, there are two possible events that may occur to the network cell performing this procedure:

■ The network may receive an ODSC_ACK containing a contention number "Nx" from "Network M" announcing an intention to occupy the same frame as one of the winner frames.

When this event happens, the network cell first compares "Nx" with its own contention number, "Nc" to determine who is in possession of a higher priority for accessing the target frames.

If the network has a higher priority (i.e., Nc < Nx), it maintains the "Winner Frames" list untouched. Otherwise, the network cell removes all the frames that are in common with the competing network cell from the list of "Winner Frames" by marking those common frames as "Loser Frames."

■ The network may receive an ODSC_REL containing a contention number "Nx" indicating "Network K" is granted to acquire the same frame as one of the winner frames.

Obviously the network cell cannot occupy the same frame with another network cell if they are within the range of mutual interference. Therefore the network checks if "Network K" is one of its one-hop neighbors.

If "Network K" is not a one-hop neighbor, the network cell will ignore the received ODSC_REL message without changing its "Winner Frames" list. Otherwise, the conflict between these two cells on accessing the target frame must be resolved. To achieve that, the network cell compares "Nx" with its own contention number, "Nc" to determine who is in possession of a higher priority for accessing the target frames.

If the network has a higher priority (i.e., Nc < Nx), it maintains the "Winner Frames" list untouched. Otherwise, the network cell removes all the frames that are in common with the competing network cell from the list of "Winner Frames" by marking those common frames as "Loser Frames."

At the time the timer to perform frame occupancy times out, the network cell first marks all winner frames as occupied frames, then schedules to occupy all the winner frames in the subsequent super-frame after the contention process is completed, and accordingly announces the super-frame MAP indicating the new frame allocation to the neighboring network cells.

9.2.3.5 Spectrum Contention Procedure at Contention Source

The Spectrum Contention Procedure at Contention Source (ODSC-SRC), as introduced in the overall procedure of the frame-based spectrum contention protocol (see Section 9.2.3.2 and Figure 9.13), is illustrated in Figure 9.17 and described in the following steps.

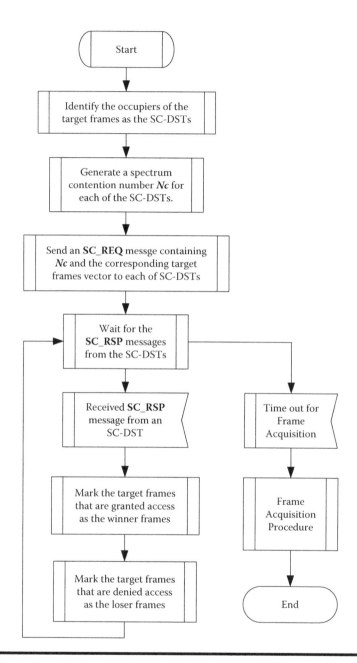

Figure 9.17 Spectrum contention procedure at contention source.

1. The contention source (ODSC-SRC) identifies the occupiers of the target frames as the contention destinations (ODSC-DSTs).
2. The ODSC-SRC generates a spectrum contention number "Nc" for each of the ODSC-DSTs.
3. The ODSC-SRC sends an ODSC_REQ message containing "Nc" and the corresponding target frames vector to each of SC-DSTs. The target frame vector is a 16-bit vector indicating the indexes of data frames within a super-frame that the ODSC-SRC requests to acquire (through the contention) for its data services, starting from the next super-frame after the contention process is completed. For each of the 16 bits of the target frame vector, the corresponding frame is requested for the contention when a bit's value is set to 1; otherwise, the bit value of the corresponding frame is set to 0.
4. The ODSC-SRC waits for the ODSC_RSP messages transmitted from the ODSC-DSTs. The ODSC-SRC will terminate the waiting process and proceed to Step 6 at the timeout of a specific timer for frame acquisition.
5. If an ODSC_RSP message is received from one of the ODSC-DSTs before the timeout of the timer, the ODSC-SRC marks the target frames that are granted access (due to a successful frame contention with the ODSC-DST) as the winner frames and, on the other hand, marks the target frames that are denied access as the loser frames (due to a failed frame contention with the ODSC-DST). The ODSC-SRC continues with Step 4.
6. When the timer for frame acquisition times out, the ODSC-SRC proceeds to conduct the Frame Acquisition Procedure, as described in Section 9.2.3.4 and Figure 9.16.

9.2.3.6 Spectrum Contention Procedure at Contention Destination

The Spectrum Contention Procedure at Contention Destination (ODSC-DST), as introduced in the overall procedure of frame-based spectrum contention protocol (see Section 9.2.3.2 and Figure 9.13), is illustrated in Figure 9.18 and described step by step as follows.

A network cell may function as a Contention Destination (ODSC-DST) when one of the following events occurs at the network cell:

- Received an ODSC_REQ message from an ODSC-SRC
- Received an ODSC_ACK message containing a contention number (SCN) "Nx" from another network cell (ODSC-SRC) announcing an intention to acquire frame "Fx," the same as one of the occupied frames

When an ODSC_REQ message is received from an ODSC-SRC, the following steps are followed by the ODSC-DST:

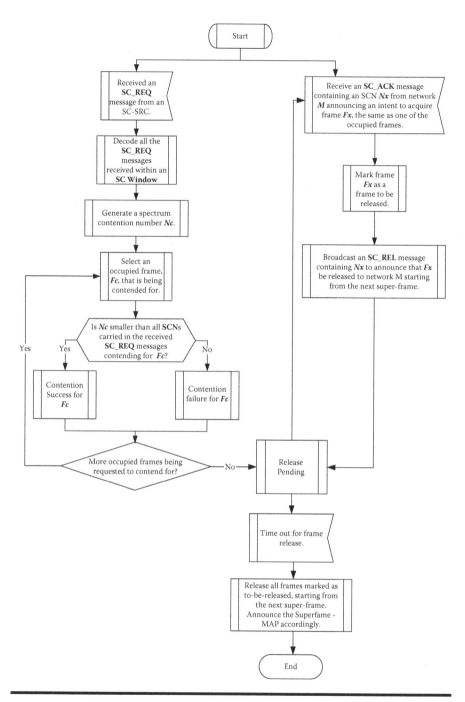

Figure 9.18 Spectrum contention procedure at contention destination.

1. Decode all the ODSC_REQ messages received within an ODSC Request Window (see Section 9.2.2).
2. Generate a spectrum contention number "Nc."
3. Select an occupied frame, Fc, that is being contended for by the one or multiple of ODSC-SRCs.
4. Conduct the Spectrum Contention Resolution Algorithm, as described in Section 9.2.2.

    If the contention for frame "Fc" is successful, the ODSC-DST constructs ODSC_RSP messages (using frame vectors) for all ODSC-SRCs indicating that frame "Fc" is not granted (see Figure 9.19).

    Otherwise, if the contention for frame "Fc" fails, the ODSC-DST proceeds with the following sub-steps (see Figure 9.20):

    i.   Identify the winner ODSC-SRC of frame "Fc" that has the smallest SCN.
    ii. Construct and transmit an ODSC_RSP message (using frame vectors) for the winner SC-SRC indicating that frame "Fc" is granted.
    iii. Construct and transmit ODSC_RSP messages (using frame vectors) for all the other SC-SRCs indicating that frame "Fc" is not granted.

5. Check if there are more occupied frames being requested to contend for. If the condition is true, go to Step 3; otherwise, proceed to Step 6 below.
6. In the "Release Pending" stage, the ODSC-DST is pending for three events:

    i.   Receiving an ODSC_ACK message from one of the Winner ODSC-SRCs for an occupied data frame. The ODSC_ACK message will confirm the frame acquisition intention of the Winner ODSC-SRC. When this event arises, go to Step 7.
    ii. Receiving an ODSC_ACK message containing an SCN "Nx" from a network cell ("Network M") announcing an intention to acquire frame "Fx," which is the same as one of the occupied frames.
    iii. Time out for data frame release. When this event arises, go to Step 8.

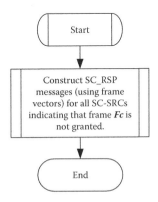

Figure 9.19   Operations of contention success of a data frame at contention destination.

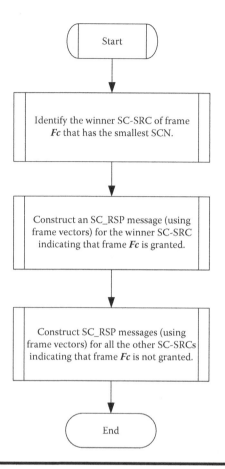

Figure 9.20 Operations of contention failure of a data frame at contention destination.

7. When receiving an ODSC_ACK message from one of the Winner ODSC-SRCs confirming the intention of acquiring one or multiple Winner data frames, the ODSC-DST first marks the winner data frame as a "Frame to be Released," then broadcasts an ODSC_REL message for the "Winner" ODSC-SRC that has acknowledged the ODSC_RSP message that grants transmission opportunities in frames currently occupied by the ODSC-DST (which sent the ODSC_RSP message). The ODSC_REL message contains the IDs of the "to-be-released" frames, the ID of the winner ODSC-SRC, and the associated Winner SCN.

Note the Winner SCN in the ODSC_REL message is used to resolve potential conflicts between the winner ODSC-SRC and other network cells (involved in another independent contention process) for accessing the same target frame. See Section 9.2.3.4 (Procedure of Frame Acquisition) for more details.

When this step is completed, go to Step 6 again.

8. When a specific timer times out for frame release, release all frames marked as "to-be-released," starting from a subsequent super-frame. Announce the super-frame MAP accordingly.

A network cell may also receive an ODSC_ACK message containing a contention number (SCN) "Nx" from another network cell (ODSC-SRC), for example, "Network M," announcing an intention to acquire frame "Fx," the same as one of the occupied frames (see Section 9.2.2.2.5 for details). In such a case, the network cell, functioning as an ODSC-DST responds to the received ODSC_ACK, following the steps below:

1. Mark frame "Fx" as a frame to be released.
2. Broadcast an ODSC_REL message containing "Nx" to announce that "Fx" be released to "Network M" starting from the subsequent super-frame.
3. Go to Step 6 above.

9.3 Beacon Period Framing Protocol

In this section we introduce the Beacon Period Framing (BPF) Protocol that enables reliable, efficient, and scalable inter-network communications in support of inter-network coordination functions, such as the ODSC protocol described in Section 9.2.

9.3.1 Synchronized Super-Frame and Date Frame Structure with Beacon Period

As depicted in Figure 9.21, the BPF Protocol adopts the synchronized super-frame and frame structure proposed in IEEE 802.22 without loss of generality. All channels are partitioned in time into synchronized super-frames, each of which consists of 16 frames with fixed frame size. Each frame is further divided into a Data Transmission Period and an optional fixed size Beacon Period (BP), which allows coexisting network cells to exchange coexistence beacons for inter-network communications. In line with 802.22, we assume that each BP allows one beacon to be transmitted, and that a network cell is only allowed to transmit coexisting beacons in BPs on its operating channel. Note that both the Data Transmission Periods and the Beacon Periods shown in Figure 9.21 can be used by any coexisting network cells.

Although BPs can be accessed using contention-based mechanisms such as CSMA or Aloha, these mechanisms cause coexistence beacons to be transmitted in nondeterministic instances that are unknown to the other coexisting cells. This nondeterministic characteristic renders unpredictable (and potentially long) delay and very low bandwidth efficiency for inter-network communications. For example, the allocated bandwidth is wasted if no inter-cell communication can

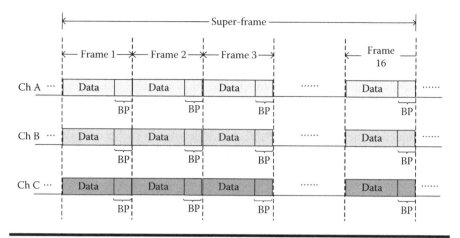

Figure 9.21   Super-frame and data frame structure with beacon periods.

be successfully conducted during a BP. In order to mitigate the nondeterministic issue of contention-based beacon transmissions, each network cell can periodically reserve on the operating channel a BP for exclusive beacon transmission. Although this reservation-based approach improves the system performance and bandwidth efficiency in static coexistence environments, it still suffers from performance limitations when the coexistence scenarios are dynamically changed, due to its lack of flexibility and scalability. The BPF Protocol, as introduced in the following, provides an efficient, flexible, and scalable method for reliable inter-network communication utilizing the beacon periods.

9.3.2 Beacon Period Frame Structure

As shown in Figure 9.22, a BP frame is a group of 16 BPs in consecutive data frames. The BP frame begins with an Announcement BP (A-BP) and ends with a BP preceding the A-BP of the next BP frame. The location of the A-BP is designed to be unique across a large number of continuous channels. To achieve that, we specify that an A-BP

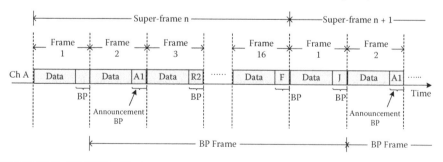

Figure 9.22   Beacon period frame structure.

for a particular channel always resides in a MAC frame with the frame index (within a super-frame) equal to the channel number of the residual channel modulo 16.

Similar to a regular data frame in a TDMA system, a BP frame consists of a MAP (the payloads' scheduling information) and the payloads, which are the 16 BPs in the BP frame. The MAP of a BP frame is carried by the announcement beacon transmitted in the A-BP. As an example, Figure 9.5 shows a BP frame for a coexistence scenario where cell 1 and cell 2 reside on channel A. The A-BP, which is reserved by cell 1 and labeled as "A1," is the first BP in the BP frame. As specified in the MAP, the second BP is reserved for cell 2 (labeled as "R2"). The other unassigned BPs are set to be Free-to-use (labeled as "F") except for the last BP, which is reserved for "Joining" (allowing a new cell to participate in transmission on the channel). The details of BP assignments will be described in the following subsection.

9.3.3 Types of BP Assignment

For the BP Framing Protocol, we define 4 types of BP assignments as follows.

1. Announcement BP (A-BP)

   The A-BP is always at the beginning of a BP frame and may be reserved by a network cell, behaving as the channel coordinator, for the transmission of the announcement beacon which aggregates the MAP and other information elements in order to allow efficient inter-cell discovery and communications.

2. Reservation BP (R-BP)

   An R-BP is reserved for a network cell, say cell x, that resides on the operating channel to perform contention-free beacon transmission. When a BP is reserved for a cell x, only cell x has the right to transmit a beacon during this BP on the operating channel of cell x. However, cell x can chose to perform any task other than beacon transmission, such as data transmission, receiving beacons on other channels, and so forth. Other network cells that intend to receive a beacon packet from cell x can tune to the operating channel of cell x during x's R-BP by referring to the MAP.

3. Free-to-use BP (F-BP)

   An F-BP can be used in many ways by all network cells residing on the operating channel: either for data transmissions, beacon transmissions (using contention-based methods) or receptions, or any other system maintenance purposes.

4. Joining BP (J-BP)

   The J-BP is used for an off-channel cell or new network cell to join the BP frame so as to participate in communication on the operating channel. All on-channel network cells in the same proximity should frequently monitor the J-BPs in order to capture the joining messages in a timely manner. Beacon transmissions in a J-BP use contention-based methods.

9.3.4 *Inter-Network Communications Using BP Framing*

In this subsection, we describe the inter-network communications procedure of the BP Framing Protocol.

For two network cells, cell A and cell B that are operating on channel Ch(A) and Ch(B), respectively, the inter-cell communications are performed using BP Framing as follows:

1. Cell A to receive beacon packets from cell B
   - Tune to the operating channel of Cell B—Ch(B), during the A-BP of Ch(B).
   - Receive and decode the BP frame's MAP of Ch(B).
   - Identify the R-BP of Cell B—R-BP(B).
   - If R-BP(B) exists, receive beacon packets from Cell B at R-BP(B).
   - Otherwise, try to receive beacon packets from Cell B during the F-BP on Ch(B).
2. Cell A to transmit beacon packets to cell B
   - If an R-BP is required for the beacon transmission, reserve one—R-BP(A).
   - Transmit the MAP of the BP frame of Ch(A), during the A-BP of Ch(A).
   - If R-BP(A) is available, transmit the beacon packet during R-BP(A).
   - Otherwise, transmit the beacon packet during an F-BP on Ch(A).

The inter-network communications method using BPF protocol as described above aims to be general enough for supporting both co-channel inter-network communications and cross-channel inter-network communications where coordination among network cells operating on different channels is necessary (e.g., synchronization of spectrum sensing quiet periods among network cells across channels). In the case of co-channel inter-network communications (e.g., for co-channel spectrum sharing), Ch(A) is set to be identical to Ch(B) in the BPF protocol.

9.3.5 *Channel Coordinator and Channel Members*

In order to facilitate efficient, flexible, and scalable management of inter-network communications, one of the network cells communicating on the operating channel may behave as the channel coordinator and is responsible for a number of coordination tasks, which include transmitting announcement beacons, managing channel membership (joining of new members and leaving of old members), and scheduling of beacon periods for all channel members by generating the MAP. By default, the network cell that occupies the channel first becomes the channel coordinator. All the other co-channel network cells behave as the channel members after successfully registered with the channel coordinator through the joining process. A channel member can request a BP reservation and shall follow the schedule in the MAP transmitted by the coordinator.

9.3.6 Flexible and Scalable Scheduling of Beacon Periods

Because there are 16 BPs available in a BP frame, up to 16 co-channel network cells can be simultaneously accommodated for deterministic inter-network communications through BP reservations, and when some of the BPs are shared using contention based access, more than 16 cells can be allowed to communicate on a channel. The assignment of BPs and scheduling of network cells to communicate on BPs can be flexibly managed by the channel coordinator. Note that the coexistence scenario could be dynamically changed over time due to, for example, channel switching or mobility of the network cells. The BPF protocol allows the scheduling of BPs to be adapted to the current coexistence scenario optimizing scalability, performance, and bandwidth efficiency for inter-network communications. For example, when the number of co-channel network cells is small, each cell can be allocated more R-BPs so that it can have more control to manage the inter-network communications or reuse some of the R-BPs for its data transmissions. On the other hand, when the number of co-channel network cells increases, each cell is allocated fewer R-BPs so as to accommodate more channel members.

9.4 Performance Evaluation and Discussion

To evaluate the performance of ODSC and BPF Protocols, we have developed a Network Simulator (NS2) [12] model for cognitive radio-based DSA networks implementing these two protocols for inter-network spectrum sharing and communications.

In our simulations, multiple network cells, synchronized to a common time source, coexist in the same vicinity sharing a SINGLE channel. We configure each super-frame to contain 16 data frames. The sizes of the frame and the BP are 10 ms and 2 ms, respectively. Each round of simulation runs for 10,000 seconds.

In order to verify the feasibility of these protocols, we conduct simulations for three basic types of coexistence scenarios modeled by contention graphs (see Section 9.2.2.1), in which vertices denote the network (e.g., WRAN) cells and edges connecting vertices represent the mutual interference between the cells:

1. Complete Graph scenarios—every pair of vertices is connected by an edge.
2. Cycle Graph scenarios—vertices are connected in a closed chain.
3. Wheel Graph scenarios—a center vertex is connected to all other vertices that form a cycle.

For each type of scenario, as shown in Figure 9.23, we vary the number of coexisting cells (up to 7 cells) to evaluate the performance scalability of the proposed protocols. Table 9.1, Table 9.2, and Table 9.3 show the simulation results that measure the channel occupancy (i.e., ratio of channel occupation time to the total operation

Table 9.1 Channel Occupancy (Utilization), Fairness, Convergence Time of Coexisting WRAN Cells in Complete Graph Scenarios

No. of Cells	Cell 1 Util.	Cell 2 Util.	Cell 3 Util.	Cell 4 Util.	Cell 5 Util.	Fairness Index	Convergence Time (Sec)	Optimal Fair Share
2	0.5009	0.4950	0	0	0	0.9999	30	1/2
3	0.3316	0.3299	0.3354	0	0	0.9999	150	1/3
4	0.2484	0.2511	0.2511	0.2462	0	0.9999	270	1/4
5	0.2023	0.1931	0.2039	0.1989	0.1981	0.9997	300	1/5

Table 9.2 Channel Occupancy (Utilization), Fairness, Convergence Time of Coexisting WRAN Cells in Cycle Graph Scenarios

No. of Cells	Cell 1 Util.	Cell 2 Util.	Cell 3 Util.	Cell 4 Util.	Cell 5 Util.	Cell 6 Util.	Cell 7 Util.	Fairness Index	Convergence Time (Sec)	Optimal Fair Share
3	0.3316	0.3299	0.3353	0	0	0	0	0.9999	150	1/3
4	0.5012	0.4959	0.4959	0.5006	0	0	0	0.9999	35	1/2
5	0.4200	0.2841	0.4197	0.4160	0.4197	0	0	0.9814	336	2/5
6	0.4979	0.4987	0.4978	0.4989	0.4976	0.4989	0	0.9999	30	1/2
7	0.4307	0.4485	0.4287	0.4369	0.4391	0.4322	0.3485	0.9946	418	3/7

Table 9.3 Channel Occupancy (Utilization), Fairness, Convergence Time of Coexisting WRAN Cells in Wheel Graph Scenarios

No. of Cells	Cell 1 Util.	Cell 2 Util.	Cell 3 Util.	Cell 4 Util.	Cell 5 Util.	Cell 6 Util.	Cell 7 Util.	Fairness Index	Convergence Time (Sec)	Optimal Fair Share
4	0.2484	0.2511	0.2511	0.2462	0	0	0	0.9999	270	1/4
5	0.3349	0.3331	0.3288	0.3346	0.3277	0	0	0.9999	505	1/3
6	0.2840	0.2779	0.3057	0.2783	0.2833	0.2789	0	0.9988	1933	2/7
7	0.3245	0.3354	0.3366	0.3308	0.3301	0.3300	0.3320	0.9999	1112	1/3

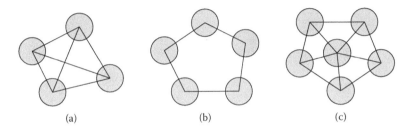

(a)   (b)   (c)

Figure 9.23   Types of coexistence scenarios.

time) of each coexisting network cell applying the ODSC protocol in three types of scenarios, respectively. The optimal occupancy as shown in each table is obtained by applying the Max–min fairness scheduling criterion [13] for each scenario.

Based on the data collected in Tables 9.1, 9.2, and 9.3, we plot the global (Jain's) fairness index [14] of the coexisting network cells as depicted in Figure 9.24. It shows that the ODSC protocol effectively enables network cells to achieve a close-to-optimal global fairness performance in all three types of coexisting scenarios without sacrificing bandwidth utilization. Moreover, the fairness performance scales very well with an increasing number of coexisting cells for a variety of scenarios.

Another important performance metric to evaluate an inter-network spectrum sharing mechanism is the time that the coexisting networks take to converge to the equilibrium of spectrum sharing activities. Figure 9.25 illustrates the convergence time of all the coexistence scenarios employing the ODSC protocol utilizing the inter-network communication channel enabled by the BPF protocol. It can be observed from Figure 9.25 that the complexity of the contention graph of a coexistence scenario determines the convergence time of spectrum sharing. Although an increased number of coexisting cells would in general increase the network convergence time, the type

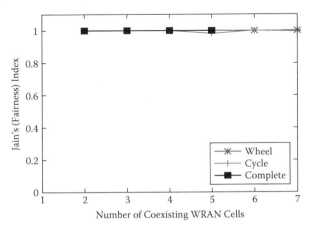

Figure 9.24   Fairness of ODSC in different coexistence scenarios.

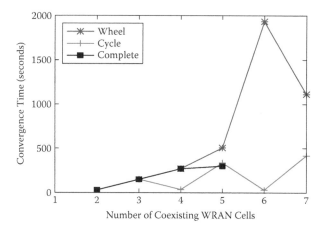

Figure 9.25 Convergence times of ODSC in different coexistence scenarios.

of contention graph of a scenario impacts the convergence performance significantly (Figure 9.25). As an example, the wheel graph scenario with 6 cells, which has an imperfect contention graph, requires almost 2000 seconds to reach the equilibrium, while the wheel graph scenario with 7 cells takes about 1100 seconds to converge. This is due to the imperfect topology of the contention graph that requires longer convergence delay to resolve more intensive inter-network contention conflicts.

Without loss of generality, the simulation results of ODSC operation presented in this section assume that the demand at each cell is the same as that for all the neighboring cells, and so the equal access time represents a "fair" system. In practice, the demands at the different cells may be different. For example, the demand at one cell is high, but its quality of service (QoS) is best effort, while the demand at another cell is low with a low-latency QoS requirement for an application such as voice. To improve QoS support of ODSC operations, spectrum contention demands at each network cell can be classified into different QoS categories, each of which may have different weight for random spectrum access number generations. That is, the higher QoS demand of a network cell has higher likelihood of winning the contention, and vice versa. On the other hand, a network cell with higher demands, which could be in low QoS categories, may attempt to acquire a shared spectrum more frequently, but could have a low likelihood to succeed for each of the acquisition attempts. Evaluations of ODSC operation with improved QoS support will be conducted in future work.

9.5 Discussions and Conclusion

In this chapter, the design challenges of inter-network spectrum sharing and communications for the emerging cognitive radio-based dynamic spectrum access (such as IEEE 802.22) networks are addressed. We present the On-Demand Spectrum Contention protocol that enables the coexisting network cells to compete for the scarce spectrum by exchanging and comparing randomly generated contention access priority numbers in an on-demand, distributed, and cooperative manner. In order to support inter-network coordination functions such as ODSC for cognitive radio-based dynamic spectrum access systems, we describe in detail the Beacon Period Framing protocol that realizes a reliable, efficient, and scalable over-the-air inter-network communication channel among coexisting cells reusing their occupied RF channels. Extensive simulations conducted on a variety of coexistence scenarios show that the ODSC protocol, enabled by the BPF protocol, provides close-to-optimal fairness, scalability, and spectrum efficiency for inter-network spectrum sharing. In the following subsections, we review prior work and discuss several system issues that emerge from comparing the concepts proposed in this chapter with other approaches. Generally, the conclusion is that the choice of protocols is closely connected to the availability of infrastructure and the purposes of the network.

9.5.1 Centralized vs. Distributed Spectrum Sharing

As proposed in [15], the centralized Spectrum Sharing networks are controlled and coordinated by a central entity, called Spectrum Broker. The service provider and users of these networks obtain time limited rights from the spectrum broker to part of the spectrum and use it to offer the network services. The management architecture proposed in [15], called DIMSUMnet, assumes that a contiguous chunk of spectrum, called coordinated access band (CAB), is reserved by regulatory authorities such as the Federal Communications Commission for centralized dynamic spectrum access. For a geographical region, the allocation of various parts of CAB to individual network operators or users is managed by the spectrum broker, who owns the CAB. The network operators or users submit spectrum lease bids to the spectrum broker and gain access to the spectrum by paying a price. Within the CAB, certain fixed frequencies are as allocated to so-called Spectrum Information Channels, which are used to deliver control information for spectrum access. Within this centralized architecture, multiple spectrum brokering servers must be redundantly deployed for each region and must maintain consistent information among them in order to ensure reliable allocation of the spectrum resources. If one of the servers fails, one of the remaining servers will continue to satisfy the spectrum lease and information request in the region. The development of a scalable mechanism in order to minimize the overhead of frequent and deterministic dissemination of spectrum information has been considered a key technical issue. Moreover, the overhead of the spectrum brokering increases as the number of network cells in a region increase. It is unrealistic to expect the base stations in the network cells to acquire spectrum lease from a spectrum brokering server located remotely; otherwise, the communication delay and overhead across multiple wireless hops could be quite significant. In order to utilize the Spectrum Information Channels for exchanging the control information for spectrum access, both the base stations and the client devices have to be equipped with dedicated RF interfaces for accessing these spectrum information channels. In certain cases, however, it may not be practical, especially if the base stations and the clients are designed to be inexpensive and simple devices that may not support the spectrum information channels.

The spectrum sharing mechanism described in this chapter is based on the distributed management architecture that is assumed by most Dynamic Spectrum Access networks employing the method of opportunistic spectrum access. In particular, the system aspects of the most distributed spectrum access-based TV white space networks (such as IEEE 802.22 WRAN) are fundamentally different from what the centralized architecture usually assumes:

- There is no dedicated spectrum reserved by FCC (or other regulatory authorities) for dynamic spectrum access in the TV bands.
- The spectrum brokering systems are not available.
- The access to the TV spectrum is opportunistic and license exempt and is not based on spectrum trading.

9.5.2 Backhaul-Based vs. Over-the-Air Inter-Network Communications

Although only over-the-air solutions are described in this chapter, inter-network communications can also be realized using the IP-based backhaul approach. The major reason that the over-the-air approach was proposed in this work is that we focus our attention on the MAC layer design. The IP-based backhaul solution involves operations at higher layers; therefore it is beyond the scope of this work. Another reason that we propose the over-the-air solution as an inter-network communications alternative is due to the following considerations on the backhaul-based approach.

The first consideration is the quality of communications offered by the backhaul solution. Latency and jitter are the major concerns. In order to connect to the IP backbone, a base station may have to route the control messages through multiple "backhaul relay radios" over the air until reaching the wired point of presence (POP) connecting to a wired backbone that is optimized to reduce latency and jitter using such technologies as a multi-protocol labeling system (MPLS) [16]. The communication latency and jitter that occur in both the wireless link and inside the relay radios accumulate as each backhaul relay is traversed. Assuming a 10 ms delay introduced by each hop of the wireless relay link and by the IP network, and 5 hops to reach the IP backbone from both sides of the communicating base stations, a 110 ms delay is required for inter-network communications. The connection to the

IP backbone could also be realized through nonterrestrial communications, such as satellite services. In such a case, however, a permanent latency in the range of 500 ms to 1000 ms could be incurred.

The second consideration is the availability of the backhaul network. Although the backhaul network is usually accessible to dynamic spectrum access networks targeting broadband services, it may not be available in emergency and public safety situations when the infrastructure is down. Moreover, when the dynamic spectrum access networks are deployed for ad hoc, infrastructureless communications, the connection to the backhaul network cannot be de facto assumed.

9.5.3 Dedicated Radio Frequency vs. In-band Radio Frequency for Inter-Network Communications

An alternative to using the in-band radio frequency (the same frequency as for the system's data service communications) for realizing the inter-network communications channel is to utilize a dedicated radio frequency (or a number of dedicated frequencies). This will facilitate coordination of the dynamic spectrum access networks in TV bands.

Unlike in the centralized spectrum sharing architecture where dedicated frequencies for spectrum coordination are reserved by the spectrum owner, it would be infeasible, or at least very difficult, to maintain such dedicated radio frequencies for inter-network communications in the opportunistic spectrum access environment. As the secondary users, the dynamic spectrum network systems must vacate the operating frequencies, including the ones allocated for inter-network communications, whenever the licensed incumbents reclaim the spectrum. In such a dynamically changing radio environment, a common frequency is not guaranteed to dynamic spectrum systems. Moreover, if one is identified, the complexity to manage it would be prohibitive.

References

1. J. Mitola, "Cognitive Radio: An integrated Agent Architecture for Software Defined Radio," Ph.D. dissertation, Royal Inst. Tech., Stockholm, Sweden, 2000.
2. FCC. ET docket no. 04-113. Notice of Proposed Rulemaking, May 2004.
3. IEEE802.22 Working Group, "IEEE802.22 Project Homepage" [Online]. Available: http://www.ieee802.org/22/, accessed on Sept. 30, 2010.
4. IEEE P802.22/D3.0 Draft Standard for Wireless Regional Area Networks Part 22: Cognitive Wireless RAN Medium Access Control (MAC) and Physical Layer (PHY) specifications: Policies and procedures for operation in the TV Bands.
5. MAC and PHY Proposal for 802.11af, IEEE 802 Doc. No.: 11-10-0258-00-00af-mac-and-phy-proposal-for-802-11af.pdf.

6. IEEE 802.19 Task Group 1 on Wireless Coexistence in the TV White Space. http://www.ieee802.org/19/pub/TG1.html, accessed on Sept. 30, 2010.

7. W. Hu and E. Sofer, "22-05-0098-00-0000 STM-Runcom PHY-MAC Outline," Technical proposal submitted to IEEE 802.22 WG.

8. W. Hu, D. Willkomm, M. Abusubaih, J. Gross, G. Vlantis, M. Gerla, and A. Wolisz, "Dynamic Frequency Hopping Communities for Efficient IEEE 802.22 Operation." *IEEE Communications Magazine*, Special Issue: Cognitive Radios for Dynamic Spectrum Access, vol. 45, no. 5, pp. 80–87, May 2007.

9. N. Abramson, "The ALOHA System—Another Alternative for Computer Communications," *Proc. Fall Joint Computer Conference (AFIPS)*, Vol, 37, 1970, pp. 281–285.

10. L. Kleinrock and F. A. Tobagi, "Packet Switching in Radio Channels, Part I—Carrier Sense Multiple Access Nodes and Their Throughput-Delay Characteristics," *IEEE Trans. Comm.*, Vol. 23, December 1975, pp. 1400–1416.

11. F. A. Toagi and L. Kleinrock, "Packet Switching in Radio Channels, Part II—The Hidden Terminal Problem in Carrier Sense Multiple Access and the Busy Tone Solution," *IEEE Trans. Comm.*, Vol. 23, December 1975, pp. 1417–1433.

12. "The Network Simulator—ns-2," http://www.isi.edu/nsnam/ns/, v2.28, accessed on Sept. 30, 2010.

13. X. L. Huang and B. Bensaou, "On Max-Min Fair Scheduling in Wireless Ad-Hoc Networks: Analytical Framework and Implementation," in *ACM MobiHOC*, 2001, pp. 221–231.

14. R. Jain, D. M. Chiu, and W. Hawe, "A Quantitative Measure of Fairness and Discrimination for Resource Allocation in Shared Systems," DEC Research Report TR-301, 1984.

15. M. J. Marcus, "Unlicensed Cognitive Sharing of TV Spectrum: The Controversy at the Federal Communications Commission," *IEEE Comm. Magazine,* May 2005.

16. J. Evans and C. Filsfils, *Deploying IP and MPLS QoS for Multiservice Networks: Theory and Practice,* San Francisco, CA: Morgan Kaufmann, 2007, ISBN 0-12-370549-5.

COEXISTENCE

Chapter 10

Spectrum Sensing in TV White Space

Rania A. Mokhtar, Rashid A. Saeed, and Borhanuddin M. Ali

Contents

10.1 Overview

The realization of recently developed wireless standards, such as the IEEE 802.22 and the IEEE 802.11af TVWS, are based on their ability to coexist with primary user and each other and to avoid transmission in the busy channels to ensure proper interference mitigation and appropriate network behavior. From the view that TV white space (TVWS) technology exploits the backyard without the intervention or assistance of the primary systems, the essential to TVWS technologies is to have a nature of spectrum sensing for the spectrum map determination, dynamic frequency selection (DFS), transmission adaptation and/or traffic scheduling. Sensing can be handled via several suboptimal and optimal techniques to extract channel information. Sensing can be used for improving power control and interference cancelations. In the United States, the Federal Communications Commission (FCC) recently announced that sensing is not required for TV band devices (TVBDs); only geo-location should be used for TVWS technology; sensing-only terminal is kept as a future option; and the report encourages research to increase sensing reliability.

10.2 Introduction

Spectrum sensing is a key to an enabling technology for white space access, because sensing is a method of measuring spectrum in a cyclical manner to detect primary transmitters for the express purpose of protecting the primary receivers and to avoid interference with the primary systems. Lately, spectrum sensing has received a lot of attention in the TVWS context. The TVWS applications, such as the 802.22 WRAN, cellular extension over TVWS, IEEE 802.11af (erroneously called TVWS 802.11af White-Fi) and TVWS 802.16 WiMAX networks, mobile TV over TVWS, emergency and public safety applications over TVWS, and some smart city applications beyond White-Fi, such as smart grid, tele-medicine, and networked robots, are to provide geographic sensing algorithms and policies basically in order to protect the primary TV broadcast service.

The secondary operation of white space devices in TV bands relies heavily on the capability of white space devices to successfully identify TVWS, suggested by regulators based on the capability of these devices to avoid harmful interference to licensed users of these bands, which in addition to the analog and the digital TV broadcast band (ATV/DTV) also includes Part 74 wireless microphones [1]. The spectrum opportunities of TVWS consist of waste and idle channels of licensed incumbent systems. The licensed incumbents that operate in the TV bands are:

■ The TV Broadcast service that comprises the analog television system, such as National Television System Committee (NTSC) (United States/Canada/South Korea/Japan/Taiwan/Philippines), PAL (Europe), and SECAM (France

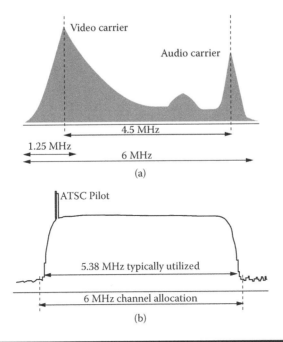

Figure 10.1 (a) NTSC channel spectrum, (b) DTV-ATSC spectrum.

analog TV), and the digital television system (DTV) including Advanced Television Systems Committee ATSC (United States/Canada) and DVB-T (Europe/worldwide). Figure 10.1 shows a channel spectrum for the analog and the digital TV [2].

■ Public/Private Land Mobile Radio System (LMRS), such as public safety and military system (380–399 MHZ in the United States) as well as commercial LMRS.

■ Licensed incumbents also include Part 74 low-power auxiliary devices or wireless microphones. These licensed incumbents do not have a standardized format, and classically use the FM modulation which occupies around 200 kHz bandwidth.

Spectrum sensing enables TVWS devices to detect the presence of TV signals and provides a smart, adaptive, and distributed solution to identify the TV white bands. However, the reliability of sensing techniques plays a major role which involves making observations of the radio frequency (RF) usage, sensing management, cooperative management, and sensing policies. Figure 10.2 illustrates the sensing functions in the physical, medium access control (MAC), and network layers, as well as a variety of methods and techniques that can be mapped to achieve the sensing stack functionalities. In the physical layer, sensing can be performed by different sensing methods; that is, spectrum sensing can be performed with wide band

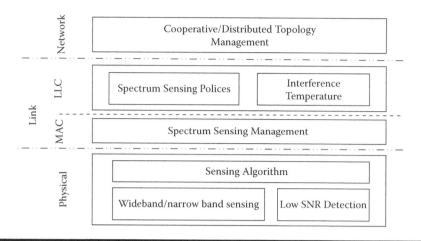

Figure 10.2 Spectrum sensing functions.

or narrow band scanning techniques in which the target band can be divided into sub-bands and scanned with a parallel or a sequential scanning method, in which case sensing may exploit the frequency hopping method for a fast scanning of the frequency band.

Sensing also requires a networking solution for coordination of a cooperative, whereby a distributed sensing management is needed for topology management, query issues, sensing load distribution, and information exchange. Further sensing is required to map with regularity and to resolve policy issues. A spectrum sensing necessitates the use of a sort of distributed and/or centralized data/decision fusion strategy for the combination of sensing results.

10.3 Overlay Spectrum Sharing and Sensing Requirements

TVWS spectrum efficiency requires domination over a number of concerns including interference constraint, end products of QoS, fulfillment of regulations, and resolving of some technical matters, such as detection of hidden nodes and white space reconfigurable radio equipment and devices development. The TVWS overlay scenario permits communications to work by [3]:

- Sensing to detect unused band or white space
- Coexisting with similar white space devices in which white frequencies will be used
- Monitoring frequency used by others
- Spectrum mobility and transmission power control as and when needed

Practical sensing of TV signal based on the FCC has two essential requirements: first, a high detection sensitivity to achieve the specified threshold which is defined as –107 dBm for the U.S. 6 MHz channel in short sensing time, and secondly, detection of a hidden incumbent that may result from strong signals from nearby device transmission sources. The challenge to achieve these two requirements is the RF front-end sensing technology that can be used to produce an acceptable probability of detection and probability of false alarm. Other challenges are out-of-band and spurious emissions due to low detection thresholds in the presence of other electronic devices functioning in the area which may affect the ability of a detector by masking the TV incumbent signal, thus resulting in a hidden node problem.

Spectrum sensing has a different aspect in TVWS systems, for example, supporting certain quality of service (QoS) in sensing-based spectrum access used for TVWS wireless communication is extremely difficult due to the absence of a full knowledge of how the primary TV services utilize the band, as well as other white device issues. A sensing threshold selection results in a tradeoff between the system throughputs and the incumbent protection.

10.4 Spectrum Sensing Techniques

There are several proposed spectrum sensing methods that enable cognitive radios to identify unused band and dynamic frequency selection (DFS). In this section, some of the most common spectrum sensing techniques in the cognitive radio literature are explicitly explained. When the structure of a primary signal is identified to the cognitive radio, the most select detector in the stationary Gaussian is matched filter, the noise is picked up in a matched-filter followed by a threshold test. However, implementing this type of coherent detector is difficult because a secondary user would need an extra dedicated circuitry to achieve a synchrony with each type of primary licensee. In addition, synchronization at such low signal-to-noise ratio (SNR) in many cases is more difficult than detection [4,9]. Moreover, there may be cases already in practice where matched-filter detector is ruled out due to lack of the associated knowledge on the primary signal's structure. In such scenarios, a general-purpose detector, in which several sensing techniques were introduced for detection of TV signal, would be much more desirable and practicable.

10.4.1 Energy Detection

A common method for the detection of unknown signals in noise is energy detection [5]. A figure in [5] depicts block-diagram of an energy detector. The input bandpass filter selects the center frequency, f_s, and the associated bandwidth, W. This filtering process is followed by a squaring device to measure the received energy and an integrator which determines the observation interval, T. Finally, the output of the integrator, Y, is compared with a threshold, λ, to determine the presence of a signal. The energy detector decision statistics can be performed as:

$$T(s) = \sum_N (s[n])^2 \tag{10.1}$$

and the number of sample is

$$N = 2[(Q^{-1}(P_f) - Q^{-1}(P_d)]^2 = O(SNR^{-2}) \tag{10.2}$$

While energy detectors have been extensively studied in the past, performance under the channel randomness has been only recently considered [4]. A noncoherent energy detector is suboptimal signal detection as compared to a coherent matched-filter or cyclostationary feature detector. A pure energy detection scheme is confounded by the in-band interference because it is not robust against spread spectrum signals, and its performance severely suffers under fading conditions [5]. The detection performance is poor under low SNR values. There is a minimum SNR referred to as SNR_{wall} below which the signal cannot be detected and the energy detector is not able to distinguish the noise, interference, and modulated signals. Therefore it is not capable of using adaptive interference cancelation techniques. The detection threshold is exposed to unknown or varying noise levels. Using adaptive estimated threshold would provide robustness and better gains; however it is affected by the in-band interference. In addition, there is no clear way to define the threshold in the presence of frequency selective fading, with regard to channel notches.

The FCC specifies the detection threshold as −114 dBm for the 6 MHz channels and −120 dBm for the 8 MHz channels, which necessitates the requirement that white space devices must be capable of sensing TV signals with a sensitivity higher than that of the original TV receivers [6]. To achieve a good sensing quality with such a high detection sensitivity level for the existence of noise is difficult and laborious with the current energy detection methods. However, the noncoherent energy-based approach does not require a prior knowledge of the signal to be detected and can give results in far fewer calculations to reach a decision, thus enabling a larger bandwidth to be surveyed at all times.

An energy detection of a TV signal can be done in a full bandwidth or a pilot detection process where the pilot detection algorithms are specific to the ATSC signals which have a DC pilot at a lower band-edge in a known location relative to the signal. Detection is achieved by setting a threshold either on the amplitude or the location of the pilot signal. Detection based on the location of a pilot is particularly robust against noise uncertainty because the position of the pilot can be pinpointed with a high degree of accuracy even if the amplitude is low due to fading [7].

10.4.2 Multi-Resolution Spectrum Sensing

A multi-resolution sensing technique produces a multi-resolution power spectral density (PSD) estimate using a tunable wavelet filter that can change its center frequency and its bandwidth [6,8]. This is performed by the filter, which sweeps over a range of frequencies, and the power at each frequency is recorded. The power spectral density estimation can be used to check the busyness of the spectrum channel, like in the previous sensing technique.

10.4.3 Segment Sync, PN511 or PN63 Sequence Detection

The ATSC sensing technique based on signature sequence correlation method involves correlating the received baseband signal with a signature sequence based on the data field sync pseudorandom noise (PN) sequences. The correlated signal is processed to a test statistic that is compared to the defined threshold. The digital TV data frame consists of 313 segments, with each segment having four data segment synchronization symbols as well as the 828 Reed Solomon (RS) encoded symbols. The data segment sync symbol pattern is represented as "1001"of a binary level. The data frame of the ATSC TV signal comprises 24.2 ms of 313 segments with the first segment being the field sync symbol pattern of 832 symbols, and this process of segmentation is repeated for every frame. The content of the 832 symbols consists of the 511 PN code sequence and the 63 PN code sequence, and this pattern is to be repeated 3 times [8,9]. From the structure of the field sync symbol pattern of 832 symbols, the detection method can be composed of three kinds of methods using (i) the data segment sync symbol pattern, (ii) using the PN511 code sequence, and (iii) using the PN63 code sequence.

The ATSC data is sent in segments of 828 symbols in which, at the beginning of each segment, a 4-symbol sequence (5,–5,–5,5) is sent. Detection of this sequence can be used in a feature detector. In the PN 511 sequence, the ATSC signal has a 511-symbol-long PN sequence that is inserted into the data stream every 24.2 ms. Because this is quite infrequent, averaging over more than one field would be necessary for detection, which leads to longer detection times.

10.4.4 Cyclostationary Feature Detection

Sensing can exploit the associated features of a TV signal to achieve a higher sensing quality via the extra sophisticated sensing methods. The cyclostationary feature detection process exploits the cyclostationary features of the received signals [10]. Cyclostationary features are caused by the periodicity in the signal or in its statistics, such as the mean and the autocorrelation. Instead of a power spectral density (PSD), a cyclic correlation function is used for detecting signals present in a given spectrum.

The cyclostationary-based detection algorithms can differentiate noise from primary users' signals. This is a result of the fact that noise is a wide-sense stationary (WSS) with no correlation, while modulated signals are cyclostationary with a spectral correlation due to the redundancy of signal periodicities [11,16]. The technique exploits the periodicity characteristics associated with the modulated signals, such as carrier wave, trains of pulse, spreading code, hopping sequences, and/or cyclic prefixes to estimate signal parameters and to use this parameter to detect signal in the presence of noise and other signals. The spectral correlation function (SCF) which is cyclic spectrum can be defined as [12]:

$$S_x^\alpha(f) = \lim_{T\to\infty} \lim_{\Delta t\to\infty} \frac{1}{\Delta t} \int_{-\Delta t/2}^{\Delta t/2} \frac{1}{T} X_T\left(t, f+\frac{\alpha}{2}\right) X_T^*\left(t, f-\frac{\alpha}{2}\right) dt \qquad (10.3)$$

$$X_T(t,v) = \int_{t-T/2}^{t+T/2} x(u)e^{-j2\pi vu} du \qquad (10.4)$$

Power spectrum density (PDS) is a special case of an SCF for $\alpha = 0$. Since the noise is a wide-sense stationary (WSS) with no correlation, the advantage of this technique is its capability to differentiate noise from the primary users' signals which are cyclostationary with a spectral correlation. The cyclic spectral density (CSD) function of a signal is defined as [13]:

$$S(f,\alpha) = \sum_{T=-\infty}^{\infty} R_y^\alpha(T)e^{-j2\pi fT} \qquad (10.5)$$

where

$$R_y^\alpha = E[y(n+T)y^*(n-T)e^{j2\pi\alpha n}] \qquad (10.6)$$

is the cyclic autocorrelation function (CAF), and α is the cyclic frequency.

Both the ATSC and DVB-T signals are cyclostationary; specifically, the means and the correlation sequences of these signals reveal the characteristics of a periodicity. Cyclostationary feature detectors were introduced as a complex two-dimensional signal processing technique for the recognition of modulated signals in the presence of noise and interference. Recently methods of resolving these anomalies have been proposed by a number of authors [14] for the detection of weak TV signals in the context of spectrum sensing for cognitive radio.

10.4.5 Spectral Correlation

Orthogonal frequency domain multiplex (OFDM) signals, including DVB-T signals, contain a special sequence called the cyclic prefix (CP), where the last D bits of the OFDM symbol are copied to the beginning of the symbol. A cyclic prefix detection is similar to an energy detection. However, the test statistic used in the algorithm is the energy contained in the cyclic prefix of each OFDM symbol, instead of the whole symbols [15]. Furthermore, due to the presence of CP the autocorrelation function of DVB-T signals shows distinct peaks at nonzero values whose amplitude and position can be used to detect the signal from the noise. In the presence of a known pattern, sensing can be performed by correlating the received signal with a known copy of itself [15,16].

The correlation method compares the received signal spectrum shape with the known shape of the target primary signal by calculating the correlations with the known values of the target signal and reveals the existence of a primary user when the correlation values are greater than the known values of a target primary TV signal. This method is only applicable to the systems with known signal patterns, and it is termed waveform-based sensing [17]. Table 10.1, describes the different spectrum sensing methods.

10.5 Cooperative Sensing

The basic problem concerning spectrum sensing is the detection of a signal within a noisy measure. The performance of the local spectrum sensing techniques is limited by the received signal strength. The spectrum sensing at a cognitive radio node (local sensing) is associated with some challenges. The channel state information (CSI) between a primary transmitter and a receiver and a receiver location are unknown to a cognitive radio [18].

Hidden node problems can arise when there is a blockage between the TVWS device and a TV station, but no blockage between the TV station and a TV receiver antenna and no blockage between the unlicensed device and the same TV receiver antenna. In such a case, a cognitive radio may not detect the presence of the TV signal and can start using an occupied channel, causing harmful interferences to the TV receiver. This problem increases the requirement in the cognitive radio sensitivity to a level that outperforms primary user (PU) receivers to the extent that it is able to detect weak signals. Local node sensing may achieve an acceptable sensing result only after a very long sensing time. Considering this limitation occurring in a local spectrum sensing, cooperation between cognitive radios is introduced for better accuracy of the primary user detection. Cooperation may solve problems of shadowing and the hidden node problem [19] as shown in Figure 10.3. The cooperation is considered as the key method for the realization of a cognitive radio.

Table 10.1   Spectrum Sensing Methods

Method	Advantages	Disadvantages
Matched filter (optimal)	Short time to achieve a certain probability of false alarm or probability of missed detection	• Requires a thorough knowledge of the primary users' signaling features • Synchronization is required and may be more difficult than detection • Implementation complexity • Large power consumption
Cyclostationarity features detection (suboptimal)	Separate target signals from noise	• Requires pre-knowledge of the primary signal • Requires considerable processing power • Performs worse than energy detection method when the noise is stationary
Waveform based sensing (suboptimal)	• Requires short measurement time • More robust than energy detector and cyclostationarity based methods	Known patterns of target signal are needed
Energy detection (suboptimal)	• Does not require *a priori* knowledge • Fast. Larger bandwidth • Low implementation complexity	• Not robust against spread spectrum signals • Confounded by in-band interference • Poor performance under low SNR values

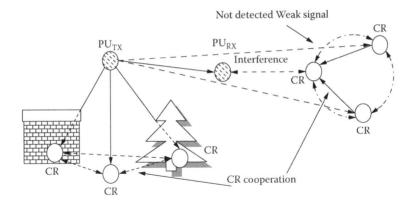

Figure 10.3 Cooperative detection.

Sensing can be improved by incorporating more observations from other cognitive radios at various locations. Network cooperation can improve a sensing performance as compared to a local sensing whereby a cooperative sensing enhances the detection and reduces the probability of interference to a primary user. A cooperative sensing is based on the fact that summing signals from two cognitive radios can improve the detection reliability by increasing the signal-to-noise ratio (SNR) if the signals are correlated [15]. A cooperative spectrum sensing exploits the broadcast nature and spatial diversity of the channel. In a cooperative sensing, the cognitive radio receivers may estimate channel variations resulting from fading, noise, and interference by relaying messages to each other that propagate redundant signals over multiple paths in the network.

Realization of a cognitive radio standard based on a cooperative radio employs a set of applications, such as relay channels, distributed antenna arrays, localization methods, data exchange, and fusion method as well as collaborative mapping.

Cooperative sensing can be implemented in different ways based on the cooperation method and network architecture. Cooperative sensing methods incorporate a set of cooperation technologies including network architecture, sensing method, data fusion algorithm, sensor selection, and data exchange protocol that characterize the performance of the method. Figure 10.4 identifies the direction of research in collaborative sensing. The chart shows cooperative technologies employed in spectrum sensing and highlights the other related cooperative methods in this area.

10.6 Spectrum Awareness in IEEE Standards

Spectrum awareness has been realized in some of the current wireless standards and has started to include cognitive awareness features which have been included in some

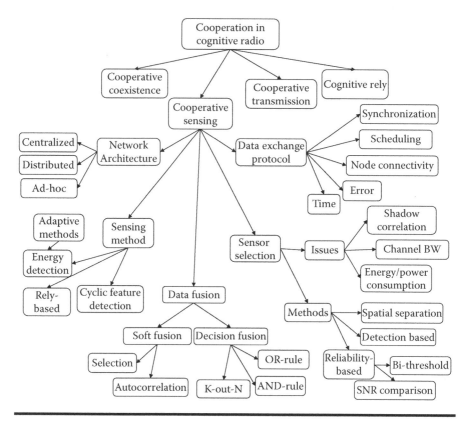

Figure 10.4 Direction of the research in collaborative sensing.

of the wireless systems. Opportunistic exploitation of spectrum has been started in many ongoing standardization efforts, while spectrum sensing and monitoring are mandatory tasks of some of the wireless standards, such as IEEE 802.11y/k, IEEE 802.22 wireless regional area network (WRAN), and IEEE 802.11af TV White-Fi, which proposes to reuse vacant spectrum in the TV broadcast bands, IEEE 802.16h cognitive radio, and other non-IEEE 802 standards, such as IEEE 1900 or standard coordinating committee 41 (SCC41) [20].

The first recognized spectrum sensing is developed from the license-exempt Radio Local Area Network (RLAN) standard for the protection of radar systems in the 5 GHz band. High sensing thresholds are considered for the detection of a radar system due to the high power nature of the radar systems (i.e., 100 mW indoor considers −62 dBm, and 1 W outdoor considers −64 dBm detection margin [21]). In what follows, some wireless standards that incorporate some kind of spectrum sensing for either adaptive purposes or opportunistic spectrum sharing will be discussed. However, the spectrum knowledge can also be used to initiate advanced receiver algorithms, such as adaptive interference cancellation.

Table 10.2 Requirements for the Physical Layer Parameters for a Typical Probability of Detection

Parameter	DTV/ATV	Part 74
Detection time	≤ 2 [sec]	≤ 2 [sec]
Move time	2 [sec]	2 [sec]
Closing transmission time	100 [ms]	100 [ms]
Detection threshold (sensitivity) at 90% probability of detection, 10% probability of false alarm	−114 [dBm] (DTV)	−107 [dBm] (Over 200 KHz)
	−94 [dBm] (ATV)	
SNR	−21 [dB] (DTV)	−12 [dB]
	1 [dB] (ATV)	

10.6.1 IEEE 802.22 WRAN

The IEEE 802.22 Wireless Regional Area Network (WRAN) standard is the first example of the TVWS application. The standard presents fixed point-to-multipoint wireless access and is set to utilize the white space of the worldwide TV broadcast bands. The FCC has formed an initiative known as the TV Band NPRM for the standards to proceed. Spectrum sensing is a major function of the 802.22 standards, and the sensing feature of 802.22 standards is the focal requirement for WRAN operation [6]. The IEEE 802.22 defined spectrum sensing requirements for both the analogue and the digital TV broadcasts and part 74 of the wireless microphones. Table 10.2 summarizes the requirements for the physical layer parameters for a typical probability of detection, with a probability of 90% and 10% of false alarm, respectively.

The spectrum sensing in WRAN network architecture is done in a distributed manner and is controlled by the WRAN base station. The client devices, defined as the customer premise equipment (CPE), employ a sort of signal measurement device which reports the measured results to the WRAN base station. The base station employs a data/decision fusion method on the reported observations and defines the local spectrum map, which is shared with the neighboring base stations, and the final spectrum map can be stored in a geo-location spectrum information database (SIB). The sensing mechanism in the IEEE 802.22 standard is triggered and controlled by the base station while the energy detection and features detection methods are used to identify idle TV channels with two sensing techniques. The first technique, known as an in-band spectrum sensing, employs two-stage detection, is fast, and the fine sensing to the frequency channels at present is being used by the base station-CPEs communication.

An in-band spectrum sensing can be performed in two phases. The first phase is carried out during a "quiet period." The quiet period sensing requires that the base station and the entire wireless devices within the designated zone are to stop all transmissions on the channel. The WRAN transmissions necessitate a break off process to allow for the in-band sensing, and therefore it is necessary for the base station to band its downlink transmission and allocate reporting time for the client CPEs to report the observation results. The sensing quiet periods are constructed into the transmission time frame. The measurement data preferred field (MDP) in the US-MAP is used by the base station to notify the CPEs which utilize a channel that they can transmit in the United States if they do not detect the presence of a primary user, and the CPEs that are not utilizing a upstream bandwidth (US BW) must make use of the UCS slots [22].

The other phase of an in-band sensing can be carried out without interrupting the WRAN operation; here the CPEs, in conjunction with a US BW, can transmit a bulk measurement report (BLM-REP) that has a priority factor over other waiting traffic, and if the band is not enough, then the CPE can exploit the fields in the MAC header, while again the CPEs that are not utilizing a US BW must make use of the UCS slots. The in-band sensing is carried out via a common sensing-eligible slot if it is not marked for use in an out-of-band sensing because the frame does not contain a US or DS traffic.

The in-band sensing is carried out on a periodic base in two stages of sensing: (i) fast sensing, and (ii) fine sensing. The in-band fast sensing form is carried out via the simple and typically fast energy detection method, which must be done in a short time period (1 ms). The observation of the fast sensing is analyzed by the WRAN base station to decide if the next step, fine sensing measurement, is required. If the base station considers there is a necessity for an additional measurement on a particular channel, a fine sensing is then performed for more accurate observations, whereupon a fine sensing stage is initiated based on the fast sensing results.

Throughout the fine sensing duration the CPEs are required to perform further detailed examinations on the specified channels. A fine sensing mobilizes more powerful sensing algorithm methods in which a fine channel sensing requires about 25 ms sensing duration, during which the CPEs scan at the signatures of incumbent signals. Fine sensing makes use of a feature detection technique for detection of incumbent modulated signal. The illustration in Figure 10.5 shows the sensing frame for the in-band spectrum sensing.

The second technique of the WRAN sensing, known as an out-of-band sensing process, is carried out on a channel sensing that is applied to channels currently not under the base station–CPEs utilization. Out-of-band sensing search for the additional white channels for spectrum band is carried out in the presence of the incumbent user in currently used channels. It also ensures that adequate guard band is available between the underutilized channels and every TV station that can be exploiting the adjacent channels. The out-of-band sensing is carried out during the system normal operation and does not require a quiet period, where the following two situations can be recognized:

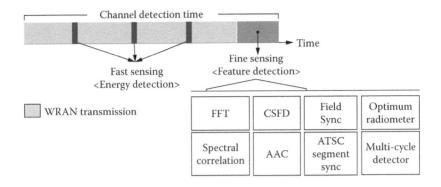

Figure 10.5 IEEE802.22 in-band sensing frame.

■ Case 1: enough guard is available between the currently utilized channel and the target sensed channel, where in this case a sufficient filtering based on the special RF circuits will be required due to the possible large variations in the signal levels between the RF transmit chain and the RF sensing chain at the RF front-end of the device.

■ Case 2: only a small parting or no particular isolation between the sensing and the transmitting/receiving RF circuits is available, whereby in this case the out- of-band sensing cannot take place within the normal system operation while the CPEs are transmitting/receiving process.

The IEEE 802.22 may possibly face a misdetection of a hidden incumbent problem which might occur when the WRAN CPEs that utilize a particular channel sense only on their active channel. Interference in the incumbent occurs for the time being while the base station is broadcasting on all aggregated channels as shown in Figure 10.6 [23]. This case can be solved by using a distributed quiet period (DQP) pattern. A hidden incumbent problem might also result due to a powerful interference from an incumbent, in which case the CPEs fail to detect a base station. The solution to this problem is done by using an off-band signaling that is periodically broadcast on the additional available channels. The CPEs can be aware of the base station to band off the current operating frequency due to a hidden incumbent case via one of the white channels [14,24].

10.6.2 White-Fi, IEEE 802.11af

One of the TVWS key applications is the IEEE 802.11af, which introduces the use of the TVWS in 802.11 technologies for the extension of the Wi-Fi range, based on better propagation characteristics below 1 GHZ, thus maximizing the range. Like any other TVWS application, White-Fi needs to incorporate a sort of spectrum sensing functionality to advance the 802.11 carriers sensing mechanism to detect

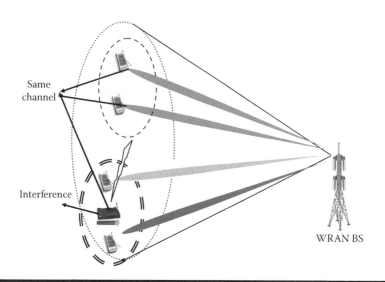

Figure 10.6   Possible hidden node scenario in WRAN.

unused channels of the TV spectrum. Potential candidates for the IEEE 802.11 extension standard that deploy sensing mechanism are the IEEE 802.11y and the IEEE 802.11k [25].

The IEEE 802.11y and the IEEE 802.11k provide suitable candidates for the TVWS technology. The IEEE 802.11y Radio Local Area Network (RLAN) standard provides broadband wireless data networks by enabling an efficient use of the spectrum in the 3.65–3.7 GHz band in the United States and also the 2.4 GHz, 5.7 GHz, and the 18 GHz bands in Hong Kong. The 802.11y RLAN allows the use of light-licensed RLAN units in the United States. This extension to the 802.11 allows detection of the existing transmissions and ensures no interference from the functions of nearby active satellite earth stations in the same frequency band by two essentials: (i) the enabling access points (licensed) have to be placed outside the omission zone (150 km) of the existing earth station, and (ii) the dependent stations are enabled by the enabling station prior to the transmission [26].

The 802.11y improvement sensing mechanism advances the original 802.11 carrier sensing protocol CSMA (Carrier Sensing Multiple Access) with the enhancements to the carrier sensing protocol and the energy detection techniques. The RLAN provides two capabilities for a dynamic spectrum sharing. First, it provides an Extended Channel Switch Announcement (ECSA) method that is used by the base stations for a reconfigurability purpose including notification of connected clients/station of such spectrum mobility request which indicates an intention to change channels, operating bandwidth, and regulatory classes. The ECSA mechanism permits the RLAN network to dynamically identify and operate on the channel with the least noise.

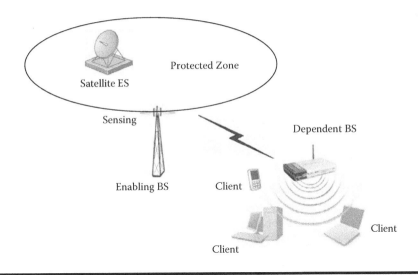

Figure 10.7   IEEE802.11y enabling mechanism.

The second capability provided by the RLAN is the Dynamic Station Enablement (DSE) mechanism, which allows an enabling station which is a high-powered registered station to send enabling beacon to the dependent remote stations in the coverage area and control of the enabling station. Figure 10.7 shows a layout of the 802.11y enabling mechanism.

The IEEE 802.11k is another extension to the IEEE 802.11 WLAN specification proposed for radio resource measurement. This standard defines and exposes radio and network information to facilitate the management and maintenance of a mobile Wireless LAN. The TIEEE 802.11k is a key standard for enabling seamless Basic Service Set (BSS) transitions in the WLAN environment. This standard incorporates different sensing reports for different measurements of channel noise, interference, load, etc. The channel energy caused by devices other than the 802.11 is measured as the channel interference level, and this measurement is presented by the noise histogram report as sensed by the client devices.

Other measurements include beacon report, which provides information about nearby base stations, and channel quality reports, such as station statistic, frame, and hidden node reports. This measurement is reported to the base station or access point where the final result is calculated. The base station then uses this sensed information to distribute the load over the network. This standard, therefore, provides a solution to the overloading problem that results from the client's device being connected to the base station with the strongest signal level and distributes the load to other base station when a base station has reached its maximum capacity. Thus this standard has a better network throughput.

10.7 Conclusions

In this chapter we discussed various spectrum sensing aspects and issues related to TV white space. Special attention has been paid to ongoing TV white space standards, i.e., IEEE 802.22 and the IEEE 802.11af. These standards were studied based on their ability to detect primary user (PU) or TV incumbent users and coexistence with other's signals, and to avoid transmission in the busy and adjacent channels to ensure proper interference mitigation within the allowable transmitted powers. Throughout the chapter, many TVWS scenarios were studied and elaborated. Sensing is important for devices that are not registered in any database, or are registered but move with uncoordinated manner and do not have geo-location capabilities. Sensing can be handled via several suboptimal and optimal techniques to extract channel information. Sensing can be used for improving power control and interference cancelations.

References

1. M. J. Marcus, "Unlicensed Cognitive Sharing of TV Spectrum: The Controversy at the Federal Communications Commission," *IEEE Commun. Mag.* vol. 43, no 5, pp. 24–25, May 2005.
2. C. Cordeiro, K. Challapali, D. Birru, and S. Shankar, "IEEE 802.22: The First Worldwide Wireless Standard Based on Cognitive Radios," in *IEEE DySPAN,* Nov. 2005.
3. Federal Communications Commission, "In the Matter of Unlicensed Operation in the TV Broadcast Bands," Notice of proposed rule making ET Docket No. 04-186 (FCC 04-113), May 2004.
4. R. Tandra, Fundamental Limits of Detection in Low SNR, Master Thesis, University of California Berkeley, Spring 2005.
5. T. Yucek and H. Arslan, "A Survey of Spectrum Sensing Algorithms for Cognitive Radio Applications," *Communications Surveys and Tutorials,* IEEE, vol. 11, no. 1, pp. 116–130, First Quarter 2009.
6. S. Shellhammer, "Spectrum Sensing in IEEE 802.22," First Workshop on Cognitive Info. Process. (CIP 2008), June 2008.
7. C. Politis (Ed.), "Spectrum Sensing and Fragmentation," White Paper, Version 1.0, WWRF WG8 on Spectrum Topics, ITU, November 2009.
8. M. Ghosh, V. Gaddam, G. Turkenich, and K. Challapali, "Spectrum Sensing Prototype for Sensing ATSC and Wireless Microphone Signals," 3rd International Conference on Cognitive Radio Oriented Wireless Networks and Communications (CrownCom 2008), pp. 1–7, 15–17 May 2008.
9. S. J. Shellhammer, S. Shankar, R. Tandra, and J. Tomcik, "Performance of Power Detector Sensors of DTV Signals in IEEE 802.22 WRANs," *Proceedings of the First International Workshop on Technology and Policy for Accessing Spectrum,* p. 4-es, Boston, MA, August 2006.
10. M. Nekovee, "A Survey of Cognitive Radio Access to TV White Spaces," *International Journal of Digital Multimedia Broadcasting,* Volume 2010, April 2010.

11. Y. Zeng, Y. C. Liang, Anh T. Hoang, and R. Zhang, "A Review on Spectrum Sensing for Cognitive Radio: Challenges and Solutions," *EURASIP Journal on Advances in Signal Processing* Vol. 2010, 2010.

12. M. Ghozzi, F. Marx, M. Dohler, and J. Palicot, "Cyclostationarity-Based Test for Detection of Vacant Frequency Bands," in Proc. IEEE Int. Conf. Cognitive Radio Oriented Wireless Networks and Commun. (Crowncom), Mykonos, Greece, June 2006.

13. A. Fehske, J. Gaeddert, and J. Reed, "A New Approach to Signal Classification Using Spectral Correlation and Neural Networks," in Proc. IEEE Int. Symposium on New Frontiers in Dynamic Spectrum Access Networks, Baltimore, Maryland, pp. 144–150, November 2005.

14. C. Cordeiro, M. Ghosh, D. Cavalcanti, and K. Challapali, "Spectrum Sensing for Dynamic Spectrum Access of TV Bands," 2nd International Conference on Cognitive Radio Oriented Wireless Networks and Communications (CrownCom 2007), pp. 225–233, Orlando, Florida, August 2007.

15. Z. Quan, S. J. Shellhammer, W. Zhang, and A. H. Sayed, "Spectrum Sensing by Cognitive Radios at Very Low SNR," in Proc. IEEE Global Communications Conference (GLOBECOM), Honolulu, Hawaii, November 2009.

16. W. Zhang, H. V. Poor, and Z. Quan, "Frequency-Domain Correlation: An Asymptotically Optimum Approximation of Quadratic Likelihood Ratio Detectors," *IEEE Transactions on Signal Processing,* Vol. 58, No. 3 (Part 1), pp. 969–979, March 2010.

17. A. V. Dandawate and G. B. Giannakis, "Statistical Tests for Presence of Cyclostationarity," *IEEE Transactions on Signal Processing,* Vol. 42, Issue 9, pp. 2355–2369 September 1994.

18. Simon Haykin, "Cognitive Radio: Brain-Empowered Wireless Communications," *IEEE Journal on Selected Areas in Communications*, Vol. 23, No. 2, February 2005.

19. X. Chen, Z. Bie, and W. Wu, "Detection Efficiency of Cooperative Spectrum Sensing in Cognitive Radio Network," *Journal of China Universities of Posts and Telecommunications,* Volume 15, Issue 3, pp. 1–7, September 2008.

20. R. V. Prasad, P. Pawelczak, J. A. Hoffmeyer, and H. S. Berger, "Cognitive Functionality in Next Generation Wireless Networks: Standardization Efforts," *IEEE Communications Magazine,* Volume 46, Issue 4, April 2008, pp. 72–78.

21. W. Lemstra and V. Hayes, "License-Exempt: Wi-Fi Complement to 3G," *Journal of Telematics and Informatics,* Volume 26, Issue 3, pp. 227–239, August 2009.

22. H. Kim and K. G. Shin, "In-Band Spectrum Sensing in Cognitive Radio Networks: Energy Detection or Feature Detection?," in Mobicom'08, September 2008.

23. S. Lim, S. Kim, C. Park, and M. Song, "The Detection and Classification of the Wireless Microphone Signal in the IEEE 802.22 WRAN System," Microwave Conference, 2007. APMC 2007. Asia-Pacific, pp. 1–4, 11–14 December 2007.

24. Hou-Shin Chen, Wen Gao, and D. G. Daut, "Spectrum Sensing for Wireless Microphone Signals," Sensor, Mesh and Ad Hoc Communications and Networks Workshops, 2008. SECON Workshops '08. 5th IEEE Annual Communications Society Conference on, pp. 1–5, 16–20 June 2008.

25. R. Ahuja, R. Corke, and A. Bok, "Cognitive Radio System Using IEEE 802.11a over UHF TVWS," in Proceedings of the 3rd IEEE Symposium on New Frontiers in Dynamic Spectrum Access Networks (DySPAN 2008), pp. 1–9, Chicago, Illinois, October 2008.

26. Xiaoyu Fu, Wenchao Ma, and Qian Zhang, "The IEEE 802.16 and 802.11a Coexistence in the License-Exempt Band," in Proc. of IEEE Wireless Communications and Networking Conference (WCNC 2007), pp. 1942–1947, Kowloon, March 2007.

Chapter 11

Distributed Spectrum Sensing

Yohannes D. Alemseged, Chen Sun,
Ha Nguyen Tran, and Hiroshi Harada

Contents

11.1 Introduction

In the last few decades, the demand for radio frequency (RF) spectrum has been increasing due to the steady fast growth of the wireless communication industry. Additional spectrum resources are needed to accommodate more users and provide broadband multimedia services.

So far, the trend of spectrum allocation has relied on assigning different portions of the spectrum for particular services, which means many existing wireless systems are regulated by a fixed spectrum assignment strategy. The regulation partitions the whole spectrum into a large number of pieces. Each piece is specified for a particular system. However, this approach turned out to be inefficient to accommodate the current spectrum demand. A large portion of frequency bands is unoccupied or only partially occupied [1]. Recent studies corroborated this fact that the assigned frequencies are not occupied all the time, which is indicative of under-utilization of the scarce spectrum [2]. The unused spectrum, also known as white space, varies temporally and spatially. The Federal Communications Commission

(FCC) has reacted to this unveiled fact by allowing an opportunistic usage of the spectrum holes with a condition that no harmful interference is induced to the licensed services or primary user (PU) signals [3]. Cognitive radio system (CRS) [4] is an appropriate and front runner technology to realize this new paradigm of spectrum allocation through dynamic spectrum access (DSA). It has a great potential to boost the evolution of wireless communication.

CRS performs spectrum sensing in the licensed band of primary users (PUs) to discover unused spectrum, which is also known as spectrum opportunity and enables unlicensed users or secondary users (SUs) to dynamically access temporally those spectrum opportunities. The spectrum sensing capability of CRS should be as robust as possible to render sufficient protection to the incumbent services or PUs while optimizing the discovery of spectrum opportunity. Hence, reliable and fast sensing becomes crucial. Identifying a PU signal is not a trivial matter, especially when its signal level is weak. Thus, obtaining reliable information using a standalone sensor, with reasonable power consumption and complexity, may not be practical.

Cognizant of these facts, a sensing method based on cooperation among multiple cognitive radios (CRs) has been suggested [5–7]. The idea is to employ multiple CRs in such a way that an aggregate of spectral information is cooperatively shared among them to maximize the reliability of the signal detection. Other similar approaches include the concept of utilizing distributed spectrum sensors for spectrum sensing [8–10]. A more specific approach of wireless sensor networks (WSN) for spectrum sensing is discussed in [8,9].

The chapter is organized aiming at exploring distributed spectrum sensing (DSS), ranging from the single sensor as a special case to the multiple sensors case. For the sake of readability and mathematical tractability, certain signal modeling and assumptions are also introduced. The second section introduces scenario and components of DSS along with the description of the architectural assumptions. Analysis of different DSS algorithms are described. DSS under different data combining such as hard information combining and soft information combining are explored. The third section describes the media access schemes in DSS. Impact of different media access schemes on DSS is analyzed. Enhancement of the commonly known media access schemes is also discussed. The fourth section introduces the concept of spectrum sensing database (SDB), where the SDB could be realized as part of a DSS system. SDB can play a key role in CRS as a sensing information source. Such database can collect sensing data from distributed sensors in a programmed fashion and make the data available to CRS after further processing. Fusion center (FC), where data fusion or aggregation of the collected sensing information can be assumed, is part the SDB. The section elaborates the architectural aspects of SDB and some performance metrics to indicate the influence of using SDB as source of sensing information for CRS. Finally, concluding remarks are given in Section 11.5.

11.2 Distributed Spectrum Sensing

We consider a distributed spectrum sensing scenario as follows. Fixed spectrum sensors are spatially distributed over an office or commercial building complex as shown in Figure 11.1. This allows secondary users to share the distributed sensors and circumvent the expenditure of installing spectrum sensors on individual secondary user devices. The distributed sensors report their sensing data to a fusion center (FC) or secondary user (SU), which is assumed by default a CR, through a common auxiliary channel. The common auxiliary channel is also known as reporting channel or common control channel in some other studies. In this chapter it is also referred to as sensor-to-SU channel. The task of individual sensors is to

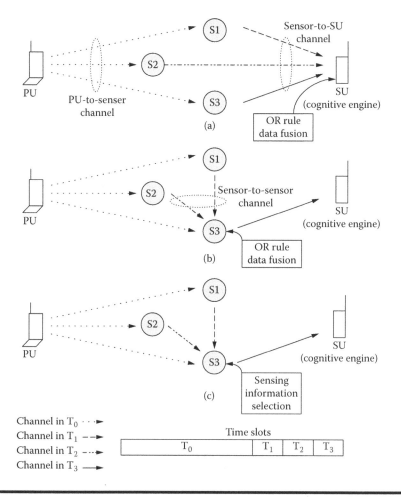

Figure 11.1 Examples of distributed sensing schemes using *N* = 3 DSSs. (a) cooperative sensing; (b) collaborative sensing; (c) selective sensing.

identify the presence or absence of primary user signal (PU). The FC combines the data collected from individual sensors to obtain a global decision on the presence or absence of PU signal. FC can be implemented in one of the sensors that serve as a gateway to further forward it to a central database or sensing database (SDB). Finally CR terminal can use the information at the FC or SDB to exploit any spectral opportunity. The actual role of FC may depend on the particular deployment scenario; for instance, FC may refer only to a fusion data function, while the control of sensors is left to another network entity. FC may also be coupled with cognitive engine and can be considered as CR. However, it is assumed here that FC, which may or may not be coupled with cognitive engine, plays a role of network controller for the sensors connected to it in addition to the data fusion capability. In general terms distributing sensing can be implemented where acquiring reliable spectrum sensing by using a single sensor becomes challenging. Especially inside a building complex, the PU signal strength can be weak due to concrete and metallic structures. Because these distributed sensors are not embedded at the SUs, we also assume the sensor-to-SU channels are influenced by multipath fading and will study the overhaul sensing, considering the multipath effects of sensor-to-SU channels that are currently considered as typical distributed sensing cases in IEEE standardization groups [11].

Distributed sensors can also be collocated and encased within wireless access points (AP) to minimize their deployment cost. Then one of the APs in the building complex can serve as FC, with the necessary application to carry out the data fusion, in addition to the WiFi service it provides. Scenarios for spatial distribution are also inherent in personal/portable devices, where spectrum sensors could be embedded in each personal/portable device. One practical application of such a setting would be enhancement of spectrum sensing capability for the use of personal/portable WiFi-type devices to be used in TV white space.* A very high sensitivity can be achieved at the FC if the distributed personal/portable devices forward their sensing information to the AP. Assuming further a cognitive engine capability in addition to data fusion capability at the AP, the AP can exploit available spectral opportunities to communicate with the personal/portable devices. We target three distributed sensing schemes for analysis, i.e., *cooperative sensing, collaborative sensing,* and *selective sensing,* for PU detection using N number of distributed spectrum sensors (DSSs), S_i, $i = 1, 2, \ldots, N$ that are distributed to N distinct locations within a service area. Here, we denote the channel from PUs to the DSSs as *PU-to-sensor channel.* And, we denote the channel from the DSSs to SUs as *sensor-to-SU channel.* The channel from a DSS to another DSS is called a *sensor-to-sensor channel.*

As shown in Figure 11.1(a), in *cooperative sensing* scheme, the DSSs provide independent (e.g., at different positions, frequencies, periods) sensing information to SUs. The DSSs perform sensing over time slot T_0. Subsequently, individual DSSs

---

* According to the FCC rule [3], the personal/portable devices which are WiFi-type devices should have spectrum sensing capability to detect TV band signals as low as −114dBm.

send the local sensing information to the SU consecutively through a sensor-to-SU channel over time slots T_i, $i = 1, 2, \ldots, N$. Then, the SU performs data fusion and makes global PU detection decisions. Schemes in [12,13] are special cases of the cooperative sensing scheme. It is also possible that the sensors exchange their information thorough sensor-to-sensor channels and perform data fusion locally before forwarding the result to the SU.

In *collaborative sensing*, DSSs perform sensing over time slot T_0. Instead of sending independent sensing information to the SU, the DSSs send the sensing information over time slots T_1, T_2, $\ldots$, T_{N-1} to the DSS that has the best sensor-to-SU channel. Subsequently, this sensor performs data fusion of sensing information and sends the SU improved sensing information in time slot T_N.

As shown in Figure 11.1(c), in a *selective sensing* scheme, sensing information is gathered at the sensor selected as a gateway. Instead of combining the sensing information, the sensing information from one of the DSSs is selected and sent to the SU by the gateway, thus the name selective sensing. The DSSs perform sensing over time slot T_0. The DSSs send sensing information over time slots T_1, T_2, $\ldots$, T_{N-1} to one DSS that provides the best sensor-to-SU channel. The scheme in [14] is a special case of the selective sensing scheme. We assume that the DSSs employ energy detection (ED) and generate local hard decision information (i.e., 1 or 0 to represent the presence or absence of the PU). This local hard decision information is sent through the sensor-to-SU channel with on–off keying (OOK) signaling to the SU. Then, the SU makes a global decision regarding the presence or absence of PUs. Both PU-to-sensor channel and sensor-to-SU are subject to Rayleigh fading. The following assumptions hold in the analysis: the sensor-to-sensor channel is assumed to be error-free, and for cooperative sensing and collaborative sensing the data fusion is based on OR rule. In addition, these schemes consume the same number of time slots in transmitting sensing information.

11.2.1 Single Device Sensing

A given CR terminal shall have at least one embedded spectrum sensor. Spectrum sensing based on such a single sensor can be viewed as a default way of obtaining sensing information for CR. The performance obtained can thus be used as a baseline performance to compare it with other schemes that utilize multiple spectrum sensors. The sensing technique we consider to identify the presence of PU signal from the observed signal is a simple energy detection.

11.2.1.1 Received Signal Model

Let us denote the sampled form of observed signal as x_i with a sampling period of T_s s. When the PU is active, which is referred to as hypothesis H_1, the observed signal at S_i is written as

$$x_i = h_{u,i}u + z_i \tag{11.1}$$

where u is a narrowband primary signal with statistics $\mathcal{N}(0,\sigma_u^2)$ propagating through a flat fading channel $h_{u,i}$. z_i is the additive white Gaussian noise (AWGN) modeled as an independent and identically distributed (i.i.d.) random variable (RV) with zero mean and variance σ_z^2. When the primary user is not active, which is referred to as hypothesis H_0, the received signal at the ith DSS is given by

$$x_i = z_i \tag{11.2}$$

The PU-to-sensor channel from the PU to the ith DSS is given by

$$h_{u,i} = \tilde{h}_{u,i}d_{u,i}^{\alpha/2} \tag{11.3}$$

where $\tilde{h}_{u,i}$ is a small-scale fading factor being modeled as a complex circular Gaussian RV with zero mean and unit variance. Furthermore, we assume that the channel is subject to block fading [15], i.e., $\tilde{h}_{u,i}$ is fixed during each block, which is longer than the total duration of sensing and sensing information exchange. Here, α is a power pathloss exponent, whereas $d_{u,i}$ is the distance from the PU to S_i. Thus, the instantaneous SNR of the PU-to-sensor channel observed at the ith DSS can be written as

$$\rho_{u,i} = \left|\tilde{h}_{u,i}\right|^2 \overline{\rho}_{u,i} \tag{11.4}$$

where

$$\overline{\rho}_{u,i} = \frac{\sigma_u^2}{d_{u,i}^{\alpha}\sigma_z^2} \tag{11.5}$$

In this study, we consider small-scale fading at PU-to-sensor channels and sensor-to-SU channels and assume that all the PU-to-sensor channels follow independent Rayleigh fading with equal average SNRs. The same goes for the sensor-to-SU channels. This is valid for situations where distributed sensors are deployed at a small area, noticing that sensing results far from the SU are not of interest to the SU, and the distances between sensors and the SU do not have much difference. Therefore, we set $\overline{\rho}_{u,i} = \overline{\rho}_u$, $i = 1, 2, \ldots, M$. The sensor-to-SU channel is used for transmitting sensing information. The hard decision information from DSSs is sent to the SU with OOK signaling.

The signal received from the ith individual DSS at the SU is given by

$$r_i = h_{s,i}s + n, \quad s \in \{0,1\} \tag{11.6}$$

where

$$h_{s,i} = \frac{\tilde{h}_{s,i}}{d_{s,i}^{\alpha/2}} \tag{11.7}$$

denotes the sensor-to-SU channel from S_i to the SU. Here, $d_{s,i}$ is the distance from S_i to SU. The power of the signal from the S_i is σ_s^2. The AWGN at the SU is modeled as an i.i.d. Gaussian RV n with mean zero and variance σ_n^2. The small-scale fading factor $\tilde{h}_{s,i}$ is modeled as a complex circular Gaussian RV with zero mean and unit variance. The sensor-to-SU channels are also assumed to be subject to block fading. The SNR of the individual sensor-to-SU channel from S_i to SU is given by

$$\rho_{s,i} = \left|\tilde{h}_{s,i}\right|^2 \bar{\rho}_{s,i} \tag{11.8}$$

where

$$\bar{\rho}_{s,i} = \frac{\sigma_s^2}{d_{s,i}^{\alpha}\sigma_n^2} \tag{11.9}$$

We also assume that all the sensor-to-SU channels have the same average SNRs, i.e., $\bar{\rho}_{s,i} = \bar{\rho}_s$, $i = 1, 2, \ldots, M$. Note that the pathloss effect in both PU-to-sensor channels and sensor-to-SU channels is taken into account in their average SNRs.

11.2.1.2 Energy Detection

The ED compares a detection statistic with a detection threshold [16] and produces hard decision information "1/0" representing the presence/absence of PU signal. At S_i the detection statistic is written as

$$T_i = \sum_{k=1}^{N_0} \left|x_i(k)\right|^2 \tag{11.10}$$

where $x_i(k)$ represents the kth sample of x_i and N_0 is the number of samples. Under hypothesis γ_0 the false alarm rate for a detection threshold γ_0 is written as [17]

$$P_f(\gamma_0, N_0, \sigma_z) = \frac{\Gamma\left(\dfrac{N_0}{2}, \dfrac{\gamma_0}{2\sigma_z^2}\right)}{\Gamma(N_0)} \tag{11.11}$$

where $\Gamma(\cdot, \cdot)$ is the incomplete Gamma function and $\Gamma(\cdot)$ is the Gamma function [18]. The detection threshold can be obtained using (11.11) for a given false alarm rate. In hypothesis H_1 the detection statistic follows a noncentral chi-square distribution [19]. We write the P_d at S_i for a given PU-to-sensor channel realization with the instantaneous SNR $\rho_{u,i}$ as

$$P_d = Q_{N_0}\left(\sqrt{2\rho_{u,i}N_0}, \sqrt{\frac{2\gamma_0}{\sigma_z^2}}\right) \tag{11.12}$$

where $Q_M(\cdot, \cdot)$ is the general Marcum Q-function [20].

11.2.1.3 OOK Receiver

In sensor-to-SU channels DSSs transmit the hard decision information to SU using OOK signaling. ED is used here to demodulate the OOK signal. Equations (11.11) and (11.12) can be used to describe the "false alarm rate" and "probability of detection" of the OOK receiver with arguments γ_1, N_1, and σ_n. In this study, we use ED-based noncoherent OOK receiver at the SU. The receiver compares the energy Z of the received signal

$$Z = \sum_{k=1}^{N_1} |r_i(k)|^2 \tag{11.13}$$

with a threshold γ_1. Here, N_1 is the number of samples of the OOK signal from DSS S_i. Based on this assumption, we can use the P_f in (11.11) and P_d in (11.12) of the ED to describe the receiver performance of the ED-based OOK receiver in AWGN. After obtaining the closed-form expression for the P_d of ED in Rayleigh fading channel, we can use that equation with (11.11) to describe the performance of the OOK receiver in Rayleigh fading channel correspondingly. Notice that P_f in fading channel remains the same, because it is independent of the SNR.

11.2.2 Analytical Models of Distributed Spectrum Sensing

Using (11.12), we first derive the generic expression for the probability of detection by ED with soft information combining (SC) in Rayleigh fading channel in this section. Using this expression, we set up analytical models of cooperative sensing, collaborative sensing, and selective sensing.

11.2.2.1 Probability of Detection by ED with SC under Rayleigh Fading

Let us assume that r_i, $i = 1, 2, \ldots, N$ are i.i.d. RVs that follow Rayleigh fading process with zero mean (i.e., r_i^2 follows exponential distribution)

$$f_{r_i^2}(x) = e^{-x}, \quad i = 1, 2, \ldots, N \tag{11.14}$$

Defining

$$r_{SC} = \max\{r_1, r_2, \ldots, r_N\} \tag{11.15}$$

as the maximum amplitude of Rayleigh fading channels and knowing that

$$f_{r_{SC}^2} = N(1 - e^{-x})^{N-1} e^{-x} \tag{11.16}$$

we can obtain the probability density function (pdf) of the maximum amplitude r_{SC} among r_1, r_2, $\ldots$, r_N through the standard RV transformation as

$$f_{r_{SC}}(x) = 2Nx(1 - e^{-x^2})^{N-1} e^{-x^2} \tag{11.17}$$

Given (11.12) and (11.17), we can write the probability of detection by ED with SC in Rayleigh fading channel into the following generic form,

$$\bar{P}_{d,SC}(N, N_0) = \int_0^\infty f_{r_{SC}}(x) Q_{N_0}(ax, b) dx$$

$$= 2N \int_0^\infty e^{-x^2} x Q_{N_0}(ax, b)$$

$$\times N - 1n = 0 \sum \binom{N-1}{n} \left(-e^{-x^2}\right)^n dx \tag{11.18}$$

$$= 2N N - 1n = 0 \sum (-1)^n \binom{N-1}{n}$$

$$\times \int_0^\infty e^{-(n+1)x^2} x Q_{N_0}(ax, b) dx$$

Making use of [20, Eq. (12)] for the integration part in (11.18), we obtain the closed-form expression for the probability of detection by ED with SC in Rayleigh fading channel as

$$\bar{P}_{d,SC}(N,N_0) = 2Np^2 \sum_{n=0}^{N-1} (-1)^n \binom{N-1}{n} \exp\left(\frac{-b^2}{2}\right)$$

$$\times \left\{ \left[\frac{p^2+a^2}{a^2}\right]^{N_0-1} \left[\exp\left(\frac{1}{2}b^2\frac{a^2}{p^2+a^2}\right)\right.\right.$$

$$\left. - \sum_{m=0}^{N_0-2} \frac{1}{m!}\left(\frac{1}{2}b^2\frac{a^2}{p^2+a^2}\right)^m\right]$$

$$\left. + \sum_{m=0}^{N_0-2} \frac{1}{m!}\left(\frac{b^2}{2}\right)^m\right\}$$

(11.19)

where

$$p = \sqrt{2(n+1)} \tag{11.20}$$

In (11.19), a and b are defined based on the number of samples N_0, detection threshold γ, average channel SNR $\bar{\rho}$, and the standard deviation of noise σ as follows:

$$a = \sqrt{2\bar{\rho}N_0} \tag{11.21}$$

and

$$b = \frac{\sqrt{2\gamma}}{\sigma} \tag{11.22}$$

Equation (11.19) provides a generic closed-form expression for the probability of detection by ED with SC in Rayleigh fading channel. Using this equation, we will formulate the analytical expressions of probability of detection and false alarm rate for the three sensing schemes. Note that the average probability of detection in Rayleigh fading channel expressed in [17, Eq. (9)] is a special case of (11.19) with $N = 1$. Also, the performance of OOK ED receiver can be described using the P_d and P_f of the ED. We will use (11.11) and the closed-form expression (11.19) to describe the performance of an OOK ED receiver with SC in Rayleigh fading channel. For simplicity of expression, in the following discussion we use $\bar{P}_{d,SC}^{(0)}$ to denote the probability of detection at DSSs which is given by (11.19) with arguments N_0, γ_0, $\bar{\rho}_u$, and σ_z correspondingly in (11.21) and (11.22). Similarly, we use $\bar{P}_{d,SC}^{(1)}$ to denote the probability of detection of the OOK ED receiver at the SU, which is given by (11.19) with arguments N_1, γ_1, $\bar{\rho}_s$, and σ_u correspondingly in (11.21) and (11.22).

11.2.3 Cooperative Sensing

For the cooperative sensing, each DSS performs ED using N_0 samples and produces local hard decision information, "1" or "0." Using OOK signaling, the hard decision information from these DSSs, S_1, S_2, ..., S_N, is transmitted consecutively to the SU through N independent sensor-to-SU channels in time slots T_1, T_2, ... ,T_N, respectively. The SU receiver uses N_1 samples to demodulate the OOK signal from each DSS. After employing data fusion based on OR rule of the demodulated local hard decision information from these DSSs, the SU determines the presence or absence of PU. Therefore, the PU detection performance is dependent on the performance of the local detection at the DSSs and the demodulation performance of OOK signal at the SU. Using (11.11) and (11.19), we can express the probability of detection after OR rule-based data fusion of the received local hard decision information from N DSSs as

$$P_{d,coop} = 1 - (1 - P_{d,i})^N \tag{11.23}$$

where $P_{d,i}$ denotes the probability that SU detects the presence of PU signal based on the hard decision information received from the ith DSSs,

$$P_{d,i} = \bar{P}_{d,SC}^{(0)}(1,N_0)\bar{P}_{d,SC}^{(1)}(1,N_1)$$
$$+ [1 - \bar{P}_{d,SC}^{(0)}(1,N_0)]P_f(\gamma_1,N_1,\sigma_n) \tag{11.24}$$

The first part on the right side of (11.24) represents the probability that the ith DSS detects the presence of PU signal and the associated local hard decision information "1" is received by the PU receiver correctly. The second part on the right side of (11.24) represents the probability that the ith DSS fails to detect the presence of PU signal but the associated hard decision information "0" is received as "1" due to the error of OOK transmission. Also, knowing that the data fusion of received hard decision information from DSSs is performed at the SU, we obtain the false alarm rate for cooperative sensing as

$$P_{f,coop} = 1 - (1 - P_{f,i})^N \qquad (11.25)$$

where $P_{f,i}$ denotes the false alarm rate based on the received local hard decision from the ith DSS as

$$P_{f,i} = P_f(\gamma_0, N_0, \sigma_z)\overline{P}_{d,SC}^{(1)}(1, N_1)$$
$$+ [1 - P_f(\gamma_0, N_0, \sigma_z)]P_f(\gamma_1, N_1, \sigma_n) \qquad (11.26)$$

The first part on the right side of (11.26) represents the probability that the ith DSS generates a false alarm and the associated local hard decision information "1" is sent to the SU receiver correctly. The second part on the right side of equation (11.26) represents the probability that the ith DSS does not produce a false alarm but the associated hard decision information "0" is received as "1" at SU due to the error of OOK transmission.

11.2.4 Collaborative Sensing

In collaborative sensing, N DSSs send their local hard decision information to a DSS that has the best sensor-to-SU channel, i.e., maximum $\left|\tilde{h}_{s,i}\right|^2$, $i = 1,2,\ldots,N$. This sensor applies the OR rule-based data fusion of the hard decision information. Knowing that the probability of detection at each DSS $\overline{P}_{d,SC}^{(0)}(1, N_0)$ is given by (11.19), we can write the probability of detection at the sensor after OR rule-based data fusion as

$$\overline{P}_{d,OR} = 1 - [1 - \overline{P}_{d,SC}^{(0)}(1, N_0)]^N \qquad (11.27)$$

The hard decision information produced after data fusion is sent to SU using OOK by the DSS that has the best sensor-to-SU channel among N DSSs. This is equivalent to the OOK transmission with SC in Rayleigh fading channel. The detection performance can be expressed by (11.19) with arguments γ_1, N_1, and σ_n. We can write the probability of detection for the collaborative sensing scheme as

$$P_{d,coll} = \overline{P}_{d,OR}\overline{P}_{d,SC}^{(1)}(N, N_1)$$
$$+ [1 - \overline{P}_{d,OR}]P_f(\gamma_1, N_1, \sigma_n) \qquad (11.28)$$

The first part on the right side of (11.28) denotes the probability that the sensor declares the presence of PU signal after data fusion and the associated hard decision information "1" is sent correctly to the SU by the DSS that has the best sensor-to-SU channel. The second part represents the probability that the presence of the PU signal is not detected after data fusion and the associated hard decision information "0" is received as "1" at the SU due to error. Similarly, we can write the false alarm rate for collaborative sensing as

$$
\begin{aligned}
P_{f,coll} &= \{1-[1-P_f(\gamma_0,N_0,\sigma_z)]^N\}\overline{P}_{d,SC}^{(1)}(N,N_1) \\
&+ [1-P_f(\gamma_0,N_0,\sigma_z)]^N P_f(\gamma_1,N_1,\sigma_n)
\end{aligned}
\tag{11.29}
$$

The first part on the right side of (11.29) represents the probability that at least one of the DSSs produces a false alarm and the hard decision information after OR rule–based data fusion "1" is sent to the SU through the best sensor-to-SU channel correctly. The second part represents the probability that none of the DSSs issues a false alarm and the associated sensing information after data fusion "0" is received as "1" at the SU OOK receiver due to error.

11.2.5 Selective Sensing

In selective sensing, local hard decision information from those N DSSs is not combined. A DSS that observes the best sensor-to-SU channel, i.e., maximum $\left|\tilde{h}_{s,i}\right|^2$ collects the hard decision information from other DSSs. Then, the hard decision information from the DSS that has the best PU-to-sensor channel, i.e., maximum $\left|\tilde{h}_{u,i}\right|^2$, is selected and sent to the SU with OOK signaling. Here we assume that the select sensing based on instantaneous gain of PU-to-sensor channel is applicable to the scenario where the PU sends paging or control signals periodically but may not necessarily transit data packets following the paging. In such situations the sensors can estimate the instantaneous gain of the PU-to-sensor channel and subsequently uses ED to check if the PU is sensing data packet. The probability of detection based on the selected hard decision information can be described by $\overline{P}_{d,SC}^{(0)}(N,N_0)$. Note that the DSS with the best sensor-to-SU channel sends the selected hard decision information to the SU with OOK signaling. The demodulation process at the SU is equivalent to the ED with SC. Thus, we can use $P_f(\gamma_1, N_1, \sigma_n)$ and $\overline{P}_{d,SC}^{(1)}(N,N_1)$ to describe the performance of the ED-based OOK receiver. The probability of detection for the selective sensing scheme can be written as

$$
\begin{aligned}
P_{d,sele} &= \overline{P}_{d,SC}^{(0)}(N,N_0)\overline{P}_{d,SC}^{(1)}(N,N_1) \\
&+ [1-\overline{P}_{d,SC}^{(0)}(N,N_0)]P_f(\gamma_1,N_1,\sigma_n)
\end{aligned}
\tag{11.30}
$$

The first part on the right side of (11.30) represents the probability that the presence of PU signal is detected based on the selected sensing information and the associated hard decision information "1" is sent to the SU OOK receiver through the best sensor-to-SU channel without error. The second part represents the probability that no PU signal is detected based on the selected sensing information and the associated hard decision information "0" is sent to the SU through the best channel but is received as "1" due to error. The probability of false alarm can be written as

$$P_{f,sele} = P_f(\gamma_0, N_0, \sigma_z) \overline{P}_{d,SC}^{(1)}(N, N_1)$$
$$+ [1 - P_f(\gamma_0, N_0, \sigma_z)] P_f(\gamma_1, N_1, \sigma_n) \tag{11.31}$$

The first part on the right side of the equation (11.31) denotes the probability that the false alarm is issued based on the selected sensing information and the hard decision information "1" is received correctly at the SU OOK receiver through the best sensor-to-SU channel. The second part represents the probability that no false alarm is issued based on the selected sensing information and the associated hard decision information "0" is sent to the best sensor-to-SU channel but is received by the SU OOK receiver as "1" due to error. Comparing (11.31) with (11.29) and (11.25), we notice that the selective sensing scheme does not directly increase the probability of false alarm. This is due to the fact that the selective sensing scheme avoids OR rule-based data fusion. We will show in the following section that the selective sensing scheme also benefits from the diversity advantage of DSSs.

11.2.6 Spectrum Sensing with Two-Stage Detection

One way of conveying the sensing information to the CR is to send directly the value of the measured signal feature; for instance, the energy of the signal and the locally estimated noise power. Strictly speaking such information is analog, but in order to transmit it to the CR it has to be quantized into discrete values and represented by finite bits. The optimum number of bits to represent the analog information is out of the scope of this chapter. At the CR side, such quantized information from each sensor is combined to make signal detection, hence the name soft information combining (SC). Once the soft information is available at the CR, we can formulate a binary hypothesis testing problem in which the aggregate information from all the sensors either corresponds to the presence of the PS (hypothesis H_1) or the absence of the PU signal (hypothesis γ_0). In a method of hard combining (HC), sensing information is conveyed in the form of decisions in short message packets. The corresponding hypothesis testing on the presence and absence of PU signal is conveyed by one-bit information (0 or 1).

Both the methods of SC and HC have their pros and cons [8,7]. From a performance point of view, the SC provides a better result for the obvious reason that it also provides more information, but the requirement of the large feedback observation makes it prohibitive [7]. Particularly this would be significant for bandwidth and power constrained transmission. On the other hand, HC provides less transmission overhead because it needs to transmit only 1 bit information, but it has inferior performance due to the loss of information during the local decision.

Following the above discussion, we propose a two-stage detection that exploits the advantages of both methods of SC and HC. The proposed scheme works as follows: at the first stage, as shown in the flow chart in Figure 11.2, the CR obtains a set of hard decision informations (p_i) from each sensor. The CR performs HC using one of the methods described in the previous section. If the outcome of the global decision is 1, a presence of PU signal is declared. But if the outcome of the global decision on the presence of PU signal is 0, the CR steps into the second level detection by requesting the sensors to provide soft information (s_i) and perform SC.

This approach will reduce the unnecessary use of the SC, especially when the PU is closer to the sensors. In other words, it is inherently adaptive to the SNR of the sensing channel. At lower SNR, however, the two-stage detector on average will have higher detection time because two detection steps are involved. The sensors should also be able to store the soft information in a temporary storage area. Implementation of such sensing strategy requires a client-server setup in which sensors provide soft information on demand. For example, it can be implemented in a distributed spectrum sensing architecture discussed in [10].

11.2.7 Numerical Results

Using the generic closed-form expression for the probability of detection, we have obtained analytical models for the three distributed sensing schemes. In this section, we will use these analytical models to investigate the performance of these schemes. First, we give numerical results to verify the derivation of the generic closed-form expression (11.19) for the probability of detection by ED with SC in Rayleigh fading channel. We consider the situation where there are N channels that are of independent Rayleigh fading processes with equal average SNRs. In each fading block, the channel that has the maximum instantaneous SNR is selected. The ED is performed in this selected channel to determine the presence or absence of signal. In Figure 11.3 we plot the probability of detection using (11.19) for different values of N when the average SNR of the channels changes from −10 to 5 dB. The detection threshold is obtained using (11.11) with a target false alarm rate P_f = 0.01. The number of samples for ED is N_0 = 10. Correspondingly, the empirical probability of detection is obtained from Monte-Carlo simulation with 10^5 channel realizations. Results show that the performance of ED is improved by using SC. For example, using the ED with SC at N = 3, we can improve the probability of detection from 0.66 to 0.94 in a 5 dB SNR environment. Moreover, the perfect match of theoretical and empirical results verifies our derivation of (11.19). In the following analysis we will use the analytical models obtained in Section 11.2.2 to evaluate the three sensing schemes.

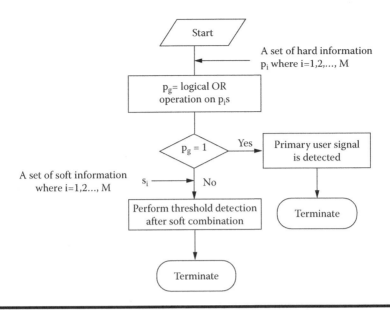

Figure 11.2 Flow chart for the two-stage detection in distributed spectrum sensing.

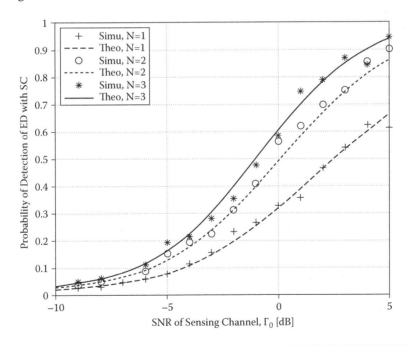

Figure 11.3 Theoretical and empirical probability of detection by ED with SC in Rayleigh fading channel. $P_f = 0.01$ and $N_0 = 10$.

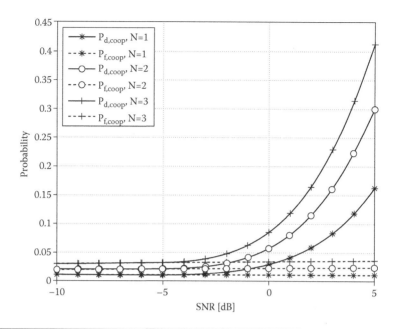

Figure 11.4 Performance of cooperative sensing at different SNRs. $N_0 = 10$ and $N_1 = 1$.

For these sensing schemes, we assume that $N_0 = 10$ and $N_1 = 1$. Furthermore the average SNRs of PU-to-sensor channels and sensor-to-SU channels are equal (i.e., $\bar{\rho}_u = \bar{\rho}_s$). The noise variances at DSSs and SU are equal (i.e., $\sigma_z^2 = \sigma_n^2$). The detection threshold at each individual DSS and that at SU ED-based OOK receiver are obtained using (11.11) with arguments N_0, $\bar{\rho}_u$, σ_z and N_1, $\bar{\rho}_s$, σ_n, respectively, whereas the local target false alarm rate is $P_f = 0.01$. Note that this P_f is different from $P_{f,coop}$, $P_{f,coll}$, and $P_{f,sele}$.

In Figure 11.4 and Figure 11.5 we plot the probability of detection and false alarm rate for cooperative sensing and collaborative sensing, respectively. Results show that both schemes improve the PU detection performance with the number of available DSSs. At 5 dB SNR the probability of detection is increased from 0.02 to 0.3 and 0.4 by cooperative sensing and collaborative sensing, respectively. However, the false alarm rate of cooperative sensing increases with the number of DSSs. This is due to the OR rule-based data fusion at the SU in (11.25), whereas the OR rule-based data fusion does have much influence on the false alarm rate for the collaborative sensing scheme as shown in (11.29).

Knowing this effect, we manually reduce the false alarm rate of cooperative sensing by changing the local target false alarm rate at individual DSSs and SU to $P_f = 0.004$. In Figure 11.6, we compare the performance of cooperative sensing and collaborative sensing using $N = 3$ DSSs when the two schemes have nearly the same false alarm rate, i.e., $P_{f,coop} \approx P_{f,coll}$. Results show that collaborative sensing

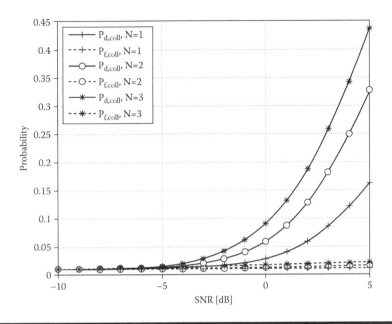

Figure 11.5   Performance of collaborative sensing at different SNRs. $N_0 = 10$ and $N_1 = 1$.

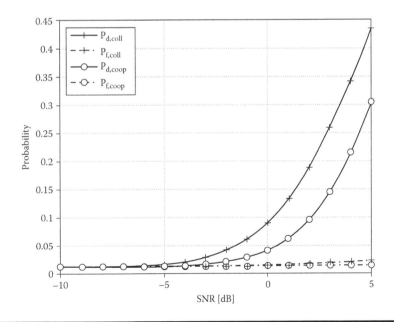

Figure 11.6   Probability of detection and false alarm rate for cooperative sensing and collaborative sensing at different SNRs. $N = 3$, $N_0 = 3$ and $N_1 = 1$.

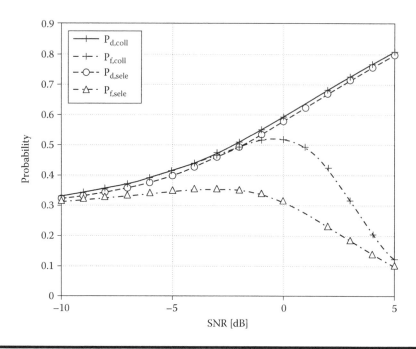

Figure 11.7 Probability of detection and false alarm rate for cooperative sensing and collaborative sensing at different SNRs. $N = 5$, $N_0 = 10$, and $N_1 = 1$.

outperforms cooperative sensing in terms of probability of detection. Finally, we compare collaborative sensing with selective sensing. In order to compare false alarm rates by the two schemes, we do not fix the detection thresholds γ_0 and γ_1 at DSSs and SU ED-based OOK receiver. Rather, they are obtained based on maximum-likelihood rule given the SNRs, noise power, and numbers of samples [16]. In Figure 11.7, we plot the probability of detection and false alarm rate for both selective sensing and collaborative sensing when there are $N = 5$ DSSs. Results show that the selective sensing scheme matches the performance of the collaborative sensing scheme. Moreover, the false alarm rate of the selective sensing scheme is much smaller than that of the collaborative sensing scheme. As described previously, the selective sensing scheme does not employ OR rule-based data fusion. Thus it avoids increasing false alarm rate.

Figure 11.8, illustrates the probability of misdetection (P_{md}) vs. P_f measured at the CR for SNR = −8 dB for cooperative sensing and single device sensing. For HC, a one-bit information on the absence and presence of PU is transmitted in an error-free channel to the SU or FC, where data is combined based on OR rule or AND rule. In soft combining (SoC), the quantized values of collected energy and threshold value which were input to the decision device for local detection are transmitted from each sensor to the SU or FC in an error-free channel. Data is combined using equal weight combining (EC) to obtain global detection. For this experiment, at

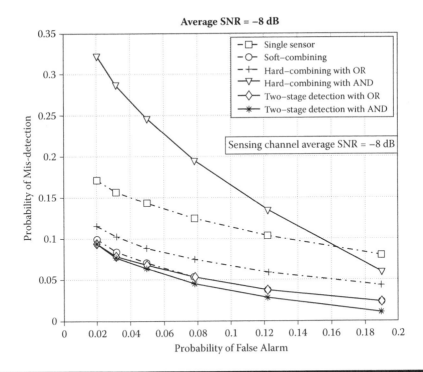

Figure 11.8 Plot showing probability of mis-detection versus probability of false alarm for SNR = –8 dB, N = 2.

each sensing period, the sensors collect 10,000 samples of the received signal to obtain smooth curve at lower P_f values.

We consider the performance obtained by a single sensor shown by "□," marker as a reference. The multiple sensing scheme with HC, indicated by "+" marker employing an OR rule, shows significant improvement compared to a single sensor for a wide range of P_f values. For comparison, a hard information combining scheme employing an AND rule is shown as indicated by a "Δ" marker. Clearly it performs even worse than the single device sensing especially at lower P_f values, because the probability that both sensors detect the PU signal under fading channel is lower than that of a single sensor. The SoC scheme, indicated by "○" marker depicts further performance improvement compared to the HC employing an OR rule. The newly proposed two-stage detection is simulated first for HC using an OR rule followed by the SoC. This plot is shown by the "◊" marker. Its performance almost converges with that of the performance obtained using SoC.

Next the two-stage detection is simulated by using an AND rule for HC instead of an OR rule. The results, shown by a "*" marker, confirm that it offers the best performance. The detectors involving the AND rule combining tend to perform better at higher P_f which should not happen under normal circumstances. This

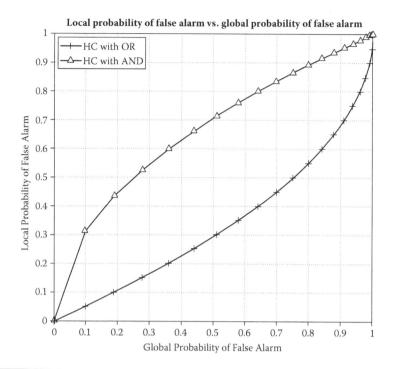

Figure 11.9 **Plot showing global P_f versus local P_f for the AND and OR rule combining.**

might be explained as follows. At higher P_f the threshold at the sensors is so low that, in some cases, the local decisions by the sensors may no longer be uncorrelated. Hence the local and global P_fs relationship assumed may not hold. For purely uncorrelated outcome of the observation by the two sensors, the relationship between the local and global P_fs is shown in Figure 11.9.

For a reasonable value of misdetection probability, e.g., $P_{md} = 0.1$, implementing the first-stage sensing by using an AND rule instead of an OR rule results in comparable performance. But the latter one has less frequency of using the second stage for successful detection, which is illustrated in Figure 11.10. For 90% detection, it is found out that using the AND rule at the first stage results in 16% exploitation of the second stage for successful detection. The use of OR rule results in only 0.6% exploitation of the second stage for successful detection. Figure 11.10 shows the snapshot (50 out of 10,000 realizations) for illustration purposes.

11.3 Media Access and Transmission

Distributed spectrum sensing can be grouped under low data rate sensor networks that are mainly event driven. Sensors start to perform the measurement and provide

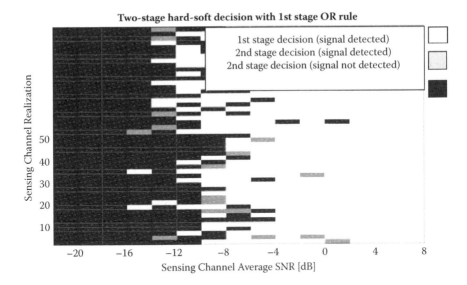

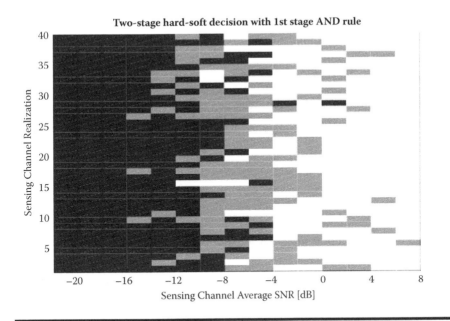

Figure 11.10 Plot showing sensing realization vs. sensing-channel SNR for two-stage detection.

measurement data upon request from FC. For instance, the measurement result may consist of information on the spectral occupancy of a certain frequency band, time stamp, and additional measurement parameters that qualify the result. A few bytes payload per frame could be sufficient to encapsulate the above information. In addition, some part of the information is needed only during an initial step and may not necessarily appear on each frame.

However, classic medium access control (MAC) and routing protocols studied for the voice and data traditional networks are not suitable for the kind of wireless sensor network described above [21]. Most traffic in the network is generated by the sensing event triggered by the FC or sending a request for sensing information. Then during a specified period determined by the FC, sensors should perform the measurement and provide the data back to the FC. Another point is, in many wireless sensor network applications, sensor nodes are battery operated. Hence, energy efficiency becomes important in designing protocols for wireless sensor networks. To do that, the protocols should address the major sources of MAC and routing levels. In this paper, we are interested in the efficiency of the MAC protocol in terms of obtaining improved sensing at FC. Although there is a strong correlation between energy consumption and particular MAC, "analysis of the trade-offs of sensing quality versus energy consumption of a given MAC in distributed spectrum sensing is out of the scope of this paper."

According to the mechanism for collision avoidance, MAC protocols for distributed spectrum sensing can be divided into two groups as follows.

11.3.1 Schedule-Based MAC

In a scheduled protocol such as time-division multiple-access (TDMA), information on a preassigned time slot is broadcasted to the sensors before commencing the spectrum sensing operation. Usually this is done through a beacon which is transmitted by the FC during beacon period. Figure 11.11(a), illustrates how sensors report the result according to a predefined time slot. The total sensing time, denoted by T_t is divided into several window time frames T_w. Each window frame begins by a time frame during which beacon is transmitted. We denote this time by T_b. The window frame also contains time T_{sen} to perform sensing by individual sensors, and Z number of time slots, each denoted by T_r, to report sensing information. Hence, T_w can be written as,

$$T_w = Z \times T_r + T_{sen} + T_b \qquad (11.32)$$

Because channels are preallocated before transmission, each sensor is guaranteed for successful transmission from the media access perspective. TDMA protocol is attractive in particular for low duty cycle (LDC) systems where transmission is done as a burst of packets followed by longer silent period. The drawback of such media access schemes is lack of flexibility. For instance, if new sensor nodes join the network, or old ones stop operating, the FC must update the slot allocation.

11.3.2 Contention-Based MAC

In pure contention based protocol, the total sensing time is divided in to T_w time frames where T_w is given by (11.32), but $Z \times T_r$ is a contention window this time. The sensor nodes attempt to transmit sensing information in any of the Z reporting time slots in a random fashion as shown by Figure 11.11(b). If two or more sensors attempt to use the same time slot at the same time, collision would occur. In contrast to the scheduled protocol, it is more flexible to entertain the change in the number of active sensor nodes. A given reporting time slot can be used by a sensor which has data to report without wasting time. In the case of scheduled protocol, sensors have to wait for their predefined time slot, and if there is an unresponsive sensor, its corresponding time slot cannot be used. Some nodes might have run out of battery or are not able to respond due to some malfunction. Contention-based protocol also does not require fine-grain synchronization like the scheduled-based protocol [21].

11.3.3 Impact of MAC on Distributed Sensing

Evaluating the impact of MAC protocol on distributed spectrum sensing requires several sets of parameters. However, for illustrative purposes, we focus on a simple contention-based MAC. We mainly emphasize the impact of packet loss due to collision on the sensing quality at the FC. Note that, in many contention-based MAC protocols, acknowledgment of successful packet reception through acknowledgment ACK signal, back-off timing, and retransmission, are introduced to reduce the collision probability. However, the collision probability cannot be totally avoidable no matter what refinements are done.

Let us assume a case where no ACK signal is expected and hence no retransmission is considered, which means the packet transmission rate of each sensor during the contention window is unity. First each sensor receives information with regard to T_{sen}, T_w, and T_r through a beacon signal in the period T_b. Then each node, after finishing sensing, transmits sensing information to the FC randomly in one of the reporting slots before the next beacon receives. Each sensor should transmit only once during the contention window T_w. Successful transmission occurs when only one sensor tries to transmit at a particular reporting slot, otherwise, packet collision occurs. The probability of a packet being lost due to collision for randomly distributed and infinite number of sensors is provided in [21], from which with minor modification we obtain P_{MAC} for M number of sensors as

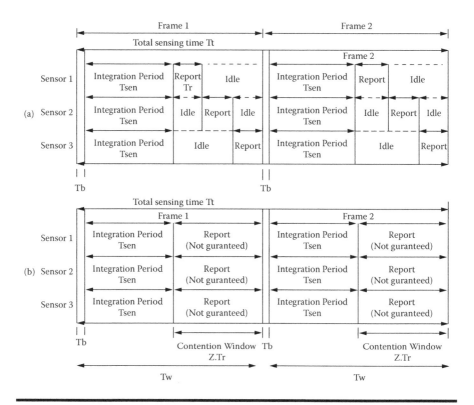

Figure 11.11   Illustration of conventional media access schemes (a) scheduled, (b) contention based.

$$P_{MAC} = \left[1 - \left(1 - \frac{1}{Z}\right)^{M-1}\right] \qquad (11.33)$$

where M is the number of sensor nodes. One can easily observe that, for increased Z, we can attain reduced collision probability. For fixed values of T_w and T_r, however, increasing the number of time slots would reduce T_{sen}, which would impact the local probability of detection. To show this, we can write P_d in terms of T_{sen} as follows:

$$P_d = Q\left[\frac{\gamma - \frac{T_{sen}}{T_{sam}}\zeta}{\sqrt{2\frac{T_{sen}}{T_{sam}}\zeta^2}}\right] \qquad (11.34)$$

where we use the notation P_d when P_D is computed for a particular sensor as a local probability of detection. ζ is a function of h and ξ and is given by $\zeta = 1 + |h|^2$

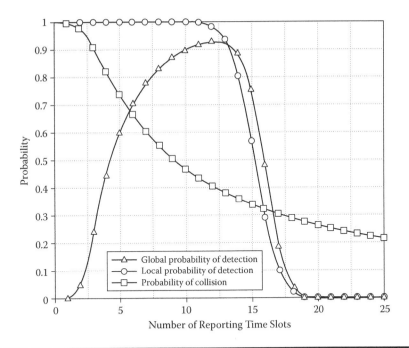

Figure 11.12 Impact of MAC collision on global probability of detection for *N* = 7 when contention based media access is used.

ξ. Equation (11.34) indicates the direct impact of optimizing Z on the local probability of detection. To compute the global probability of detection P_g we assume that the fusion center employs an OR rule to combine the binary sensing from M identical sensors. Then P_g as a function of individual sensor P_d is given by

$$P_g = 1 - (1 - P_d)^M \quad \text{for OR counting rule} \tag{11.35}$$

We further look at the effect of packet collision that accounts for unsuccessful delivery of sensing results from some of the sensors. Whenever collision occurs, the FC receives no sensing information; hence to reflect this effect, we rewrite (11.35) as

$$P_{gc} = 1 - (1 - P_d[1 - P_{MAC}])^M \tag{11.36}$$

where P_{gc} is the new global probability of detection, taking into account the probability of having collision during the reporting period. By substituting P_d and P_{MAC} in (11.36) by (11.34) and (11.33) respectively, and writing Z as a factor of T_{sen} from (11.32), we obtain

$$P_{gc} = 1 - \left[1 - Q\left(\frac{\gamma - \frac{T_{sen}}{T_{sam}}\zeta}{\sqrt{2\frac{T_{sen}}{T_{sam}}\zeta^2}} \right) \right.$$

$$\left. \times \left(1 - \frac{T_r}{T_w - T_b - T_{sen}} \right)^{N-1} \right]^M$$

(11.37)

From (11.37), one can easily notice that, for a given T_w and T_r, if we increase T_{sen}, which is the same as reducing Z, less collision and hence less packet loss is achieved. On the other hand, reducing T_{sen}, which is the same as increasing Z, would result in reduced probability of detection. By solving (11.34), one might be interested in seeking for an optimum T_{sen}. This is a cross-layer problem where we are optimizing the sensing period at physical layer and optimizing the MAC parameter, which is the number of reporting slots in this case. Without going for further elaboration, an illustrative example is provided by plotting P_{gc} against Z (Figure 11.12) that indicates the optimum number of reporting slots to achieve the maximum global probability of detection.

11.3.4 Enhanced MAC for Distributed Sensing

The enhanced MAC approach we propose is based on the scheduled MAC approach. The main idea is to exploit any idle time, from MAC perspective, to maximize the sensing quality from individual sensors. The method works as follows.

11.3.4.1 Level One Enhancement

In a similar manner to the scheduled sensing described in Section 11.3.1, local sensors are triggered and controlled by the FC. First they receive a beacon frame carrying timing information such as the time slot in which they should report the result. Report of spectrum sensing information from all sensors has to be completed before the next beacon frame arrives. The illustration in Figure 11.13(c) shows three sensors performing sensing and reporting their result. The sensing duration or integration interval of each sensor is different, unlike the ordinary scheduled bases sensing where all sensors employ identical T_{sen}. The enhancement allows each sensor to continue sensing until a specific time slot assigned to that particular sensor arrives. After reporting, a sensor may or may not have idle time depending on its position.

As a result, sensors with more sensing interval provide a result with higher sensitivity, which maximizes the overall global detection probability. Modifying equation (11.35) to incorporate the extended sensing period, we obtain

$$P_{g'} = 1 - (1 - P_{di})^M \quad \text{for OR counting rule} \tag{11.38}$$

$$P_{di} = P_d(T'_{sen}) \tag{11.39}$$

where Pd_i represents the ith sensor probability. It is obtained by substituting T_{sen} by T'_{sen} and computing (11.34). The equation for T'_{sen} is given by

$$T'_{sen} = T_{sen} + T_r \times (i - 1) \tag{11.40}$$

11.3.4.2 Level Two Enhancement

In level two enhancement, sensors are allowed for extended sensing both before and after reporting time slot, depending on their order of reporting. As shown in Figure 11.13(d), the first sensor will have an extended sensing period only after reporting sensing information, while the last sensor to report will have an extended sensing period only before reporting. The rest of the sensors shall have extended sensing periods both before and after their respective reporting slots.

In level two enhancement, sensors have to store the signal energy they gathered after their reporting period. In the next window frame the collected data from the previous window frame will be augmented into the newly collected energy to

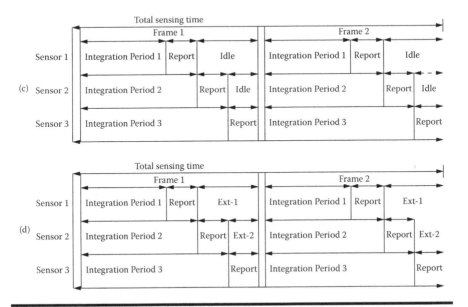

Figure 11.13 Illustration of enhanced scheduled media access schemes (c) with idle time, (d) with no idle time.

enhance the detection probability. This scheme, like the ordinary scheduled sensing, could yield identical probability of detection for each sensor but with higher quality. This approach has both pros and cons. On the pro side, it achieves the highest efficiency as it exploits all the idle time slots. On the con side, it is required to store signal energy collected from the previous window frame, and the primary user signal under consideration should be more or less stationary during two window frame intervals. In addition, in level two enhancement, the total lack of idle time may drain sensors' energy. Hence, energy requirement of the whole system should be taken into account when choosing level two enhancement over level one enhancement.

11.3.5 Computer Simulation

Two sets of computer simulation are performed. The first one is to study the global probability of detection under the different schemes discussed in the preceding sections against the primary user signal *SNR*. The second one portrays the global probability of detection improvement against the number of sensor nodes in the network.

For the first case, three sensors, sensor A, sensor B, and sensor C are assumed to observe a given primary signal independently. A Rayleigh fading channel is assumed for the PU signal propagation. At each sensor, an i.i.d. additive white Gaussian noise is generated and added to a PU signal modulated with quadrature phase shift keying (QPSK) and convolved with the PU signal channel. All the channel paths are assumed uncorrelated. For the sensors to transmit their sensing information to the FC, a perfect reporting channel is considered. This assumption is reasonable for a low data rate reporting channel and transmission scheme with sufficiently low bit error rate performance. Each sensor reports one-bit information with regard to the presence or absence of the PU signal. Finally a global decision is effected at the FC by combining the information from different sensors.

For each sensing period, a minimum of 1,000 samples of the received signal are collected to derive the test statistic. This number would be reasonable for the Gaussian model approximation of the test statistic. To obtain the performance curves under each scenario, the simulation is repeated for 10,000 realizations.

For the conventional media access schemes as in Figure 11.12, T_{sen} is selected in such a way that $T_{sen} = T_p$, and the frame duration would be $4T_{sen}$ for three sensors. T_b is omitted in the simulation because its impact is identical to both enhanced media access schemes and conventional media access schemes. Each sensor collects 1000 samples of signal energy during sensing period T_{sen}. For the enhanced media access schemes as in Figure 11.13, T'_{sen} of an individual sensor is computed by using (11.40). Based on the above setup, four simulation plots are generated. In all cases a 5% constant false alarm rate or probability of false alarm is employed. The first two plots represent a contention-based protocol. To emulate a packet loss due to collision, we considered two cases where, in the first, a maximum of one packet loss is anticipated due to collision, while in the second case, a maximum of two packets, loss is anticipated during the contention window. The third plot indicates a traditional TDMA

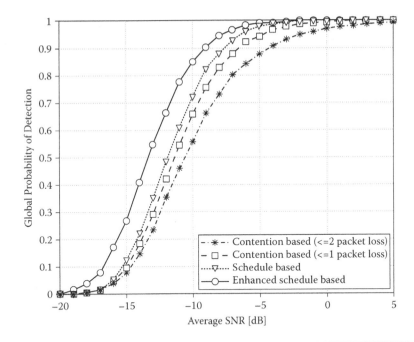

Figure 11.14   Global probability of detection at the FC vs. average SNR of PU signal channel.

protocol where the sensing information is reported in the scheduled manner. In this protocol all sensors are configured to implement equal integration interval T_{sen}. Then a simple counting rule, "OR," is used for data fusion. The fourth plot demonstrates the enhanced MAC approach where idle waiting times are exploited for further signal energy collection. Note that only the level one enhancement is addressed here because the basic merit of exploiting idle times can be displayed either by level one or level two enhancements. In addition, level two enhancement requires extra assumption of primary user signal stationarity during two frame durations.

Figure 11.14 illustrate the merits of the enhanced MAC approach. For 90% of global probability of detection, the enhanced MAC approach shows 2 dB gain compared to that of the traditional approach of schedule-based MAC and 5 dB gain compared to that of the contention-based MAC, where a max of two packets' loss is anticipated. These gains are expected to be pronounced if more sensors are used.

Figure 11.15 shows the global probability of detection against the number of sensors. An *SNR* value of –10 dB is selected. The global probability of detection obtained under each MAC scheme, when the number of sensors becomes 3, matches with that of the plot shown in Figure 11.14. It can be seen that the enhanced MAC approach achieves 100% detection when the number of sensors is 6, while the other schemes achieve the same performance only when the number of sensors is about 12.

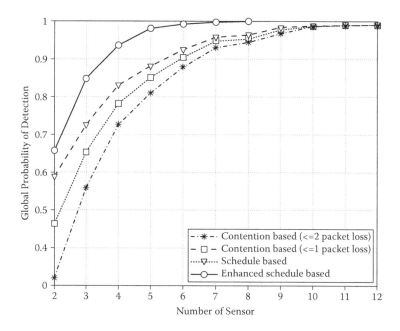

Figure 11.15 Global probability of detection at fusion center vs. number of sensors.

11.4 Sensing Database (SDB) and DSS

Apart from the need for immediate use in CR, the sensing information obtained from DSS may also be needed for other purposes, for instance, spectrum monitoring that provides a view on long-term spectrum usage. In this case, the sensing information can be collected on a regular basis and uploaded to some predefined locations, for example, a local information server, for data processing, storage, and analysis [22]. We denote such server as "sensing database" (SDB).

Introducing the SDB server that can provide similar information for the CRs as the regulatory database, but located closer to the CRs than the regulatory database, would reduce the latency and reduce the traffic to the global network. Since each SDB server is only responsible for a relatively small geographical area, it can also provide local information (for instance, sensing information obtained from distributed sensors), and it is also easy to do maintenance. This section shows the architecture design of SDB and the analysis and performance evaluation results on the efficiency of SDB with respect to traffic to global network, sensing cost, and sensing quality.

11.4.1 Regulatory Database

The regulatory database approach was developed based on an assumption that the protected operating areas of primary (or incumbent) users are clearly defined and can be calculated from certain information. Therefore, after receiving geo-location of a CR, the database can determine whether the CR is sufficiently far from the protected service areas of primary users or not, in order to ensure the CR will not introduce harmful interference to those primary users.

Following this approach, the regulatory organizations in several countries have been considering regulations and technical requirements on accessing the regulatory database in order to support secondary operations in TV white spaces. The summary about U.S. FCC regulation and U.K. Ofcom consultation is given as follows.

11.4.1.1 Database Consideration by the US FCC

The United States Federal Communications Commission (FCC) released its second report and order, and memorandum opinion and order, in November 2008 for devices to operate in TV white spaces [23]. According to the released notice, the secondary devices can be classified into two types: fixed device and personal/portable device. Personal/portable devices are further classified into three categories: Mode I, Mode II, and sensing-only devices. Fixed devices can operate on channels 2 to 51, except channel 37 (which is reserved for radio astronomy and medical telemetry) and channels which are adjacent to the channel operated by a TV station. The fixed devices must have geo-location mechanism to provide location information to the accuracy of ±50 m, and must access the regulatory database at least once a day.

Personal/portable devices can operate on channels 21 to 51, except channel 37. A personal/portable device that can operate independently is called a Mode II device. The Mode II devices must have a geo-location mechanism with accuracy of ±50 m, must access a regulatory database at least once a day, and must access the database when it is powered on or changes location. The personal/portable device that operates as a client under the control of a fixed or Mode II device is called a Mode I device. Mode I devices do not need to have any geo-location mechanism.

According to this regulation, fixed devices and Mode II devices must have the capability to access the regulatory database. These devices will submit their identifications and geo-location information to the database and obtain a list of available channels for the current location.

Information stored in the database includes regulatory information, registered information by incumbent users (e.g., location, power, identification, channel of the broadcast station), and other information and algorithms to determine the protected service contours of licensed users.

11.4.1.2 Database Consideration by the U.K. Ofcom

The U.K. Office of Communications (Ofcom) is an independent regulator and competition authority for the broadcasting, telecommunications, and wireless communications sectors in the United Kingdom. Ofcom issued its consultation document in February 2009 about license-exempting cognitive devices using interleaved spectrum, and outlined several key points for cognitive devices to access the interleaved spectrum (i.e., channels 21–30 and 39–60, where the bandwidth of each channel is 8 MHz) [24]. The document was updated several times and is currently still seeking external feedback. The document discussed the practicality of different approaches for identifying available spectrum and concluded that both geo-location database and spectrum sensing can be deployed. Regarding the database approach, the detailed requirement about frequency of database access is not available yet; however, the suggested value for location accuracy is ±100 m.

11.4.2 Differences with RDB and the Needs for an SDB

There are differences in the purpose and the design of regulatory database (RDB) and SDB. The goal of the RDB approach is to protect incumbent users. Its development is based on an assumption that the service areas of incumbent users do not change often, and their operating parameters (for instance, channel numbers, transmitter geographic coordinates) are relatively static and can be stored in a database. Information in each database must be consistent with information in the others (if there are multiple operators that provide the database service) and will be updated on a daily basis. Even if sensing must be performed, there is no requirement about managing that sensing information (e.g., to upload to the database) [23,24]. Also, because the sensing information changes dynamically and it is valid for a small range, it is not suitable for centralized management.

Sensing information is not only for immediate use or internal use within a cognitive radio.* Recent studies show that by sharing sensing information among secondary users the sensing quality can be improved [25,26]. It would be difficult to use a single standalone sensor to obtain high quality of sensing. However, a sensor network consisting of spatially distributed sensors can provide high-quality sensing by overcoming effects of noise, interference, fading, and limitation of the sensing method. The distributed sensing also refers to time, frequency, and function distribution; it can be used where each individual sensor may or may not be collocated with each cognitive device. Sensing information includes raw measurement data, and local decisions can be exchanged directly between sensors and cognitive devices. A local information server is needed to store information collected by distributed sensing, then process and provide the results to other devices in the area in order to assist secondary operations. The SDB is suitable for this purpose.

---

* A cognitive radio is a type of radio in which communication systems are aware of their environment and internal state and can make decisions about their radio operating behavior based on that information and predefined objectives.

Table 11.1   Comparison between Two Databases

	Regulatory Database (RDB)	*Sensing Database (SDB)*
Scope	Global	Local
Main purpose	Protection of primary users	Assistance for secondary users
Regulatory information	Included	Included (e.g., cache and relay)
Sensing information	None	Included
Information update	Static (daily)	Dynamic (1 or several seconds order)
Information processing	Determining service contours	Distributed sensing algorithms
Deployment place	Internet	Local server (e.g., can be colocated with AP)
Service provider	Specified by regulatory bodies	Any

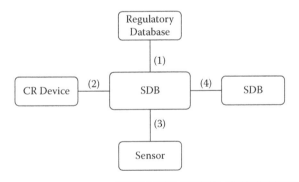

Figure 11.16   SDB and interfaces to other entities.

The differences between RDB and the proposed SDB are given in Table 11.1.

11.4.3 Architecture Design

The architecture shown in Figure 11.16 defines SDB and other entities, with their high level functions and exposed interfaces for information flows among the entities. There are four interfaces as follows:

■ Interface (1) between regulatory database and SDB: This interface is used by the SDB to obtain regulatory information corresponding to the locations of devices which are using SDB service.

■ Interface (2) between SDB and CR devices: This interface is used to exchange sensing results, and to update the operation status of CR devices to SDB.

■ Interface (3) between SDB and sensors: This interface is used to collect sensing information from sensors, and to control the sensing activities of the sensors.

■ Interface (4) between two SDBs: This interface is used to exchange sensing related information.

11.4.3.1 Functional Requirements

1. Information acquisition capability: to obtain sensing information from spectrum sensors, CR devices, SDBs or other sources, and to obtain regulatory and policy information from regulatory repositories.
2. Information storage capability: to store sensing related information, regulatory/policy information, and the results of data processing.
3. Information provision capability: to provide raw and/or processed information to CR devices or other SDBs.
4. Information organization and data processing capability: to improve the performance of information exchange, obtained data should be classified, sorted, indexed, and reformatted. Also, the SDB may provide obtained information in the original form, or it may provide the results of data processing of information from different sources according to some algorithms. One of the results could be reliable sensing information collected over a long period of time, possibly obtained from multiple sensors.

11.4.3.2 Information Contents

Information obtained by the SDB can be classified into the following categories.

1. Regulatory/policy information
2. Sensing related information
3. Other information
   - **Regulatory/policy information**: Information obtained from (regulatory) geo-location database, which includes information on license holders, facility operation parameters (e.g., frequency, location), and any special conditions that apply. This information is relatively stable; it does not change frequently for a relative large area. It can be some lists of available channels for secondary usage corresponding to a small area. It can be the same as the register information in the geo-location database, for example, transmitter location or geographic area of operation, effective radiated power, antenna height, call sign, etc.

- **Sensing related information**: Information obtained from spectrum sensors, information to be provided to CR devices or SDBs, and information used to control the sensing activities of sensors.
  - Sensing information: spectrum measurement information with time-stamp, parameters indicating the accuracy (e.g., confidence level), local decisions, etc.
  - Sensing control information: start/stop frequency, sensing method, report method, etc.
- **Sensor information**: sensor specifications, identification, etc.

  Note that sensing information changes dynamically, and that it is valid for a short period of time and for a relatively small geographic area.
- **Other information**: from the secondary users' perspective it would be beneficial if they could obtain further information which reduces sensing costs (time, power consumption, calculation complexity etc), the location of other secondary users, or information about whether a channel is being used by primary users or by secondary users, how the channel is used, etc.

11.4.3.3 Deployment Example

Figure 11.17 shows an example for the deployment of the SDB. In this deployment the SDB obtains regulatory information from a regional regulatory repository, stores the information in local storage devices for a certain duration (caching), and provides this information to the CR devices under its service area. Note that the SDB can be colocated with the access points and base stations, or it can be distributed among the CR devices.

11.4.4 Analysis on the Effect of Sensing Database

Sensing operation of a CR device that adopts the distributed sensing method can be divided into four steps as follows: database access, signal measurement, information exchange, and decision making about sensing results.

In database access duration, each CR device connects to the database and obtains a list of available channels for secondary operation. There are fixed and mobile devices in the system. The fixed device is required to check the database at least once every T_{DB}^{fix}, and the mobile device is required to check their geo-location after moving a certain period of time, which is denoted as T_{DB}^{mob}. Assume that the area is divided into a mesh with size D×D, and the mobile device must obtain new information from the database for available channels when it moves to another cell of the mesh.[*]

---

[*] Note that mesh-type geo-location database is not indicated by the FCC, and the accuracy of geo-location which is ±50 m is not equivalent to cell size D.

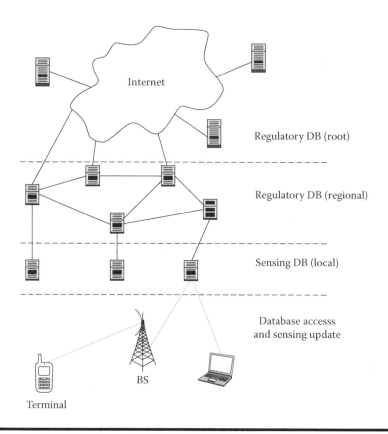

Figure 11.17 Deployment example of SDB.

In measurement duration, each CR device performs spectrum sensing for all available channels. Let us assume that the sensing time for each channel by each device is the same and fixed. As a result the measurement time depends on the latency between CR devices with the database, and the number of channels to be sensed by each CR device.

In information-sharing duration, assume that the CR devices use a contention based scheme to access the common channel for sharing sensing related information, e.g., Carrier Sense Multiple Access (CSMA) [27]. In decision making duration, each CR device makes a final decision based on its own measurement result and the results that have been received from other CR devices or sensors during the information-sharing phase.

The effect of SDB is analyzed from the following aspects: the amount of signaling traffic to and from global network (e.g., the Internet) introduced by database requests and responses, the sensing cost, and the sensing quality based on the number of independent results obtained from the distributed sensing process.

11.4.4.1 Amount of Signaling Traffic to/from Global Network

Assume that the numbers of fixed and mobile CR devices which are required to obtain the list of available channels from database are N_{fix} and N_{mob}, respectively. The access frequency to the database is adopted from the system model as follows:

- Fixed device accesses the database once per $T_{\text{DB}}^{\text{fix}}$ duration.
- Mobile device updates its geo-location after travelling for $T_{\text{DB}}^{\text{fix}}$ duration and accesses the database if it moves to another cell, where the cell size is D×D.

Without SDB, all devices must access the RDB on the Internet. Assume that the amount of traffic to/from global network for each request/response is G. Therefore the total traffic to/from global network over a unit time for database lookup, which is denoted as F_{RDB}, can be described as follows:

$$F_{\text{RDB}} = G \times \left(\frac{N_{\text{fix}}}{T_{\text{DB}}^{\text{fix}}} + \frac{N_{\text{mob}}}{T_{\text{DB}}^{\text{mob}}} \sum_{i=1}^{N_{\text{mob}}} C_i \right) \tag{11.41}$$

where C_i is the number of times that mobile CR i changed its geo-location to another cell during $T_{\text{DB}}^{\text{mob}}$.

When there is SDB, all devices can obtain the list of available channels via SDB. Assume that information from RDB is cached in SDB and the cache hit rate is h, where $0 \le h \le 1$. Assume that the amount of traffic to/from local network for each request/response is L. Therefore the total traffic F_{SDB} over a unit time is as follows:

$$F_{\text{SDB}} = hL \times \left(\frac{N_{\text{fix}}}{T_{\text{DB}}^{\text{fix}}} + \frac{N_{\text{mob}}}{T_{\text{DB}}^{\text{mob}}} \sum_{i=1}^{N_{\text{mob}}} C_i \right)$$
$$+ (1-h)G \times \left(\frac{N_{\text{fix}}}{T_{\text{DB}}^{\text{fix}}} + \frac{N_{\text{mob}}}{T_{\text{DB}}^{\text{mob}}} \sum_{i=1}^{N_{\text{mob}}} C_i \right) \tag{11.42}$$

The first term in Eq. (11.42) corresponds to local traffic (when cache hits), and the second term corresponds to global traffic (when cache misses).

Both the hit rate h and the number of geo-location changes C_i depend on the mobility pattern of mobile CR devices. Numerical results for a popular mobility model are given in Section 11.4.5.

11.4.4.2 Sensing Cost: Number of Sensing Activities

A channel being used by a CR device is called an operating channel. To reduce service interruption due to the presence of primary users, each CR device also maintains a list of backup channels, and the CR devices perform spectrum sensing on these channels regularly [28].

Assume that the operating channel must be sensed at least once every t_{op} seconds, and each backup channel must be sensed at least once every t_{bk} seconds. Let us denote N as the number of CR devices and B as the number of backup channels for which each CR device must perform spectrum sensing. In addition to these backup channels, each CR device must perform spectrum sensing for the operating channel.

After receiving the list of available channels from the database, all devices start to perform sensing according to the above rules. Sensing costs (in term of number of sensing activities per unit time) for primary users' protection, which is denoted as C_{RDB}, can be described as follows:

$$C_{RDB} = N\left(\frac{1}{t_{op}} + \frac{B}{t_{bk}} \right) \tag{11.43}$$

In addition to the information obtained from RDB, the SDB may provide further information to indicate whether one or some of the available channels are available for a much longer time compared to other channels. Such results are based on long-term monitoring, possibly with some learning algorithms. An example for this case is reserved channels for future TV transmitters. There would be no primary users in these channels in the near future, and the CR devices do not need to perform regular sensing on the channels. Let us denote such channels as $\hat{r}$.

After receiving $\hat{r}$ as one of the available channels, assume that $\hat{r}$ will be selected by each CR as an operating channel at probability α, and as a backup channel at probability b, where $0 \le \alpha, b \le 1$, and $\alpha + b = 1$. Sensing costs C_{SDB} (in terms of number of sensing activities per unit time) for primary users' protection, when there are i CRs that select $\hat{r}$ as operating or backup channel, can be estimated as follows:

$$C_{SDB}^{(i)} = i\left(\beta\frac{B-1}{t_{bk}} + (1-\alpha)\frac{1}{t_{op}} + (1-\beta)\frac{B}{t_{bk}} \right)$$
$$+ (N-i)\left(\frac{1}{t_{op}} + \frac{B}{t_{bk}} \right) \tag{11.44}$$

11.4.4.3 Additional Time for Information Exchange

According to the system model, sensing time for each channel is the same and is denoted as τ_{ch}. Measurement time when the CR device uses information provided by the RDB is as follows:

$$T_{RDB}^m = \frac{1}{N} \times C_{RDB} \times \tau_{ch} \tag{11.45}$$

When the CR device utilizes information provided by the SDB, the measurement time is as follows:

$$T_{SDB}^m = \frac{1}{N} \times C_{SDB} \times \tau_{ch} \tag{11.46}$$

Therefore the SDB can help CR devices to reduce the following amount of time for signal measurement operation:

$$\Delta T = T_{RDB}^m - T_{SDB}^m \tag{11.47}$$

The time duration ΔT can be used to share sensing information among the CR devices. Hence, reducing time required for signal measurement operation would increase time that can be allocated for information exchange operation.

11.4.4.4 Number of Successful Transmissions

Focusing on any time slot of the contention-based channel, assume that each CR device transmits with probability x. The probability that all devices do not transmit is $(1-x)^N$; therefore the probability that at least one device transmits in the considered time slot $P_{(1+)}$ is described as follows:

$$P_{(1+)} = 1 - (1 - x)^N \tag{11.48}$$

The probability that exactly one device sends on the channel and the other (N–1) devices do not send is

$$P_{(1)} = C_1^N x(1 - x)^{N-1} = Nx(1 - x)^{N-1} \tag{11.49}$$

where

$$C_k^n = \frac{n!}{k!(n-k)!}$$

is the k-combinations of n.

The successful transmission P_s occurs when exactly one device transmits, with the condition that at least one device transmits. Therefore the probability of successful transmission is

$$P_s = \frac{P_{(1)}}{P_{(1+)}} = \frac{Nx(1-x)^{N-1}}{1-(1-x)^N} \qquad (11.50)$$

If N devices transmit during information-sharing time frame, the number of sensing results that are successfully being shared with other nodes is described as follows:

$$N' = N \times P_s = N \times \frac{Nx(1-x)^{N-1}}{1-(1-x)^N} \qquad (11.51)$$

where transmission probability on a duration which is divided to S time slots is defined as

$$x(S) = \frac{1}{S}$$

Let us denote the time allocated for information-sharing phase as T_{ex} and the slot duration as τ; then S is determined as follows:

$$S_{RDB} = \left\lfloor \frac{T_{ex}}{\tau} \right\rfloor \qquad (11.52)$$

$$S_{SDB} = \left\lfloor \frac{T_{ex+\Delta T}}{\tau} \right\rfloor \qquad (11.53)$$

where $\langle a \rangle$ is the largest integer which does not exceed a.

11.4.4.5 Improvement in Sensing Quality

Sensing quality is defined as the number of successful transmissions during information-sharing phase. The number of successful transmissions when the CR devices only obtain information from the RDB is described as follows:

$$N_{RDB} = N \times \frac{N \dfrac{1}{S_{RDB}} \left(1 - \dfrac{1}{S_{RDB}}\right)^{N-1}}{1 - \left(1 - \dfrac{1}{S_{RDB}}\right)^{N}} \qquad (11.54)$$

When the CR devices obtain additional information provided by the SDB which help to reduce measurement time, the number of successful transmissions is as follows:

$$N_{SDB} = N \times \frac{N \dfrac{1}{S_{SDB}} \left(1 - \dfrac{1}{S_{SDB}}\right)^{N-1}}{1 - \left(1 - \dfrac{1}{S_{SDB}}\right)^{N}} \qquad (11.55)$$

where S_{RDB} and S_{SDB} are determined by Eq. (11.52) and Eq. (11.53), respectively.

11.4.5 Numerical Results and Performance Evaluation

11.4.5.1 Global vs. Local Traffic Rate

Because the fixed devices will send the same geo-location information and always get hit from the second request to the database, this evaluation will be focused only on mobile devices. When a mobile CR device moves to a new cell, it needs to access the database. If information on the available channels for the same cell has been requested by other devices within a certain duration, it can be used to respond for the current request. In order to determine the cache hit rate h of the SDB, the number of database requests, and the traffic rate between global/local network, we adopt the Random Waypoint Mobility Model, because the model is flexible and it appears to create realistic mobility patterns for many scenarios [29,30].

According to the Random Waypoint Mobility Model, each mobile device stays in one location for a certain period of time. After the staying period, the device chooses a random destination in the area and a speed that is uniformly distributed between [min,max]; then it travels toward the newly chosen destination at the selected speed. The parameters described in Table 11.2 are used in the simulation.

Figure 11.18 shows the cache hit rate and global vs. local traffic rate when the simulation times change in two cases: 50 and 200 mobile devices, respectively. Figure 11.19 shows the cache hit rate and global vs. local traffic rate when the number of mobile devices change in two cases: 1000 sec and 5000 sec, respectively.

Table 11.2   **Mobility Parameters**

Item	Value
Number of mobile nodes	10–200
Area size	1000 m × 1000 m
Mesh size D	50 m
Speed mean and variance	1.5 m/s, 0.5 m/s
Pause time mean and variance	10 s, 5 s
Simulation time	100–18,000 s
Number of CRs	10
Number of available channels	10 channels
Number of operating channels	1 channel
Number of backup channels	1–4 channels
Sensing step for operating channel	every 2 seconds
Sensing step for backup channel	every 6 seconds

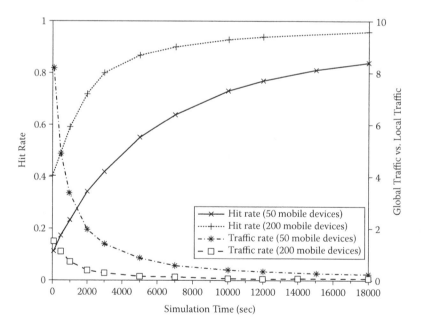

Figure 11.18   Cache hit rate and global/local traffic rate (50 and 200 mobile devices).

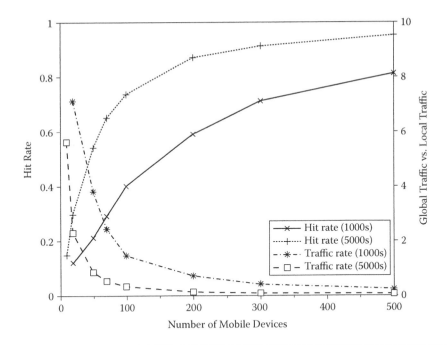

Figure 11.19 Cache hit rate and global/local traffic rate (1000 and 5000 seconds).

Note that information on available channels of each cell remains unchanged during simulation time; therefore simulation time is equivalent to the time-to-live (TTL) of the information. The following results are observed:

■ Cache hit rate is higher than 0; therefore by deploying SDB the amount of traffic to global network is reduced.
■ If user density is fixed, the amount of traffic to global network reduces when simulation time (or TTL of information in the database) is longer. This result can be used to design the refresh time for cache information.
■ If simulation time (i.e., TTL) is fixed, the amount of traffic to global network reduces when the user density is higher. This result can be used to decide the necessity of deploying SDB for each service area.

11.4.5.2 Sensing Cost and Sensing Quality

Table 11.2 shows the parameters used to evaluate the measurement time and the number of successful transmissions. The sensing step for each channel is adopted from IEEE 802.22 draft standard [28].

Figure 11.20 shows the sensing cost (in terms of number of sensing activities per unit time) when using information from RDB and SDB, respectively. It can be

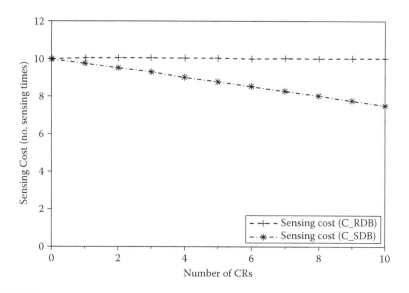

Figure 11.20 **Sensing cost (number of sensing times).**

observed that the number of sensing activities per unit time when the CR devices utilize information provided by the SDB is a maximum of 20% lower than when using information provided by the RDB.

Figure 11.21 shows the reduction rate of sensing cost for RDB case compared to SDB case corresponding to the number of CR devices in the system. When the number of required backup channels is small, the effect of using information from the SDB is higher. The rate varies from 0 to 0.5; however, a middle value which is 0.2 is selected for the evaluation on sensing quality.

From formulas (11.43), (11.44), and (11.47) we obtain

$$\frac{\Delta T}{T_{\text{RDB}}^{\text{m}}} = \frac{C_{\text{RDB}} - C_{\text{SDB}}}{C_{\text{RDB}}} \approx 0.20 \tag{11.56}$$

Assume that $T_{\text{RDB}}^{\text{m}}$ and T_{ex} are in the same scale; therefore the time which can be allocated for information sharing is also increased as follows:

$$\frac{\Delta T}{T_{\text{ex}}} \approx 0.20 \tag{11.57}$$

Figure 11.22 shows the improvement of sensing quality in terms of the number of successful transmissions during information-sharing phase. It is clear that the sensing quality can be achieved a maximum of 1.3 times higher by using information provided by the SDB.

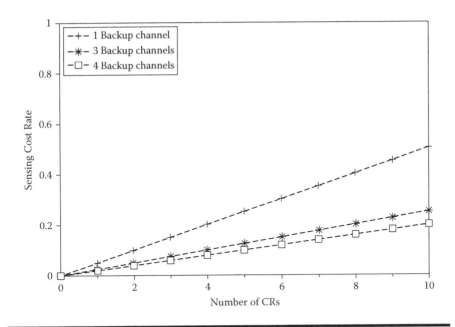

Figure 11.21   Reduction rate of sensing cost.

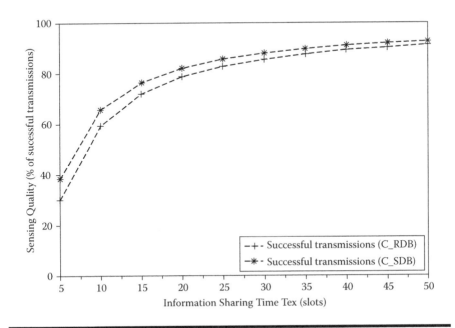

Figure 11.22   Successful transmissions during information-sharing phase.

11.5 Conclusion

In this chapter we explored energy detection-based distributed spectrum sensing (DSS). In particular we have studied three distributed sensing schemes, i.e., cooperative sensing, collaborative sensing, and selective sensing. To theoretically analyze the performance of these sensing schemes, we derived a generic closed-form expression for the probability of detection by energy detection (ED) with selective sensing in Rayleigh fading channel. Using this expression, we built up the analytical models for the three sensing schemes. Analytical study showed that all schemes improve the probability of detection by using DSS. In the case study provided, the collaborative sensing scheme outperforms the cooperative sensing scheme. The study also showed that the selective sensing scheme can achieve a good performance in terms of the probability of detection. Furthermore, the selective sensing scheme provides a lower false alarm rate than the collaborative sensing scheme.

The theoretical analysis of the DSS schemes is performed assuming that sensors provide one-bit information to represent the detection outcome, and a hard combining (HC) is performed at the fusion center (FC). Alternatively, sensors also could forward directly the soft information, or quantized version of the captured signal energy, to the FC, and soft combining (SoC) is performed at the FC. To investigate effects of SoC and HC, global detection at the FC based on SoC and HC are compared under perfect user-channel. For the method of SoC, an equal gain combining rule over the soft information is applied. For the method of HC, the logical operators AND or OR and *minimum probability of error criterion* are applied to combine the hard information. Simulation results show that, for a case of two identical sensors having independent observation of a primary (PU) signal, the SoC provides better performance in terms of global P_{md} compared to the HC. Finally, we investigated a hybrid scheme that performs two-stage detection that uses the SoC only when the first-stage detection based on HC results in "No detected signal, or 0." The hybrid scheme achieves comparable performance with that of the SoC while saving the transmission resource needed to transmit the soft information. In particular, the two-stage detection that uses OR logical rule is attractive because it requires fewer second-stage detections and hence results in a lesser overall sensing time compared to the two-stage detection that uses an AND logical rule.

The last portion of this chapter also introduces a sensing database (SDB). Without loss of generality, the SDB can be assumed as a logical entity that can be realized along with the FC. We discussed a potential architecture design of the SDB which supports the DSS scheme and can be deployed in local network of cognitive radio systems. The SDB contains regulatory information obtained from regulatory database, and sensing information obtained through the DSS scheme. A case study is introduced to analyze the influence of SDB and to evaluate the performance of SDB under certain assumptions. Performance evaluation results show that the number of sensing activities in a unit time is a maximum of 20% lower than when using SDB, and the sensing quality can be achieved at a maximum of

1.3 times higher than when using information provided by the SDB. By providing available information and the results of preliminary information processing to CR devices, the SDB can help CR systems to reduce the amount of traffic to global network (e.g., the Internet), to reduce the sensing cost by reducing the necessary number of sensing activities per unit time, and to improve the sensing quality by providing more time for the information-sharing phase that increases the number of successful transmissions of distributed sensing information.

References

1. G. Staple and K. Werbach, "The End of Spectrum Scarcity," *IEEE Spectrum*, vol. 41, pp. 48–52, Mar. 2004.
2. FCC, "FCC Spectrum Policy Task Force: Report of the Spectrum Efficiency Working Group," Tech. Rep. 02-135, FCC, Nov. 2002.
3. FCC, "FCC Second Report and Order and Memorandum Opinion and Order: In the Matter of Unlicensed Operation in the TV Broadcast Bands," Tech. Rep. 08-260, FCC, Nov. 14, 2008.
4. J. Mitola, III, and G. Q. Maguire, Jr., "Cognitive Radio: Making Software Radio More Personal," *IEEE_M_PCOM*, vol. 6, no. 4, pp. 13–18, 1999.
5. G. Ganesan and Y. Li, "Cooperative Spectrum Sensing in Cognitive Radio Networks," in IEEE Symposium on New Frontiers in Dynamic Spectrum Access Networks, DySPAN (Atlanta, GA), Nov. 2005.
6. S. M. Mishra, A. Sahai, and R. W. Brodersen, "Cooperative Sensing among Cognitive Radios," in IEEE International Conference on Communications (Istanbul, Turkey), Jun. 2006.
7. J. Ma and Y. G. Li, "Soft Combination and Detection for Cooperative Spectrum Sensing in Cognitive Radio Networks," in IEEE Global Telecommunications Conference, GLOBECOM (Atlanta, Georgia), Nov. 2007.
8. R. Thobaben and E. Larsson, "Sensor-network-aided cognitive radio: On the optimal receiver for estimate-and-forward protocols applied to the relay channel," in IEEE Asilomar Conference on Signals, Systems, and Computers (Monterey, California, USA), Nov. 2007.
9. G. Ole, B. Frode, H. Vegard, L. Markku, O. Bogar, T. Isabelle, H. Aawatif, M. Bertrand, and M. L. Christof, "Sensor Network for Dynamic and Cognitive Radio Access: Scenario Descriptions and System Requirements," Tech. Rep. ver. 1.0, Mar. 2008.
10. C. Sun, Y. D. Alemseged, H. N. Tran, and H. Harada, "Cognitive Radio Sensing Architecture and a Sensor Selection Case Study," in IEEE Vehicular Technology Conference, VTC (Barcelona, Spain), Apr. 2009.
11. "IEEE p1900.6/draft1, Draft Standard for Spectrum Sensing Interfaces and Data Structures for Dynamic Spectrum Access and Other Advanced Radio Communication Systems," Apr. 2010.
12. E. Peh and Y.-C. Liang, "Optimization for Cooperative Sensing in Cognitive Radio Networks," in IEEE WCNC'07 (Hong Kong), pp. 27–32, Mar. 2007.
13. W. Zhang, R. K. Mallik, and K. B. Letaief, "Cooperative Spectrum Sensing Optimization in Cognitive Radio Networks," in IEEE ICC'08 (Beijing), pp. 3411–3415, May 2008.

14. C. Sun, Y. D. Alemseged, H. N. Tran, and H. Harada, "Cognitive Radio Sensing Architecture and a Sensor Selection Scheme," in VTC'09 (Spain), Apr. 2009.

15. L. H. Ozarow, S. Shamai, and A. D.Wyner, "Information Theoretic Considerations for Cellular Mobile Radio," *IEEE_J_VT*, vol. 43, pp. 359–378, May 1994.

16. S. M. Kay, *Fundamentals of Statistical Signal Processing Volume II Detection Theory*. New Jersey: Prentice Hall, PTR, 1993.

17. F. F. Digham, M.-S. Alouini, and M. K. Simon, "On the Energy Detection of Unknown Signals over Fading Channels," *IEEE_J_COM*, vol. 55, pp. 21–24, Jan. 2007.

18. I. Gradshteyn and I. Ryzhik, *Table of Integrals, Series and Products*. Elsevier, 7 ed., 2007.

19. J. G. Proakis, *Digital Communications*. McGraw-Hill, 3rd ed., 1995.

20. A. H. Nuttall, "Some Integrals Involving the Qm Function," *IEEE_J_IT*, vol. 21, pp. 95–96, Jan. 1975.

21. R. Verdone, D. Dardari, G. Mazzini, and A. Conti, *Wireless Sensor and Actuator Networks: Technologies, Analysis and Design*. London: Elsevier, 2008

22. H. N. Tran, Y. D. Alemseged, C. Sun, and H. Harada, "Whitespace Sensing Database for Cognitive Radio Systems," Tech. Rep. SR-2009-99, IEICE, Mar. 2010.

23. FCC Rules No.08-260, "In the Matter of Unlicensed Operation in the TV Broadcast Bands Additional Spectrum for Unlicensed Devices Below 900 MHz and in the 3 GHz Band," Reports & Order, September 2008.

24. Ofcom, "Digital Dividend: Cognitive Access—Consultation on License-Exempting Cognitive Devices Using Interleaved Spectrum," Feb. 2009.

25. A. Ghasemi and E. S. Sousa, "Collaborative Spectrum Sensing for Opportunistic Access in Fading Environments," New Frontiers in Dynamic Spectrum Access Networks, First IEEE International Symposium on New Frontiers in Dynamic Spectrum Access Networks (DSPAN 2005), pp. 131–136, November 2005.

26. S. M. Mishra, A. Sahai, and R.W. Brodersen, "Cooperative Sensing among Cognitive Radios," *IEEE*, Jun. 2006.

27. H. Takagi and L. Kleinrock, "Throughput Analysis for Persistent Csma Systems," *IEEE Trans. Comm.*, vol. 33, pp. 627–638, Jul. 1985.

28. IEEE, "Draft Standard for Wireless Regional Area Networks Part 22: Cognitive Wireless RAN Medium Access Control (Mac) and Physical Layer (PHY) Specifications: Policies and Procedures for Operation in the TV Bands," May 2009.

29. T. Imelinsky and H. Korth, *Mobile Computing*. Kluwer Academic Publishers, 1996.

30. T. Camp, J. Boleng, and V. Davies, "A Survey of Mobility Models for Ad Hoc Network Research," *Wireless Communications Mobile Computing (WCMC)*, vol. 2, pp. 483–502, 2002.

Chapter 12

Leveraging Sensing and Geo-Location Database in TVWS Incumbent Protection

Rashid A. Saeed and Rania A. Mokhtar

Contents

12.1 Overview

Due to the transition of digital TV (DTV), recently several countries have begun to allow TV white spaces (TVWS) to be used by unlicensed devices. TV white spaces are considered an important step toward providing broadband access to millions of digital dividend household around the world and enabling a wide range of innovative wireless devices and services. The key challenge needed by the TV white space systems is that it is highly necessary for them to have the extra capabilities to avoid causing harmful interference to the licensed services in the TV band as well as to the unlicensed devices.

This requires that the TV band devices (TVBDs) must be aware of the presence of each other and of incumbent devices in the TV band. Specifically, TVBDs should have the ability to determine whether a TV channel or a frequency band is unused before it could transmit. Due to the difficulties associated with the sensing process and in order to ensure high protection to the broadcaster devices, a geo-location database has been proposed recently, where each is required to contact the database device before transmission—at least once per day—and get the idle frequency carrier.

In this chapter we present TVWS coexistence by leveraging database and sensing technologies. Coexistence is usually a term that has been used for devices/techniques of using the same unlicensed spectrum (i.e., Bluetooth, WiFi, and Zigbee at 2.4 GHz). However, the TVBDs need to share/coexist with incumbent devices as well as other TVBDs. In this chapter we try to introduce new system architecture for coexistence for both incumbent and TVWS devices. Most of the work that has been done or that is ongoing is running under one of two categories:

- Coexistence between TVBDs, i.e., IEEE802.19.1
- Spectrum sharing with incumbent devices, i.e., white space database (WSDB) group

12.2 Introduction to TVWS Coexistence

Due to the spectrum scarcity and spectrum efficient utilization, coexistence has been discussed widely in wireless communications circles, and many coexistence standards of activities are going on concurrently. For example, the IEEE802 has extensive activities in the coexistence protocols, i.e.:

1. IEEE802.15.2 wireless personal area network (WPAN) coexistence at ISM band
2. IEEE 802.16h wireless metropolitan area network (WMAN) coexistence for cognitive radio operation
3. IEEE802.22/11af is TVWS standards [1]
4. IEEE802.19.1 is a coexistence standard in TVWS

In addition to that, there have been other similar standards, i.e., IEEE 1900 and ECMA-392 [2].

TVWS coexistence should be performed for peaceful working with incumbent users in TV bands as well as other TVWS license-exempt technologies. There are three coexistence methods (i.e., RF sensing, beacon, and geo-location database). RF sensing or incumbent detection is listening to TVWS channel (i.e., if it is free, the TV band device can use it). Sensing is usually associated with many challenges, for example, hidden problems [1]. Consequently, in the United States the Federal Communications Commission (FCC) recently announced that sensing is not required for TVBDs, that only geo-location should be used for TVWS technology, that sensing-only terminal is kept as a future option, and the report encourages research to increase sensing reliability [1]. However, sensing still can play an important role in enhancing spectrum sharing among TVBDs.

In the beacon method, TVBDs transmit based on a control signal (beacon) identifying vacant channels within their service areas. The beacon signal can be sent by a TV station, an FM broadcast station, or a TVBD transmitter. However, of late the beacon method has experienced a lack of support from the relevant industry due to the expensive Capax needed, because it needs a separate infrastructure to build upon. Finally, the TVBD gets information on the vacant channels from a geo-location database. TVBD needs connections to the database and to update its information status at least each time of power-on or when it changes its location. Finally, the TV white space database (WSDB) is a database of authorized services in the TV frequency bands that is used to determine the available channels at a given location for use by TVBDs.

According to FCC part 15.700, the TVBDs can have one of three modes of operations [3]. Figure 12.1 shows the basic deployment example of TVWS systems architecture:

■ **Fixed Mode:** An operating mode in which the TVBD has the capability to transmit without receiving an enabling signal. The TVBD is able to select a channel itself based on a list provided by the database and to initiate a network by sending enabling signals to other devices. Fixed-mode TVBDs are usually working with a maximum transmission power of 4 W. According

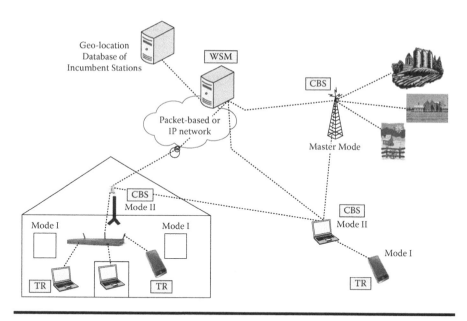

Figure 12.1   Basic deployment example of TVWS systems architecture.

to FCC rules, fixed TVBD needs to access the WSDB at least once a day. Figure 12.2 depicts the master/fixed-mode WSDB access [4].

■ **Mode I operation:** Also called sensing, it is the only mode used in operation of a personal/portable TVBD operating only on the available channel identified by either the fixed TVBD or Mode II TVBD that enables its operation. Mode I operation does not require the use of a geo-location capability or access to the TV bands database, nor does it require operation in a client mode. Mode II TVBDs are usually working with a maximum transmission power of 50 to 100 mW [5].

■ **Mode II operation:** In the operation of a personal/portable TVBD, its device determines the available channels at its location using its own geo-location and TV bands database access capabilities. Devices operating in Mode II may function as master devices. According to FCC rules mode II TVBD needs to access the WSDB each time it is activated from a power-off condition and/or after location change during operation. Figure 12.3 shows mode II WSDB access.

Figure 12.4 shows the TV white space devices mode of operations with white space database. The TVWS communication system contains three main parts; (1) WS air interface for data and management planes which include the PHY and MAC specifications, (2) spectrum sensing, and (3) geo-location. The system also may contain a coexistence mechanism. The TVWS communication system general structure is as

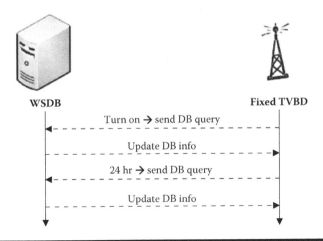

Figure 12.2   Fixed-mode WSDB access.

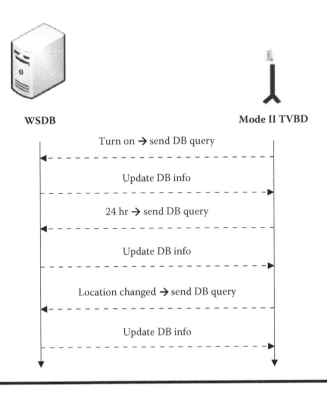

Figure 12.3   Mode II WSDB access.

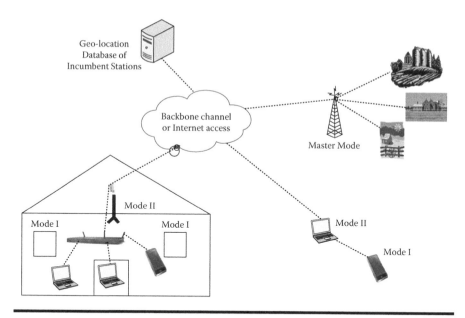

Figure 12.4   TV white space devices operation mode with white space database (WSDB).

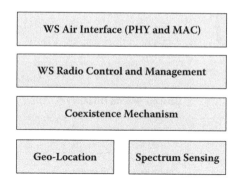

Figure 12.5   White space communication system.

shown in Figure 12.5. In the two next sections we will discuss white space sensing and database as individual coexistence solutions.

12.3 White Space Sensing Only

Sensing is the process of detecting the presence of incumbent TV devices, and it identifies the white space spectrum. RF sensing has been extensively used for coexistence between license-exempt devices, i.e., carrier sensing multiple access collision avoidance (CSMA/CA) in IEEE802.11 WLAN. The TV incumbent devices (i.e., the TV receivers and the low power auxiliary station devices, including wireless microphones) are usually much more sensitive to the interference than to other unlicensed bands systems (i.e., the 2.4GHz ISM band). However spectrum sensing in carrier sensing in IEEE802.11 is quite different. IEEE802.11 carrier sensing works at a high enough power level (maximum threshold −72 dBm) which only requires sensing of the channel for a very short time (up to 16 μs) using clear channel assessment (CCA), while spectrum sensing in cognitive radio applications works at very low power levels and hence needs much longer sensing time. As a result, CSMA can check the channel before each packet transmission, which cannot be done in spectrum sensing because the sensing time is typically longer than the packet duration.

There are plenty of spectrum sensing methods in the relevant literature, which explained that these can be performed by using the optimal likelihood ratio test, energy detection, matched filtering detection, cyclostationary detection, eigenvalue-based sensing, joint space-time sensing, and robust sensing methods [6]. However, these can be classified into three main categories, namely, matched filter sensing, energy sensing, and cyclostationary feature sensing. For reliable sensing, each TVBD can perform a two-step sensing process to increase the probability of detection (P_{det}) and reduce the probability of collision.

1. In the first step, a fast and rough spectrum sensing using energy detection (ED) sensing is used to receive signal strength indication (RSSI) or a mean received signal strength (MRSS) for initial spectrum evaluation.
2. In the second step a process of fine/feature sensing with extensive computation process using cyclostationary sensing is used to identify if the selected channel is free (H1: hypothesis 1) or occupied (H0: hypothesis 0) by other TVBDs. If the channel is occupied, then medium access control (MAC) layer needs to select another channel and repeats the fine sensing step procedure. Figure 12.6 shows the architecture of this method.

Generally, sensing-only TVBDs must not only determine the primary users' detection, but they must also perform a power control mechanism based on the primary users working in the adjacent channels (if allowed). RF sensing challenge, in contrast to a database approach, is that the spectrum sensing technology usually cannot distinguish between the licensed operations entitled to protection and those systems that operate illegally [7]. Moreover, detection of low-power auxiliary signals at extremely weak levels will add substantial TVWS development time and expense. Another challenge is the interference that may occur from the TVBDs to the TV receivers at the edge of protected service contours, even though the TVBDs are outside the protected contour. In this case, the TVBD will be sensing the incumbent transmitter, which is far enough for consideration of the channel which is free for use by TVBDs. However, the receiver may be affected by TVBDs

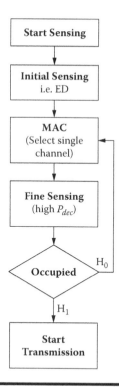

Figure 12.6 Coexistence scenario for incumbent sensing only.

transmission (i.e., the hidden receiver node). This is shown in Figure 12.7. For such a case, the FCC has set common interference protection ratios for both the co-channel and the adjacent channel interference for DTV and analog TV [8].

12.4 White Space Database (WSDB)

Sensing is important for determining the quality of TV band channel relative to real and potential interference sources and enhancing spectrum sharing among TVBDs; however, the database approach is the most reliable solution for incumbent detection and interference avoidance. In many circumstances, sensing only is not reliable for protecting incumbent users, where two experimental tests conducted by Motorola's and Microsoft's devices in 2008 had failed to detect the incumbent devices. In the United States several parties—including broadcasting and wireless microphone industries—had made recommendations to the Federal Communications Commission (FCC) by submitting ex parte or petitions for adoption of white spaces database (WSDB) [9]. In general, TV white space bands database should provide all the following functions [10]:

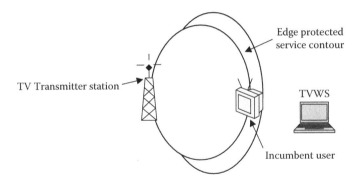

Figure 12.7   Hidden receiver node problem.

1. Detect, validate, and register incumbent devices from the authorized regulatory database (i.e., FCC) every 24 hours. Registration should include all incumbent associated information, including incumbent location, and convert all location data to NAD83 where necessary.
2. The registration includes any other protected radios not already in FCC databases.
3. Enable the registration of fixed TV band devices.
4. Enable the verification, correction, or removal of records in the database.
5. Synchronize all sub-databases with other authorized administrators.
6. Calculate protected services contours for each respective protected entity.
7. Perform calculation for available channels for any given location, time, and device parameters.
8. Provide policies for accept/reject channel update request.
9. Authentication and validation for the certification and enforcement status of TVBDs.
10. Provide transaction capability TVBD channel-queries from transmitted/nontransmitted devices.
11. Security implementation that protects the TV bands administration system from any unauthorized access.

The components that perform these functions are discussed in the next few sections.

12.4.1 Database Manager/Administrator

The database manager or database administrator has several functions that include data repository, registration, and determination of available channels/query process. The database manager interacts with all external interfaces and protocols with the TV band devices (TVBDs) or other database entities. Database managers should provide lists of available channels. Data analysis is required to select the best one among these channels; this analysis may be performed in the TVBDs as well. Some database managers may choose to provide full database services or may choose to provide only a repository function or "look-up" service. Database administrators may perform any other value-adding functions, such as tracking active channel use if reported by the TV bands device, or sending additional information to a TV bands device to enable it to determine the best available channel to use, or any other functions that can improve the spectral sharing efficiency and/ or TVBD performance.

12.4.2 Database Repository

White space database is a repository database that contains data about the incumbent and nonincumbent devices (i.e., the registered fixed devices) and their identification and location. Mode II devices just access the database; their information is not registered with the database As a result, FCC decided in the Second Report and Order (R&O) [11] to designate one or more database administrators from the private sector to create and operate TV band database(s), which will be a privately owned and operated service. Database administrators may charge fees to register fixed TV band devices and temporary broadcast auxiliary fixed links and to provide lists of available channels to the TV band devices.

First, a database repository should consist of all the registered incumbent devices in the FCC Consolidated DataBase System (CDBS): digital television (DTV) stations; digital and analogue Class A television stations; low-power television (LPTV) stations; television translator and booster stations; broadcast auxiliary service (BAS) stations (including received only sites) other than low-power auxiliary stations; private land mobile radio service (PLMRS) stations; commercial mobile radio service (CMRS) stations; and offshore radiotelephone service (ORS) stations [2]. Second, it should be able to register and store all the nonregistered TV band devices: the unlicensed microphone, telemetry, and medical devices. And finally, it should be able to register and update the location of all fixed, portable, and mobile TVBDs [12].

One of the challenges for TVWS geo-location database is that many wireless microphones are not registered in the FCC database, which can be used any time and anywhere by the broadcasters, i.e., news gathering (NG), charges, theatres, stadiums, etc. In this case the RF sensing only can be used to locate, register, and update the nonregistered wireless microphone operations to the WSDB server. According to the FCC rule, the device using sensing only must work with the of maximum transmission power 50 mW.

12.4.3 Devices Registration

Fixed and Mode II TVBDs with geo-location capabilities need to connect to the WSDB repository directly or through master devices to get the required information (i.e., available channels and maximum power allowed). This information is based on the existing incumbent in the surrounding environment entitled to interference protection, including, for example:

1. The licensed devices registered in the regulatory database (i.e., the FCC), which includes full power (DTV and analog class A TV stations) and low-power TV stations (LPTV), TV translator and booster stations, broadcast auxiliary point-to-point services (BAS), private land mobile radio service (PLMRS)/commercial mobile radio service (CMRS) operations on channels 14–20, and the offshore radiotelephone (ORS) service. These radio frequencies' location may vary from country to country.
2. The unlicensed devices that are entitled for the protection but not registered in the regulatory database, including cable TV, TV translation receivers, licensed wireless microphones, and the low-power auxiliary stations (LPASs).
3. Unlicensed fixed and Mode II TVBDs.

In addition, a TV bands database will be required to contain the locations of registered sites where wireless microphones and other low-power auxiliary devices are used on a regular or scheduled basis. For irregular events using wireless microphone registration, this can be solved by advanced registration, if there is an event for a certain time during the day (i.e., 3:00 p.m.). Consequently, the channel will be available for TVBDs until 3:00 p.m. only; after that the channel will not be free. In this case, these wireless microphones should be registered in the database far enough in advance. Because fixed TVBDs recheck the database for available channels on a daily basis, interference may happen if the microphones had not been registered at least one day before the event. At the border between countries, all TV stations in the neighbor country should be registered and included in WSDB to ensure the incumbents' protection.

12.4.4 Query Functions

Upon request from the fixed or the Mode II TVBDs, the database repository will provide a list of available white spaces spectrum/channels which allow power levels using the applicable interference protection requirements. The query can be periodical or upon request (i.e., when the interference occurs). Billing and charges for query service can be determined by the WSDB service providers or government regulatory agencies.

12.4.5 Information Exchange

With each database query, the TVDB devices need to send their particulars to the WSDB, i.e., the FCC identifier (FCC ID), the serial number (SN) assigned by the manufacturer, and the geo-location (Geo) information which includes the

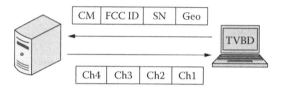

Figure 12.8 Data flow for WSDB query functions.

latitude and the longitude with other required control messages (CM). The WSDB should respond to the available channels if allowed to be accessed and with valid information, or deny the services if the TVBD is not allowed to be accessed for the service (blacklisted TVBD). Figure 12.8 and Figure 12.9 depict the data flow chart between the WSDB and the TVBD, respectively.

12.4.5.1 TVBDs Transmitter IDs

Because rapid identification of the interference source is a crucial process, each fixed TVBD should have unique identification. An identification signal/message can be transmitted by fixed devices each time the device needs to access the database. This message can carry much identification information about the TVBDs (i.e., ID number, location, vendor ID, operator ID). This information should be standardized by industry for the database to understand it. The disadvantage of identification standardization process is that it may delay time-to-market (TTM) for TVBDs; however, it ensures the consistency and compatibility between vendors. Mode I personal/portable devices should not require identification because they use low power and their potential for interference is low. Also Mode I devices do not have localization capability, which makes transmitting an identification signal meaningless.

In addition to TVBDs, Google, Inc. proposed [13] a public access interface to the repository database, which allows any individual or entity to access and review the data. The company believes that making the database accessible to the public will help ensure continued innovation in the unlicensed spectrum and transparent nature of the unlicensed TVWS.

12.5 White Spaces Database Group Proposal

In November 2009, the FCC Office of Engineering and Technology (OET) announced a public notice inviting proposals from entities seeking to be designated white space database (WSDB) manager(s). The proposals must address how the basic components of WSDB(s) as required by the FCC's rules will be satisfied, and whether the proponent seeks to provide all or only some of these functions, and must affirm

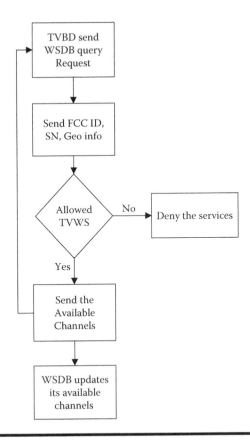

Figure 12.9 Flowchart for WSDB query function.

that the database service will comply with all of the applicable rules [14]. The white spaces database group stopped meeting in early 2010. However, we believe that their proposal is useful, and it gives a platform to many other works in WSDB.

Google Inc. for white space database (WSDB), along with the Consumer Electronics Association, Comsearch, Dell, Phillips, Microsoft Corporation, Motorola, and many other industries and associations [13], form a group which aims to provide a proper and compliant TV white space database as the basic functional architecture database management solution for the TVWS systems. The group has agreed on the three main entities for the database basic architecture (i.e., a data repository, a data registration process, and determination of available channels/query process) and a description of how each of these key functions will operate and interact. References include interfaces for external access and connection with other databases and TVBD for purposes of refreshing and updating the database. However, there are many scenarios and proposals of how these three entities will be combined. Here are some of the combination scenarios:

1. Type A: The WSDB as (i) a data repository, (ii) a data registration process, and (iii) a query process separately can be provided as shown in Figure 12.10.
2. Type B: The WSDB as (i) a data registration process and (ii) a query process under one service provider.
3. Type C: All three entities come under one service provider, where the TVBDs should talk directly to the integrated information system.
4. Type D: Clearinghouse model is distributed system architecture. It works as an independent entity to provide the repository services for the multiple geo-location databases, which can register the devices, calculate the channel availability, and respond to queries through the multiple TVWS database service providers.
5. Google, Inc. proposes a clearinghouse for its database proposal. A clearinghouse is considered a highly efficient structure for a large number of database service providers for the provision of TVWS database services.

The main challenge in the geo-location database coexistence is not to sense for the presence of incumbents, because they already know the geographic location of the incumbents. This method is efficient for a fixed analog and digital TV. However, this may not work for other devices entitled to protection which needs to be installed in a variety of places in a random manner—for example, licensed and unlicensed wireless microphones used by TV news gathering (NG) and public talks/festivals, TV translation receptions, and low-power auxiliary stations (LPASs). A TV bands database will be required to contain the locations of registered sites where wireless microphones and other low-power auxiliary devices are used on a regular or scheduled basis. In the next section we will discuss in detail a leveraging database and sensing coexistent solution.

12.6 U.S. FCC Database Regulation Exercise

On 26 January 2011 [3], the FCC designated nine companies—Comsearch, Frequency Finder Inc., Google Inc., KB Enterprises LLC and LS Telcom, Key Bridge Global LLC, Neustar Inc., Spectrum Bridge Inc., Telcordia Technologies, and WSdb LLC—as TV bands device (TVBDs) database administrators for a five-year term. The TV bands databases will be used by fixed and personal portable unlicensed devices to identify unused channels that are available at their geographic locations. TVWS will be used to provide broadband data and other services for consumers and businesses. The designation of these nine companies as database administrators is only on a conditional basis where designated companies should comply with the rule adopted by FCC in [3] and satisfy the following criteria:

1. Should have the technical expertise to be a TV band database administrator.
2. Should identify database functions it intends to perform and how it would synchronize data between multiple databases.

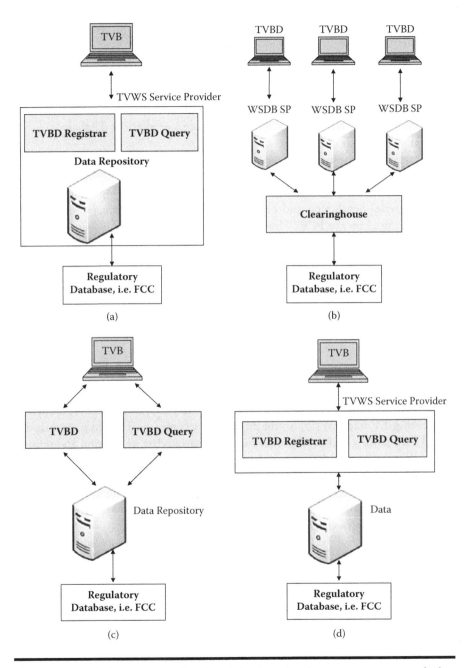

Figure 12.10 Reference functional architecture database management solution for TVWS systems (a) type A, (b) type B, (c) type C, (d) type (D).

3. Should provide diagrams of the architecture of the database system and a detailed description of how each function operates and interacts with the other functions.
4. Should identify the business relationship between itself and these other database administrators.
5. Other information and services like communication method, verifications, validations, and security will be used between database and TVBDs, to ensure that unauthorized parties cannot access or alter the database.

The operation of multiple database administrators in the United States is good in the sense that it prevents monopolies and will clearly benefit from the number and variety of mass market and niche services due to growing competition between multiple parties developing and providing channel assignment services as a new mechanism used by the FCC. However, this mechanism presents some coordination challenges between different parties, which may lead to some sort of coalition or forum between all winners to coordinate, standardize, and build the database structure models between various interfaces. Especially, the FCC mentioned that this exercise later may extend beyond databases for the TV bands, and it will consider employing similar database approaches in other spectrum bands.

12.7 Leveraging Sensing and Geo-Location Database

In this section, a dual-mode approach of leveraging database and sensing is discussed to prevent interference with wireless incumbent users (i.e., microphones, telemetry, and existing TV stations) [15]. In the dual-mode scheme, the TV white space device should have sensing and geo-location capabilities to update its location and sensing access to the database each time it wishes to use the spectrum. The motivation behind using a dual-mode approach is that connecting a database through a backhaul-based coexistence through a channel assignment is somewhat inefficient and slow for the time sharing. Over-the-air coexistence through sensing is also necessary, especially, for the short-range coexistence or when the backhaul is absent.

U.S Google has submitted a proposal to the FCC to run a free trial WSDB [13]. A database approach is usually combined with a geo-location capability where the TVBDs should send their sensing data along with their location to the WSDB to determine the suitable vacant TV channels at that location, with some other parameters like maximum transmission power. The prescribed location accuracy is usually about 50 m [16]. Each fixed TVBD will access the database at their geographic coordinates prior to their initial service transmission at a given location and at least once a day to verify that the operating channels continue to remain available. For portable and mobile devices, in addition to the two access requirements above, the device should connect each time it is activated from a power-off condition and if it changes location during an operation. Figure 12.11 shows a database accessed by

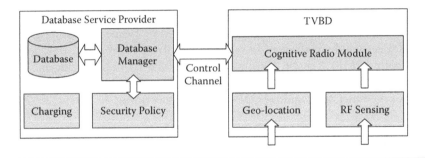

Figure 12.11   Leveraging white space geo-location database and RF sensing.

the TVBD, combined with a real-time spectrum sensing and geo-location capabilities where all the TVBDs must report their collaborative or non-collaborative sensing measurements periodically, where each device must be able to sense active TV spectrum channels, at least initially when powered on or changing their location.

12.7.1 RF Sensing

TVWS devices required a sensing capability for the incumbent devices with a threshold detection level of −114 dBm. Even though only sensing is not sufficient for protecting the incumbent users, it is compulsory for all TVBDs, according to the FCC rules. Sensing using the TVBDs alone is still under debate at the FCC level due to the challenges and problematical issues associated with the sensing alone, such as the hidden node problem. In addition, sensing of low-power auxiliary and wireless microphone signals at extremely weak levels will add substantial TVWS development time and expense, which leads IEEE802 and WiFi Alliance to file petitions to the FCC for reconsideration in this issue.

12.7.2 Geo-Location

All TVBD's personal/portable devices should have a geo-location capability to access the WSDB channels that may be used by incumbent users at that location. The accuracy of the geo-location should be 50 m to 100 m, depending on the application [17]. For outdoor applications (4 W devices), GPS can be used to support these requirements, but in the case of indoor applications (100 and 50 mW devices) there are issues with the accuracy inside buildings due to the GPS line-of-sight nature. However, a relative location can be determined by the help of other devices/ base stations using the triangular location technique.

12.7.3 Cognitive Radio Module

There should be cognitive radio capability in each device for decision making in the cases that the database connection is not available or the sensing is not reliable.

Cognitive radio module is connected with the geo-location and RF sensing parts for data analysis and decision, where a near real-time frequency planning among local devices will be performed to refine the available channel list. In a point-to-multi-point (PMP) scenario (i.e., master and Mode II) these intelligent entities can be shifted to the base station (BS), which reduces cost and size of the TVBDs. Some of the functions cognitive radio module can perform:

■ Adjustment of the current transmission of power of the radio response from sensing to database, to get information
■ Select from channel list given by the database
■ Stop transmission in one channel and switch to other channel

12.7.4 Control Channel

Each device needs a dedicated control channel to the database for refresh and update. This connection can also be an Internet connection. The TVBD communicates with the database for channel assignment and database update. The communication channel between cognitive radio modules and white space database manager includes:

■ Provision of information on frequency bands currently available for secondary usage to CR module in TVBD.
■ Provision of TVBD location and registration information to white space database.
■ For Mode II TVBDs, it provides all Mode I TVBDs location and registration information to white space database.

12.7.5 Security

A security policy should be applied for protecting the database from unauthorized access or any unintended activity. This shall include user/device authentication, integrity and confidentiality of open exchanges, and data privacy and policy correctness attestation and enforcement [18]. In addition, a fixed or Mode II device must check with its database that the Mode I device has a valid FCC Identifier before providing a list of available channels.

12.8 Synchronization with Other White Space Devices

Synchronization is the process that ensures the same data content among several participating entities, possibly having different sets of content. TVWS database synchronization can be performed in two ways:

1. Unidirectional synchronization or centralized synchronization: This scheme is working only with the authorized government database, i.e., the commission's Consolidated DataBase System (CDBS) and the universal licensing system (ULS). In a unidirectional synchronization, all of the contents from the CDBS/ULS will be placed in other clearinghouse/databases, which also implies any content in the clearinghouse/database entities that do not exist in the CDBS/ULS database will be deleted. This is applied only to the registered incumbent devices in the CDBS/ULS. This scheme also is suitable with the hierarchical database within the same database service provider.

2. For nonregistered TVBDs a bidirectional database synchronization scheme is suitable, the contents of which will be merged with all the participating databases. However, the security module should make sure that no unauthorized data will be injected into the database.

There are several other factors that need to be taken care of when implementing the database synchronization, i.e., how frequently this synchronization of data will be conducted, speed of implementation, security, performance, and the immediate update of data [19].

12.9 Database Fees

Database administrators/managers can charge fees for registering fixed devices, spectrum sensing, and providing lists of available channels to fixed devices, personal/portable devices, or any other nontransmitted entity. Charges also can be identified by policy in case of a monopoly provider.

12.10 Collaborative TVWS Databases

Collaborative technology is a new scheme that has been applied in several telecommunication industry applications, which is intended to enhance the overall system throughput. For example, in wireless communication, a cooperative antenna system, also known as MIMO (multiple input multiple output), has approved the system capacity almost m times by using m antennae. Collaborative databases also can be useful to improve spectrum utilization and efficiency. In scenarios of deploying very large scalable networks such as smart grid networks, it would be useful if Mode II and fixed-mode devices could summarize/optimize the information from different databases.

A cooperative system can be centralized or distributed. Centralized cooperatives can be realized by defining centralized topology that connect TVDB with TVWS operators and service providers. A spectrum broker scenario is proposed by COGnitive

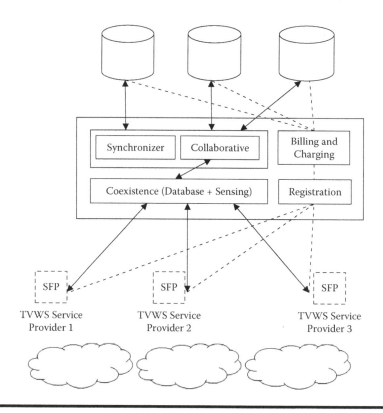

Figure 12.12   Reference model for centralized collaborative scenario.

radio systems for efficient sharing of TV white spaces in EUropean context (COGEU) [20] in their reference model. COGEU introduces a new entity called spectrum broker to control spectrum flow from the spectrum information supplier (TVDB) to the TVWS operators (e.g., WiFi, WRAN, safety and emergency networks). Figure 12.12 shows the illustration for a centralized collaborative database scenario. Spectrum broker or spectrum virtual network operators (SVNO) can have several modules [21]. Leveraging coexistence of sensing and database, this module gets the collaborated information from databases through the collaborative module. It also gets the sensing information from sensing focal point (SFP) from each operator. Sensing focal point can serve as a fusion node for the operator to report all sensing observations in different locations for the same operator. SFP can also process all gathered sensing information from different nodes through a cooperative sensing scheme [22].

12.11 Conclusions

The TV white spaces (TVWS) have opened the door for numerous innovative technologies to be initiated (i.e., IEEE802.22, IEEE802.11af, and IEEE1900.4a). The success of TVWS basically depends on the peaceful coexistence between analog and digital TV and other technologies using the same band. Coexistence is an old task that has been utilized by many technologies.

However, the sophisticated methods and the diversity of technologies that exist in this band make coexistence a problematical task. Due to the difficulties associated with sensing and ensuring high protection of the incumbent devices, many parties called for using geo-location database for coexistence. However, the database in many scenarios may not be accessible to reduce frequency agile (efficiency of spectrum allocation and sharing) and spectrum utilization efficiency. In this paper we present the TVWS coexistence by leveraging the database and the sensing technologies. The system architecture and design is discussed and compared with existing/ongoing standards.

References

1. R. Ahuja, R. Corke, and A. Bok, "Cognitive Radio System Using IEEE 802.11a over UHF TVWS," 3rd IEEE Symposium on Dynamic Spectrum Access Networks, 2008. (DySPAN 2008), pp. 1–9, Oct. 2008.
2. Ecma International, "Standard ECMA-392: MAC and PHY for Operation in TV White Space." Available http://www.ecma-international.org/publications/standards/Ecma-392.htm.
3. Mark Gibson, "TV White Space Geo-Location Database," TV White space coexistence 802.19.1 Workshop, IEEE 802 Plenary meeting, San Diego, July 16, 2010.
4. FCC TV Service Contour Data Points http://www.fcc.gov/ftp/Bureaus/MB/Databases/tv_service_contour_data/, updated on Aug. 28, 2010. Accessed on August 29, 2010.
5. D. Gurney, G. Buchwald, L. Ecklund, S. L. Kuffner, and J. Grosspietsch, "Geo-Location Database Techniques for Incumbent Protection in the TV White Space," 3rd IEEE Symposium on New Frontiers in Dynamic Spectrum Access Networks, 2008. (DySPAN 2008), pp. 1–9, 2008.
6. M. Gosh, V. Gaddam, G. Turkenich, and K. Challapali, "Spectrum Sensing Prototype for Sensing ATSC and Wireless Microphone Signals," Proc. Third International Conference on Cognitive Radio Oriented Wireless Networks and Communication (CrownCom'08), May 2008.
7. IEEE1900.4, "Joint Session of 1900.4, 1900.5, 1900.6 WGs and WS Radio Overview of 1900.4a," document id scc41-ws-radio-10/0018r0, July 2010.
8. FCC, "47 CFR Section 73.333, Figure 1 and Section 73.699, Figure 9," updated Dec. 2008, accessed on August 29, 2010.
9. Mark Bruner et al. "Ex Parte Comments of Shure Incorporated in the Matter of Unlicensed Operation in the TV Broadcast Bands ET Docket No. 04-186," May 2008. Available http://www.shure.com/idc/groups/public/@gms_gmi_web/documents/web-content/us_pro_pr_ws_2008_exparte_may.pdf.
10. Jessy Caulfield, "Proposal administer a TV band Database: Summary and Review," Key Bridge Global, Jan. 15, 2010.

11. Federal Communication Commission, In the Matter of Unlicensed Operation in the TV Broadcast Bands, ET Docket No. 04-186, Second Report and Order, 23 FCC Rcd. 16807, 2008. ("TV White Spaces Second R&O").

12. Ha Nguyen Tran et al., "Requirements and Amendments Regarding TVWS Database Access," IEEE 802.11-10-0262-00-00af, February 2009.

13. Google Inc., "Proposal to Provide a TV Band Device Database Management Solution," FCC, January 4, 2010. http://www.scribd.com/doc/24784912/01-04-10-Google-White-Spaces-Database-Proposal.

14. IEEE 802.19.1 TVWS Coexistence group, "System Description Document," IEEE 802.19-10/0055r3, March 2010.

15. D. Gurney, G. Buchwald, L. Ecklund, S. Kufner, and J. Grosspietsch, "Geo-Location Database Techniques for Incumbent Protection in TV White Space," *3rd IEEE International Symposium on New Frontiers in Dynamic Spectrum Access Networks, DySpan,'08*, Chicago, IL, pp. 1–9, Oct. 2008.

16. Federal Communication Commission, Public Notice, "Office of Engineering and Technology Invites Proposals from Entities Seeking to be Designated TV Band Device Database Managers," ET Docket No. 04-186, released November 25, 2009 at http://fjallfoss.fcc.gov/edocs_public/attachmatch/DA-09-2479A1.pdf.

17. Federal Communication Commission, R-6602 (F-) "FCC TV Contour Data," http://www.fcc.gov/ftp/Bureaus/MB/Databases/tv_service_contour_data/.

18. Linda K. Moore, "Spectrum Policy in the Age of Broadband: Issues for Congress," Congressional Research Service, R40674, July 1, 2010.

19. Thomas C. Jepsen, *Distributed Storage Networks: Architecture, Protocols and Management*, John Wiley and Sons Ltd, 2003.

20. COGEU "D3.1 Use-Cases Analysis and TVWS Systems Requirements," Report FP7 ICT-2009.1.1 July 2010.

21. Rania Mokhtar, B. M. Ali, and N. K. Noordin, "Distributed Cooperative Spectrum Sensing in Cognitive Radio Networks with Adaptive Detection Threshold," Proceedings of the Asia Pacific Advanced Network (APAN), Aug. 2010.

22. Federal Communication Commission, "Second Memorandum Opinion and Order (SMO&O) in ET Docket No. 04-186, FCC 10-174, adopted Jan. 26, 2010 (amendment). http://static.arstechnica.com/DatabaseOrder.pdf.

OTHER ASPECTS

Chapter 13

Elements of Efficient TV White Space Allocation Part I: Acquisition Principles

Joseph W. Mwangoka, Paulo Marques, and Jonathan Rodriguez

Contents

13.1 Overview

This chapter reviews the roles of a number of elements that may lead to reliable and efficient allocation of the TV white spaces. These elements have to be coordinated such that the goals of multiple stakeholders are considered. This is in contrast to an approach where only the goals of a single stakeholder are considered. For example, a regulator with the goal of curbing interference might use a command and control approach of allocating spectrum. However, satisfying the goal of controlling interference inadvertently leads to inefficient spectrum usage. Holistic consideration of the goals of other stakeholders can be achieved through market-based spectrum allocation. The market-based approach provides incentives for efficient usage of spectrum such that the resource goes to the most valuable user. Moreover, it is challenging for large or small organizations to consistently achieve all their goals in a timely and efficient manner [1]. Therefore, by coordinating different elements, each contributing its strength in a value chain, efficient exploitation of the TV white spaces can be achieved.

The elements discussed consist of: actors that provide reliable information on the availability of the TV white spaces; intermediaries that allow the exchange of information between demand and supply sides by providing a platform which brings them together and facilitates the transaction of that information; repositories to provide information on the TV white spaces availability based on location, time, reliability levels, etc.; as well as wireless service operators that wish to provide or extend their services through the TV white spaces. This chapter looks at these elements as role players in the value chain of an ecosystem that enables efficient and noninterfering exploitation of the TV white spaces to provide reliable wireless services while complying to regulatory requirements. This will in turn mitigate high transaction costs, interference risks, possibilities of conflicts between actors, and lack of investments in the TV white spaces.

The chapter proceeds as follows: Section 13.2 introduces the content covered; Section 13.3 spells out the motivations for the need of intermediaries to facilitate the functioning of the TV white spaces systems from a market perspective and presents cognition as an enabling technology; Section 13.4 gives the factors affecting the sharing of TV white spaces; Section 13.5 presents spectrum acquisition models in the TV white spaces, where the evolution of the methods to acquire secondary spectrum resources is given. Finally, Section 13.6 concludes the chapter.

13.2 Introduction

Broadcast television services operate in licensed channels in the VHF/UHF portion of the radio spectrum. Currently, there is a global move to convert TV stations from analog to digital transmission. This is called the digital switch-over (DSO) or in some cases the analogue switch-off (ASO), referring to the time when digital transmission effectively starts, or when analogue transmission effectively stops operation, respectively [2].

Due to the frequency efficiency of digital TV (DTV), some of the spectrum bands used for analog TV will be cleared and made available for other usage. Moreover, DTV spectrum allocation is such that there are a number of TV frequency bands which are left unused within a given geographical location so as to avoid causing interference to co-channel or adjacent channel DTV transmitters; that is to say, the spectrum bands are geographically interleaved. The geographic interleaved spectrum bands are generally called the Television White Spaces, TV white spaces, or simply TVWS. Therefore, because not all of this spectrum in any particular location is used by the DTV stations, what remains is available for other services on a shared basis. The cleared bands and the unused geographical interleaved spectrum bands provide an opportunity for deploying new wireless services. These opportunities create what is called the *digital dividend* in the literature [2,3]. In other words, the digital dividend refers to the *leftover* frequencies resulting from the change of TV broadcasting from analog to digital.

The focus of this chapter is on the efficient usage of the TV white spaces. In order to further demonstrate the uniqueness of the TV white spaces, an analogy is used to contrast them with the basic traditional models for spectrum management, namely, the command and control, the commons, and the market models. Briefly, the traditional models work as follows:

1. *Command and Control:* Under this model, the spectrum regulator issues a license to the operator with fixed specification on the frequency band, type of technology, transmit power levels, operation location, etc. for using license. In general, the license does not allow any change of usage in its validity duration. Although this model is effective in preventing harmful interference among users, lack of flexibility makes it wasteful, especially when more efficient technologies are available, or when new ventures could use some of the band, in areas where or times when the license holder does not operate, for provision of new wireless services.

2. *Spectrum Commons:* In this model, there are no exclusive usage rights, and multiple users can share access to a single frequency band, like the ISM (Industrial, Scientific, and Medical) band. There are no service guarantees to its users, and hence this model may not be suitable for services requiring quality-of-service (QoS) guarantee. Users have to comply with technical parameters or standards that define the power limits and operational restrictions for unlicensed devices to control interference within the spectrum band. Otherwise, this mode of spectrum usage runs the risk of suffering the *tragedy of the commons* resulting from overuse of a limited set of resources that are held in common. However,

it is a very flexible way to assign and allocate spectrum, as usage varies dynamically with technology capability and deployed applications.

3. *Market Model:* In this model, market-based mechanisms within a regulatory environment allow a dynamic trading of a spectrum segment, enabling the assignment of spectrum to the users which value it the most. In this context, trading is the transfer of spectrum usage rights between parties in a spectrum market. Therefore, the market decides spectrum assignment to users. With liberalization, restrictions on services and technology associated with spectrum usage rights are relaxed; that is, the market will dictate the type of services and technology to be deployed. However, the efficiency of the market model will depend on low transaction costs and on the presence of an arbitration system for dispute resolution.

Having briefly seen the traditional models of spectrum management, let us look at spectrum as a tradable commodity, like an apple, for example. Then the command and control spectrum management model is like selling whole apples. They are free of contamination; however, because sometimes people do not eat them all, there is wastage associated with them. The commons model is like eating the half-bitten apples; in this case, the leftovers of the command and control model. In this case, the leftovers are generally obtained for free; but because they are half bitten, they run the risk of being contaminated; and there is no guarantee of their availability whenever one is "hungry." In this context, in order to trade off the reliability of whole apples and the flexibility of half-bitten ones, it might be plausible if apples were supplied in the form of slices. The slices are uncontaminated, avoid wastage, and may be obtained at prices lower than whole apples. Slices of apples represent granularity in spectrum license parameters, where usage rights are traded while giving flexibility on type of usage. In the apple analogy, the TV white spaces are a special case where sliced, sanitary apples are leftovers! Therefore, there are two ways to acquire TV white spaces: free or paid, as shown in Figure 13.1, and detailed in Section 13.4. Free acquisition can be done through spectrum sensing, the use of beacons, or free access to geo-location database. On the other hand, paid acquisition can be done through paid access to the geo-location database or a broker. The paid approach is a bit tricky, as to who will be selling what? This chapter tries to provide an answer to this question.

The purpose of exploiting the TV white spaces is to achieve efficiency in the VHF/UHF bands. Spectrum usage efficiency is a multifaceted term that reflects the gain that the users, economy, and society obtain from its usage. It therefore has different implications within different contexts. For example, technically it could mean the amount of data that can be transmitted per unit bandwidth; economically it could mean the added value per unit bandwidth; functionally it could mean the reliability of the service deployed per unit bandwidth [4]. Achieving efficiency in one context does not necessarily imply efficiency under other criteria. For example, the command and control mode of spectrum usage enables the deployment of

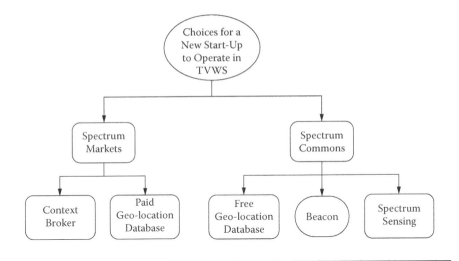

Figure 13.1 Spectrum commons ("free") and market-based ("paid") spectrum acquisition models in the TV white spaces.

reliable services due to exclusive usage of a given frequency band, hence achieving functional efficiency. However, the holder of a spectrum license under the same regime is not allowed to change transmission technology within the license period. Therefore, by hindering the replacement of old systems with better technologies or services in the band, this constraint leads to poor economic efficiency. As a result, a number of enabling factors at the technical, regulatory, and market levels have to be weighed simultaneously to achieve efficient usage of spectrum and, in this context, the TV white spaces.

Spectrum efficiency is more an economic concept than an engineering metric [5]. This is because the goal of spectrum management is to achieve economic efficiency through the proper utilization of spectrum resources in conjunction with others such as workforce, technology, end users, etc. Moreover, alternative applications such as national defense, safety, scientific exploration, as well as alternative technical approaches for similar applications in the same frequency band, such as broadband wireless access (which can be equally achieved by WiMAX, LTE, etc. over VHF band), have to be determined so that the most valuable usage is allocated.

Spectrum usage efficiency can be achieved through market forces of demand and supply [6–8]. The existence of a free market for spectrum as a resource forces system designers to use it in a cost-effective way. This can be achieved when other aspects of a wireless communication system (besides spectrum) are considered in the design, and the most economically efficient combination is determined. Therefore, several enablers have to be in place before efficient TV white space usage is achieved. In our perspective, these enablers should function in such a way that *whenever there is a need for spectrum resources on the user's side, and there are unused*

TV white spaces resources somewhere, the enablers are able to match the demand and supply sides in real time. In other words, the functioning of these elements removes the obstacles that hinder the movement of TV white spaces to more valuable users. These obstacles are mostly related to a lack or bad functioning of an intermediary entity between the regulator and the spectrum users [9–11].

The absence of an intermediary entity leads to the lack of information on available spectrum. Access to this information is an important aspect for enabling a healthy TV white space ecosystem. This information can be available through a query database of DTV stations' spectrum usage located in a centralized place and accessible upon demand. Moreover, the information can be obtained through sensing mechanisms, or by collaborating with other users—to increase the reliability of sensed information [12]. This information reflects the correct value of spectrum at any particular location and time, and hence could lead to effective TV white spaces utilization.

The availability of information solves many challenging problems for the enabling of TV white space efficient usage. These include:

1. Lowering the *transaction costs* of acquiring information on available TV white space.
2. Reduced *risk of interference.* Service providers use the information to identify alternative bands for communication, and hence mitigate interference in congested bands. This information is in conjunction with operating thresholds set by regulators or technology standardization bodies.
3. Avoidance of *conflicts between service providers.* Clear TV white spaces usage rights lead to straightforward conflict resolution procedures.
4. *Attracting more investment* in the TV white spaces. Uncertainty on the economic value of the white spaces at a given place and for a future time discourages investment, whereas reliable information encourages investment.

The TV white spaces, although conducive to more decentralized methods of spectrum usage, present risks and challenges which necessitate the introduction of centralized coordination or enabling intermediaries. That can be realized through government agencies or private firms providing TV white space information services, respectively. Arguments based on regulatory policy, market efficiency, as well as technological development support their existence [11,13–15]. From a regulatory perspective, intermediaries would ensure conformance to policies and regulations and help in achieving economically efficient usage of spectrum. From a business perspective, intermediaries would provide a mechanism for interaction between TV white space supply and demand sides, hence lowering transaction costs by centralizing information. From a technology perspective, intermediaries would reduce the complexity of the device by eliminating the need for sensing modules for spectrum acquisition (which will also lower battery consumption) [16].

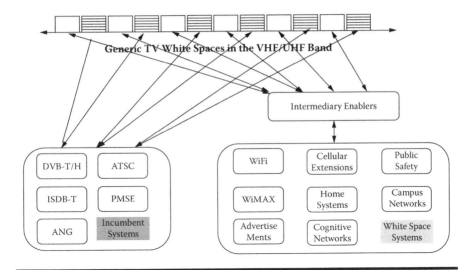

Figure 13.2 **The position of intermediary enablers for TV white space systems relative to incumbent systems.**

13.3 An Intermediary for the TV White Spaces

The TV white spaces offer a lot of opportunities for the provision of new wireless services. The challenge is how to acquire reliable spectrum availability information for wireless service providers to operate with QoS guarantee in the TV white spaces. This is because, even if there is no interference, the pure engineering solution may misallocate resources [17]. Reality necessitates combined administrative, market, and technological approaches to achieve an effective TV white space market environment, which not only allows diversity of spectrum usage models, but also provides incentives for innovation. Therefore, apart from focusing on ways to utilize the TV white spaces from a technical point of view, an intermediary for an efficient TV white space eco-system has to be established. Figure 13.2 shows the position of intermediary enablers relative to the regulators, incumbent systems, and white space systems. Incumbent systems have guaranteed and direct access to the spectrum resources, whereas secondary white space users do not have guaranteed access to the TV white spaces. Therefore, the intermediary enablers add value to the TV white spaces by providing reliable white space availability information to its users. This section focuses on the motivation for the intermediary market enablers and on cognition as an enabling technology to achieve reliable operation of wireless services in the TV white spaces.

13.3.1 Motivations

The motivations for the introduction of intermediaries in developing a viable TV white space spectrum market are: (1) maximizing spectrum utilization through trading; (2) counter-marketing experiences in other countries; (3) using market forces to resolve conflicting views in the U.S. experience; (4) lowering transaction costs, and (5) congestion measurement. These motivations are detailed as follows:

1. *Maximizing spectrum utilization through trading:* In spectrum usage, market models have been identified as a key to achieving the goal of efficient spectrum usage [10]. An effective market can be established if trading transactions' costs are low. In such a market, traders can find a trading partner, agree on terms, settle the trade, and monitor performance of the contract without incurring substantial costs or delays in the process [18]. Spectrum context brokers are like real estate agents in the residential property market, and can therefore achieve the function of matching and assigning TV white spaces resources to users who value them the most, hence maximizing the social value of the spectrum. In Chapter 14, a business model for the provision of TV white space availability information will be presented with the goal of establishing a viable spectrum exchange market.

2. *Counter-marketing experiences in other countries:* Countries implementing ambitious spectrum liberalization policies observed 61% lower wireless license sales prices compared to countries without similar policies [19]. It is argued, therefore, that even though allowing flexible license exchange may increase profits by allowing deployment of innovative technologies, the increase in the competitiveness of some licenses can lower the targeted profits. This insight cautions against developing and implementing TV white spaces exploitation mechanisms. As pointed out in [7], allowing spectrum license trading alone is not enough; other micro-level factors need to be considered to make the market produce the intended social benefits. In this context, increasing the sales volume of the TV white spaces will require an intermediary system which provides accurate and reliable information on the resource availability and is able to match the willingness-to-pay of users, or attract new users for that purpose. This will lead to a dynamic market that will maximize the economic efficiency of the TV white spaces.

3. *Using market forces to resolve conflicting views in the U.S. experience:* There are differing views on whether spectrum should be licensed or unlicensed among competing firms or stakeholders. A stakeholder who promotes a given usage model is motivated to present the model promoted as having greater value than is actually the case. In [8], the merits of employing market forces to address the issues of wireless spectrum overcrowding and the allocation of spectrum between stakeholders with differing views are examined. When unlicensed spectrum is assigned to all competing users during periods of excess demand, an inefficient outcome related to the tragedy of the commons is likely to result. It is argued that this inefficiency can be substantially reduced when the assignment of users to unlicensed spectrum is based on the minimum system requirements that are needed to be operational, such as

bandwidth and latency tolerance levels of the competing users. Alternatively, further efficiency gains can be obtained when the users are required to pay based on the way they value the resource. The efficiency gains mainly depend upon the auction mechanism's ability to solve a collective action problem in which firms desiring unlicensed spectrum have an incentive to pay less or consume more than their fair share of the common resource [17,20]. It is therefore evident that an intermediary is needed to set a stage for the competing parts to interact and let the market decide who is the most valuable user of the band in question, in this context the TV white spaces.

4. *Lowering transaction costs:* Transaction costs include the costs of planning, adapting, executing, and monitoring a task, for example, the use of the TV white spaces. A transaction occurs when a good or service is transferred across a separable interface. Transaction cost economics identifies transaction efficiency as a major source of value. Nevertheless, the emphasis on transaction efficiency is not meant to overlook other fundamental sources of value such as technological innovation and the reconfiguration of resources, as can be seen in the DSO. In general, the cost structure of most service businesses, including mobile services, is characterized by a high ratio of fixed to variable costs [21] and by a high degree of cost sharing (such that the same facilities, equipment, and personnel are used to provide multiple services) [22]. The high fixed costs typically lead to economies of scale as increased production lowers the average production costs. Similarly, the high degree of cost sharing leads to economies of scope as the provisioning of a number of different services together leads to reductions in cost [22]. A functioning intermediary value chain will therefore lower the transaction costs of the TV white spaces through increased demand and resources sharing.

5. *Congestion measurement:* Because of differences in service needs, users also vary in the extent to which they can tolerate spectrum congestion. Congestion in the TV white space context is taken to mean excess (unmet) demand for the spectrum to offer higher than current grade of service at a given price. Congestion is deemed to exist if any of the following holds [8,18]:
   - Physical congestion exists, for example, in the form of interference.
   - More users would like equivalent spectrum to run similar services.
   - There are potential users who would like to use the spectrum for something else.

An understanding of where congestion exists is necessary in order to identify areas where there is unmet demand for spectrum and therefore allocate the resources according to the service requirements. Moreover, the degree of congestion determines the ease or difficulty in accommodating new users without causing harmful interference. For example, a user requiring high quality of service (QoS) level might find the quality unacceptable whenever there is high congestion, but users requiring lower QoS levels would find the quality acceptable under the same conditions [8]. The need for correct measurement of congestion limit in relation to users' willingness to pay further articulates the need for an efficiently functioning intermediary system.

From a technological perspective, the growth of intermediaries will be facilitated by cognitive technologies, which we present in the next section.

13.3.2 Cognition as an Enabling Technology

Cognitive technology is an emerging direction in wireless connectivity to enable intelligent wireless communications and efficient spectrum usage. Cognitive technology provides a mechanism for efficient spectrum band utilization through dynamic radio resource allocation and adaptive transmission with advanced methods of spectrum acquisition. The main advantage of cognitive technology is flexibility. Through cognitive technology, the regulator can set flexible spectrum policies; the network operator can deploy cost effective services; and the end user can gain access to preferred services without worrying about the complexities underlying wireless technologies and spectrum availability. Thanks to cognitive technology [23], the regulator, network operator, and end users become equipped with the capability to perceive current network conditions and then plan, decide, and act on those conditions. Furthermore, they can learn from past experiences and use them to make future adaptations in ways that lead to consistent and holistic goals, thus optimizing the whole ecosystem.

So far, most research in spectrum sharing approaches enabled by cognitive radios (CR) has mainly concentrated on mechanisms for reclaiming *free* spectrum through opportunistic usage, while the *paid* spectrum alternative has largely been delegated to the spectrum regulating bodies, and hence remains unexplored from the automated management perspective [24]. Previous works have been dedicated to spectrum sensing, especially by considering dynamic and opportunistic spectrum access methods [25,26]. These works and others surveyed in [27] and [28] cover a wide range of problems related to secondary spectrum usage. Challenges of device coordination and the risk of the *tragedy of commons* in opportunistic spectrum access methods lead to unreliable communications and hence discourage investment. Alternatively, a market approach is a strong candidate for efficient spectrum usage in which cognitive technology can play an important role.

Market usage leads to efficient spectrum usage and provides incentives for innovation. Hence, the market model of spectrum usage is more technically and investment friendly than the "free" TV white space usage. Moreover, through cognitive technology, the flexibility in trading TV white space information and in changing the type of radio access technology will be achieved automatically. Therefore, by tapping into the flexibility of cognitive technology [29], TV white spaces can be efficiently and dynamically utilized.

Further efforts have to be directed toward integrating cognitive technology with the communication protocol stack to enable radios to function well in competitive market environments. This will require a holistic approach for efficient TV white space utilization and optimal radio resources allocation from policy, engineering, and end user perspectives. In this chapter, we assume the operation of cognitive technology in the network, end user devices, and intermediaries.

13.4 Factors Affecting TV White Spaces Sharing

This section presents the basic factors affecting the sharing of the TV white spaces. Three points will be discussed from a technical perspective, namely: (1) interference protection ratios, (2) keep out distances, and (3) general factors that influence the sharing of TV white spaces. The understanding of these principles is important for efficient exploitation of the TV white spaces.

13.4.1 Interference Protection Ratios

The field strength of radio waves is the intensity of the received electromagnetic field which will excite a receiving antenna and thereby induce a voltage at a specific frequency in order to provide an input signal to a radio receiver for radio-related applications, such as cellular, WiFi, WiMAX, broadcasting, etc. [30]. The intensity of the radio waves (received field intensity) from a transmitting station decreases with distance until reaching the sensitivity limit of the receiving device. For example in Figure 13.3 [31], broadcasts from Station *E* can be received between Points *E* and *G*. Similarly, broadcasts from Station *F* can be received between Points *F* and *H*.

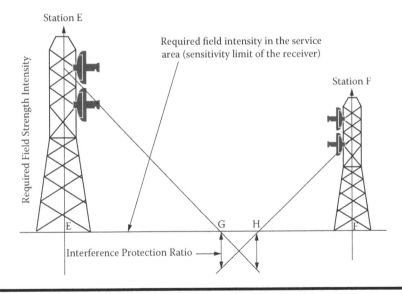

Figure 13.3 **Field intensity condition for enabling shared use of the same channel (adapted from [31]).**

Assume that both Stations *E* and *F* are transmitting in the same frequency band. Potentially, a receiving device at Point *H* could be interfered by signals from Station *E*, the same way a device at point *G* would be interfered by signals coming from Station *F*. However, if receiving devices at point *G* and *H* are sufficiently far from each other, no interfering signals will be experienced. This is because the transmitted signals have decreased to a level that is below the sensitivity level of the device, or to the level at which the receiving device is able to extract the signal successfully. When the transmitting *power* of either station is increased, the distance between Points *G* and *H* can be narrowed infinitely. The same is the case when the *distance* between the two stations using the same channel is narrowed without changing the transmission power. However, in practical cases, service providers avoid interference by imposing a protection margin called the *interference protection ratio*. The margin ensures that the field strength of the radio waves from Station *F* received at point *G* or the field strength of the radio waves from Station *E* received at point *H* is limited to a lower level than the sensitivity limit of the receiver, so that there is no interference experienced.

Table 13.1 shows the protection ratios of the Advanced Television Systems Committee (ATSC) used in the United States, the Digital Video Broadcasting–Terrestrial (DVB-T) standard used in Europe, and the Terrestrial Integrated Service Digital Broadcasting (ISDB-T) adopted in Japan.

When the desired wave to be received at Point *G* is the analog broadcast wave from the antenna of Station *E,* the digital broadcast wave from the antenna of Station *F* interferes with the desired wave at Point *G*. The technical condition for the DVB-T scenario, as specified in Table 13.1, requires that the field intensity of the radio wave from Station *F* must be 34–37 dB smaller than that from Station *E;* on the other hand, the interference protection ratio when both the desired and interference waves are digital is specified as 19 dB, which allows the field intensity of the interference wave to be 15–18 dB larger than that in the case of analog versus digital waves [31]. When the desired wave is analog, the same interference protection ratio value is required for both analog and digital interference waves [31].

The upper adjacent and lower adjacent mean the cases where the interference wave uses an upper channel and a lower channel, respectively, immediately adjacent to the channel used by the station transmitting the desired wave. When Station *E* uses Channel *N* and Station *F* uses its upper adjacent (i.e., Channel *N*+1), the required interference protection ratio for digital versus digital waves in the DVB-T scenario is −30 dB, as shown in Table 13.1. A negative value of the interference protection ratio means that the desired wave can be properly received without interference even if the field intensity of the interference wave is larger than that of the desired wave [31].

As described above, digital broadcasting has a stronger tolerance for interference than analog broadcasting, making it possible to narrow the distances between broadcast stations using the same channel and facilitating the effective use of adjacent channels. Thus, digital broadcasting can make more effective use of radio waves than analog broadcasting. This is the technical justification of the DSO

Table 13.1 Interference Protection Ratios for Frequency Planning in the ATSC 8-VSB, the DVB-T COFDM, and the ISDB-T BST-OFDM Digital Terrestrial Television Broadcasting Systems

Station E (Desired Signal)	Station F (Interfering Signal)	Interfering Frequency Band	ATSC (USA)	DVB–T (EU)	ISDB–T (Japan)
Analog	Digital	Upper adjacent	–11.95 dB	–1 to –10 dB	–5 dB
		Same channel	+34.44 dB	+34 to 37 dB	+38 dB
		Lower adjacent	–17.43 dB	–5 to –11 dB	–6 dB
Digital	Analog	Upper adjacent	–48.71 dB	–36 to –38 dB	– 37 dB
		Same channel	+1.81 dB	+4 dB	+4 dB
		Lower adjacent	–47.33 dB	–34 to –37 dB	–37 dB
	Digital	Upper adjacent	–26 dB	–30 dB	–29 dB
		Same channel	+15.27 dB	+19 dB	+19 dB
		Lower adjacent	–28 dB	–30 dB	–28 dB

Note: Data extracted from [32] and its references.

move, which reduces the frequency range by using the range of only 40 channels (geographic interleaved bands) for DTV transmission instead of the 62 channels for analogue transmission, making remaining range (cleared spectrum) available for new applications [31].

13.4.2 Keep Out Distance

In order for white space devices to use TV white spaces without interfering with incumbent users, especially when they are operating in the same channel as the TV service, there must be a *keep out distance* between the TV transmitter service contour and the white space device operating area. To elaborate the *keep out distance* concept, consider a white space system whose coverage area is located close to the co-channel TV service contour, as shown in Figure 13.4.

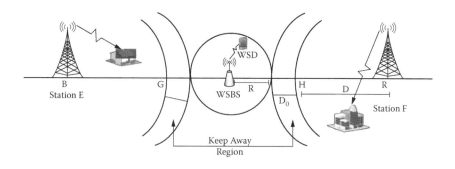

Figure 13.4 Keep away distance illustration where WSBS stands for white space base station and WSD for white space device.

Path loss in TV broadcasting band is modeled by F-curves, which are statistical propagation models derived from actual measurements and are fully specified by operating band, effective radiated power (ERP), antenna height above average terrain (HAAT), location (*l*), and time (*t*) percentage reliability in the format of F(*l, t*) [33]. F(*l, t*) denotes the propagation characteristics of the TV broadcasting in terms of spatial and temporal characteristics, and it represents the actual field strength that would exceed a certain threshold at *l*% of locations for *t*% of time.

Each TV station has a commonly regulated protected service area that is determined by its Grade B Contour for analog broadcast operations or its Noise Limited Contour (NLC) for digital broadcast operations. In general, it is assumed that operation of TV white space devices (WSDs) is not allowed co-channel within these predefined service areas [33]. The FCC has specified the protected service contour level in terms of minimum TV signal E-field strength based on F-curve propagation models. For example, the FCC has specified the use of F(50,50) curves when computing analog Grade B service contour levels, and F(50,90) curves when computing digital NLC levels. It can be noticed that digital (ATSC) modulations are afforded a higher time reliability because service degrades much more abruptly (at the signal Threshold of Visibility, or TOV) compared to analog (NTSC) modulation. Table 13.2 shows protected service contour levels as proposed by the FCC [38].

Furthermore, signal propagation in a mobile environment is complicated and is affected by deterministic path-loss variation with distance, random slow shadowing, and random fast multipath fading. For simplicity of illustration and without loss of generality, a two-ray propagation model is used to model path-loss [33–35]. According to this model, the average received signal power P_{RX} is given by

$$P_{RX} = K \frac{d_B^2}{r^4} P_{TX} \tag{13.1}$$

Table 13.2 Protected Service Contour Level

Type of TV Station	Band/Channel	Protected E-Field Level (dBu)
Full-power analog TV	Low VHF (2–6)	47
	High VHF (7–13)	56
	UHF (14–69)	64
Low-power analog TV	Low VHF (2–6)	62
	High VHF (7–13)	68
	UHF (14–69)	74
Full-power digital TV	Low VHF (2–6)	28
	High VHF (7–13)	36
	UHF (14–51)	41
Low-power digital TV	Low VHF (2–6)	43
	High VHF (7–13)	48
	UHF (14–51)	51

where K is constant, r (in meters) is the distance between the mobile and base station (BS), P_{TX} (in watts) is transmitted signal power, and d_B is a break-point (in meters) between square-law and fourth-law path loss curve and is given by $d_B = 2h_{TX}h_{RX}/\lambda_c$, where h_{TX} is the transmit antenna height, h_{RX} is the receive antenna height, and λ_c is the wavelength of the carrier frequency.

Consider the white space system in Figure 13.4. The system operates in the same band as Station E. In the Figure, D and R represent TV service contour radius and cell radius of a white space system, respectively. D_0 represents the width of the keep away region.

Transmit power of the mobile station (MS) in the cell for white space service can be controlled through accessing white space database (WSDB). WSDB stores radio station permission records including location, carrier frequency, transmit power, etc. In order to protect wireless microphones or cameras used in arena or venue, sporting, or performance events in case of electronic news gathering (ENG), their information is also filled in the database. Based on this information, WSDB can determine maximum transmit power level so as not to increase carrier to interference ratio (CIR) level of incumbent user above required interference protection ratios which is specified in terms of desired-to-undesired (D/U) signal levels [34].

Adopting the notation in [34], D_p is the signal level received from TV broadcasting transmitter with power P_T, and U_p is the signal level received from MS with power P_{TX}. We see that the WSDB controls P_{TX} so as to satisfy criterion

$$D_P / U_P > \beta \qquad (13.2)$$

where β is protection ratio. Using the propagation model of (13.1) and referring to Figure 13.4, D_p and U_p are calculated as

$$D_P = KP_T \left(\frac{\sqrt{d_{B,T}}}{D} \right)^4$$

$$\qquad (13.3)$$

$$U_P = KP_{TX} \left(\frac{\sqrt{d_{B,W}}}{D} \right)^4$$

where $d_{B,T}$ and $d_{B,W}$ are break-points corresponding to propagation channel of TV transmitter and MS, respectively, Thus, maximum transmit power is given by

$$P_{TX,\max} = \alpha P_T \left(\frac{s}{D} \right)^4 \qquad (13.4)$$

where $s = \sqrt{r^2 + (R + D + D_O)^2 - 2r(R + D_O + D)\cos\theta} - D$ and $\alpha = (1/\beta)(d_{B,T}/d_{B,W})^2$, whereas θ is the angle made by the line between the TV station white space base station (WSBS) and the line between the WSBS and the white space device (WSD).

In practice, it may be difficult for WSDB to control transmit power precisely because CIR estimated from the propagation model may be imperfect. Thus, additional margin $\Delta\beta$ should be added to the protection ratio, i.e., $D_P / U_P > \beta + \Delta\beta$. Using this additional margin, the effect of other co-channel interferences can also be evaluated indirectly [34]. This information is important in determining the availability of the TV white spaces in a given location. It will allow the sharing of different empty channels in different areas, by allowing wireless systems to emit appropriate power levels to maximize the efficiency of using the TV white spaces.

13.4.3 General Factors That Influence the Exploitation of the TV White Spaces

The interference protection ratios and the keep out distances are technical parameters to be considered when deploying a wireless system to operate in the TV white spaces. Furthermore, this information can be helpful in quantifying the availability of white spaces. The result of quantifying the available TV white spaces is information to help a secondary-market or commons spectrum user to operate without causing interference to incumbent systems. Specifically, in cases where channels are available in a given geographical location, the results help the cognitive user to determine how much transmit power, and also what kind of modulation to use, depending on the closeness or sparseness of available channels, i.e., whether contiguous or noncontiguous, respectively. Under such environments, obtaining TV white spaces is subject to several additional factors. The following are some of the other general factors to consider when estimating the number of TV white spaces in a given geographic location.

1. *Low-power CR transmission:* If the ratio of the power of CR transmitter to power of DTV station > 1, that means estimation of threshold at the DTV station can be approximated to negligible because the CR device transmits at low power. Hence, the pattern of the TV white spaces geographic area can be approximated to be bounded by the DTV station outer contour.

2. *High-power CR transmission:* In case of high-power CR devices, the range is much higher, and hence the potential for causing harmful interference to incumbent devices increases. Therefore, finding the opportunity region becomes more challenging. Specifically, this is the case when considering the fact that the constraints set by the regulators do not reflect the real amount of available TV white spaces. This is due to the observation that the protective thresholds set by the regulators are static in nature, which means that, even beyond the protective boundaries of incumbents, they are still in effect. This forces the white space user to transmit with low power levels, unnecessarily. If the protection thresholds were to be dynamic, the white space users would transmit at higher power levels whenever located beyond the protection boundaries, and hence increase the efficiency of utilizing the TV white spaces. Therefore, determination of dynamic transmit power levels for white space devices is important for successful exploitation of the TV white spaces.

3. *Adjacent channel interference:* The limit of the amount of interference to adjacent channels has to be considered. The limits set by the regulators are important. Standards should have equal or lower levels.

4. *Contiguity of channel bands:* The TV white spaces are fragmented by nature, while most technologies depend on contiguous band transmission. In case of noncontiguity of available bands, the number of applicable spectrum opportunities may suffer considerably. Otherwise, noncontiguous modulation technologies have to be considered for cognitive devices.

5. *Irregularity of DTV transmission contours:* Since the geographic interleaved spectrum opportunity patterns are not radial in shape, intelligent antennas should be encouraged in order to maximally cover the opportunity region.

These factors make the efficient usage of the TV white spaces a challenging task. Simple solutions for TV white space acquisition may not be efficient enough. Therefore, it is important to consider this problem in a holistic way, by bringing together its technological, regulatory, and business model aspects, to be able to efficiently utilize these resources. One of the key technological enablers is cognition technology, which is presented above. On the regulatory aspect, the requirements for protecting the incumbents have to be more dynamic in nature, even eliminating them whenever possible. On the business model side, there has to be a comprehensive operator-like business model for the provision of TV white space availability information. A candidate business model will be presented in Chapter 14 to illustrate this point.

13.5 Spectrum Acquisition Models

Figure 13.5 shows the evolution of methods for acquiring spectrum information. The evolution starts with autonomous sensing, where a single device collects information on spectrum availability from the surrounding RF environment. Then, the evolution develops to multiple devices cooperating in obtaining more accurate spectrum sensing information. From there, the sensing and the communication parts are separated. A specified spectrum sensor network collects spectrum information, which can be used by interested stakeholders to evaluate spectrum opportunities. The fourth stage involves a standardized dissemination of spectrum information through a cognitive pilot channel (CPC). The CPC broadcasts information on spectrum bands availability to users within its vicinity. In Chapter 14, a comprehensive business model where a spectrum broker provides intelligent spectrum context information will be presented.

Furthermore, TV white spaces acquisition can be achieved directly (actively) or indirectly (passively). The difference is in how the information on a white space is acquired or created. In the direct approach, the device uses sensing methods to discover the *white spaces*. On the other hand, indirect methods exploit past or stored information of a given band and geographical location to discover available spectrum resources. The two approaches can further be matched to a communication network perspective categorization, that is, device-centric and network-centric,

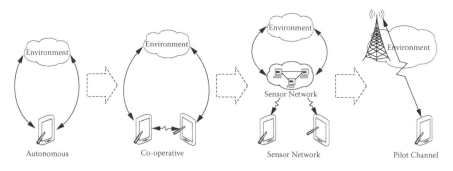

Figure 13.5   Evolution of methods for acquiring secondary spectrum availability information.

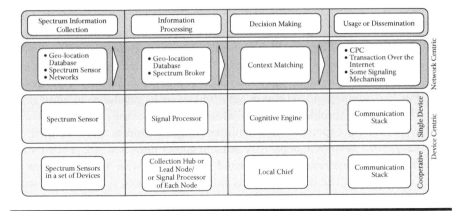

Figure 13.6 The relationship between network- and device-centric methods where the process for spectrum acquisition is the same except that it is performed by different entities, the former by the network, while the latter by the device itself.

respectively. In device-centric approaches, the device itself, by its own resources, collects spectrum availability information and uses it to support decision making on whether to use the spectrum band(s) or not—which can also be seen as a direct or an active method. In network-centric approach, spectrum availability information is centralized and made available to intended users through a predefined access mechanism—which can also be seen as an indirect or a passive method. The device-centric and network-centric approaches are illustrated in Figure 13.6.

13.5.1 Device-Centric Approaches

In device-centric approaches, the spectrum sensing ability is bundled into the cognitive device. There are two main approaches, namely (single-device) spectrum sensing and cooperative sensing. In this context, spectrum availability is mostly free of charge, except for the device's complexity and power consumption costs.

■ *Autonomous spectrum sensing:* Autonomous spectrum sensing is the conventional way for getting spectrum information. It is a well-researched approach. The main idea is to *listen-before-talk*-ing. In this approach, it is required that each individual device autonomously scan and identify unused TV channels before transmitting. This allows the establishment of peer-to-peer, home or business networks without depending on external sources for spectrum information availability. However, the main challenges for this approach include low accuracy and long delay time before acquiring reliable information.

The accuracy of autonomous sensing may be enhanced through co-operative sensing where sensing data is shared with other devices in range as discussed below. Acquisition delay may be mitigated by employing different sensing strategies, for example, the higher the rate of change of the target channel for sensing, the higher the chances of obtaining a vacant channel faster [36]. However, the presence of a sensing functionality in a cognitive radio increases the complexity and cost of the device and shortens battery life. Further, autonomous sensing is very limited in its capability to address the "hidden node" problem in the TV white spaces, hence the need for alternative ways to acquire spectrum in a noninterfering and more accurate manner.

■ *Cooperative sensing:* Cooperative sensing allows devices to share their spectrum discovery resources and avoid interference with the incumbent. It can also solve the hidden node terminal problem and reduce the effect of channel impairments, thanks to increased information reliability through cooperation. However, despite increasing sensing accuracy, the challenges of increased device complexity and cost, and shorter battery life prevalent in autonomous sensing remain. Besides, cooperative sensing requires effective device coordination mechanisms, which is in itself a challenge.

13.5.2 Network-Centric Approaches

This section presents network-centric approaches for acquiring spectrum availability information. In network-centric approaches, the spectrum information acquisition function is not embedded in the end-user device (the cognitive radio), but is located in a centralized agent. The complexity of the task is shifted from the device to the network. The information end user may pay for the service or obtain it for free. That will depend on the business model of the service provider.

■ *Use of beacons:* In this approach, a beacon acts as a source of spectrum information as well as a channel *traffic light.* Users acquire spectrum bands usage information from the beacon. The information contains *green* bands and red bands, that is, free and occupied bands, respectively. Hence, a TVWS device is not allowed to transmit in the red bands unless a permission beacon or a *green light* signal indicating that the specific channel is free to use has been given. The spectrum beacon is generated by base stations or generic service providers operating in a given locality. Nevertheless, the use of beacons consumes bandwidth and may not be effective if users cannot find the signal. This approach will be detailed when presenting the cognitive pilot channel in Chapter 14.

■ *Geo-location database:* In this approach, a database provides frequency band information to TVWS devices based on their particular geographical location. In this scheme, the TVWS device (such as a broadband Internet access point) sends its location information to a centralized (online) database to retrieve information on available white spaces in its vicinity. Location information can be obtained through a GPS receiver. This is the approach anticipated by the evolving IEEE 802.22 working group, which is developing a standard for fixed-location broadband networks using TV white spaces [26]. The use of geo-location database overcomes most of the problems associated with sensing, but leads to other issues related to obtaining the white space information.

■ *Spectrum sensor network:* In this approach, a specific spectrum sensor network (SSN) for collecting spectrum usage information is deployed. The purpose of the SSN is to provide spectrum information to operators, network planners, regulators, etc. Furthermore, the information can be capitalized [37] to populate the geo-location databases or more comprehensive entities such as spectrum brokers with more detailed or accurate spectrum usage information.

■ *Spectrum context broker:* In this approach, a spectrum context broker adds value to spectrum information provided by regulators, the geo-location database, or spectrum sensor networks by processing and reselling it as business intelligence information. User habits information such as time of appearance could be extracted based on statistical reasoning or others. This forms a paid spectrum context information provision service model. Under this model, the cognitive device becomes lighter or less complex, with better energy efficiency. Device complexity is traded off for service costs. The device is still cognitive by virtue of operating in a cognitive network where seamless spectrum handover mechanisms enable roaming in spatial, temporal, and spectral domains. Extra elaborations of a spectrum context broker will be given in Chapter 14 discussing the TV white space information provision business model.

13.6 Conclusions

The TV white spaces offer a lot of opportunities for new services provision in the wireless communications sector. The challenge is on how to acquire reliable spectrum availability information in order to operate with quality of service (QoS) guarantee in the TV white spaces with the goal of achieving economic efficiency. Spectrum usage efficiency can be achieved through market forces of supply and demand, which require functioning intermediaries to lower the cost of transactions and enable the matching of the two sides. In this chapter, elements that will enable a viable resource exchange mechanism have been reviewed.

The presence of intermediaries leads to accessibility to information on the availability of spectrum resources, hence enabling a healthy TV white space ecosystem. The technical aspects that could affect the availability of the TV white spaces and the principles behind noninterfering sharing of the resource have been presented. Moreover, a classification of methods for the acquisition was made into device-centric and network-centric approaches, where the former is based on the resources of an autonomous device, and the latter relies on information from the network to obtain white space availability information.

The intermediary elements identified in this chapter are important in that they reflect the correct value of spectrum at any particular location and time, and the potential business models lead to economic efficient utilization of the TV white spaces in an organized way in contrast to chaotic free access. In this way many challenging problems for enabling efficient TV white space usage are solved. These include: (i) the lowering of *transaction costs,* (ii) the reduction of *interference* in TV white spaces, (iii) the avoidance of *conflicts* among potential wireless service providers due to interference, and the (iv) attraction of more *investment* into TV white spaces, to mention a few.

13.7 Acknowledgments

The research leading to these results has received funding from the European Community's Seventh Framework Programme [FP7/2007-2013] under grant agreement No. 248560 [COGEU]. We thank anonymous reviewers for detailed comments, which contributed to improving the presentation of this chapter. We also thank Cláudia Barbosa for line editing the final manuscript.

References

1. Phillip Olla and Nandish V. Patel. "A Value Chain Model for Mobile Data Service Providers." *Telecommunications Policy,* 26(9), pp. 551–571, October 2002.
2. Hogan & Hartson, Analysys Mason, and DotEcon. *Exploiting the Digital Dividend—A European Approach.* Report to the European Commission, August 2009.
3. Office of Communications—UK. Digital dividend: Cognitive Access—Statement on Licence-Exempting Cognitive Devices using Interleaved Spectrum. OFCOM, September 2009.
4. J. W. Burns. "Measuring Spectrum Efficiency—The Art of Spectrum Utilisation Metrics," Proc. of IEE Conf. on Getting the Most Out of Radio Spectrum, October 25, 2002, London, UK.
5. C. Bazelon. "Licensed or Unlicensed: The Economic Considerations in Incremental Spectrum Allocations." *IEEE Communications Magazine,* 47(3), pp. 110–116, March 2009.
6. R. H. Coase. "The Federal Communications Commission." *Journal of Law and Economics,* 2, 1–40, 1959.
7. Pietro Crocioni. Is allowing trading enough? Making secondary markets in spectrum work. *Telecommunications Policy,* 33(8), pp. 451–468, 2009.
8. Mark M. Bykowsky, Mark Olson, and William W. Sharkey. "Efficiency Gains from Using a Market Approach to Spectrum Management." *Information Economics and Policy,* 22(1), pp. 73–90, 2010. Wireless Technologies.
9. Pieter Ballon and Simon Delaere. "Flexible Spectrum and Future Business Models for the Mobile Industry." *Telematics and Informatics,* 26(3), pp. 249–258, August 2009.
10. Patrick Xavier and Dimitri Ypsilanti. "Policy Issues in Spectrum Trading." *INFO,* 8, pp. 34–61, 2006.

11. S. Sengupta and M. Chatterjee. "An Economic Framework for Dynamic Spectrum Access and Service Pricing." *IEEE/ACM Transactions on Networking,* 17(4), pp. 1200–1213, August 2009.

12. K. Ben Letaief and Wei Zhang. "Cooperative Communications for Cognitive Radio Networks." *Proceedings of the IEEE,* 97(5), pp. 878–893, May 2009.

13. Junjik Bae, E. Beigman, R. Berry, M.L. Honig, Hongxia Shen, R. Vohra, and Hang Zhou. "Spectrum Markets for Wireless Services." In *DySPAN'08: IEEE Symposium on New Frontiers in Dynamic Spectrum Access Networks,* pp. 1–10, October 2008.

14. R. Berry, M. L. Honig, and R. Vohra. "Spectrum Markets: Motivation, Challenges, and Implications." *IEEE Communications Magazine,* 48(11), pp. 146–155, November 2010.

15. John W. Mayo and Scott Wallsten. "Enabling Efficient Wireless Communications: The Role of Secondary Spectrum Markets." *Information Economics and Policy,* 22(1), pp. 61–72, 2010. Wireless Technologies.

16. Federal Communications Commission. FCC 10-174: Second Memorandum Opinion and Order. Online: http://www.fcc.gov, September 2010.

17. Mark Bykowsky, Kenneth R. Carter, Mark A. Olson, and William W. Sharkey. "OSP Working Paper #41: Enhancing Spectrum's Value through Market-Informed Congestion Etiquettes." SSRN, February 2008.

18. Office of Communications—UK. SRSP: the Revised Framework for Spectrum Pricing—Proposals Following a Review of Our Policy and Practice of Settling Spectrum Fees. OFCOM, March 2010.

19. Thomas W. Hazlett and Roberto E. Muñoz. "Spectrum Allocation in Latin America: An Economic Analysis." *Information Economics and Policy,* 21(4), pp. 261–278, 2009.

20. Mark Bykowsky, Mark A. Olson, and William W. Sharkey. "OSP Working Paper #43: A Market Based Approach to Establishing Licensing Rules—Licensed versus Unlicensed Use of Spectrum." FCC, February 2008.

21. Carl Shapiro and Hal R. Varian. *Information Rules: A Strategic Guide to the Network Economy.* Boston, MA: Harvard Business School Press, 1998.

22. Joseph P. Guiltinan. "The Price Bundling of Services: A Normative Framework." *The Journal of Marketing,* 51, pp. 74–85, April 1987.

23. R. W. Thomas, L. A. DaSilva, and A. B. MacKenzie. "Cognitive Networks." In *DySPAN'05: IEEE International Symposium on New Frontiers in Dynamic Spectrum Access Networks,* pp. 352–360, November 2005.

24. Joseph Wynn Mwangoka. Resource Allocation in Cognitive Radios. PhD thesis, Tsinghua University, Beijing, China, January 2009.

25. T. Weiss and F. Jondral. "Spectrum Pooling: An Innovative Strategy for the Enhancement of Spectrum Efficiency." *IEEE Communications Magazine,* 42, pp. 8–14, March 2004.

26. C. Cordeiro, K. Challapali, D. Birru, and S. Shankar. "IEEE 802.22: The First Worldwide Wireless Standard Based on Cognitive Radios." In *DySPAN'05: IEEE International Symposium on New Frontiers in Dynamic Spectrum Access Networks,* pp. 328–337, November 2005.

27. S. Haykin. "Cognitive Radio: Brain-Empowered Wireless Communications." *IEEE Journal on Selected Areas in Communications,* 23(2), pp. 201–220, February 2005.

28. Ian F. Akyildiz, Won-Yeol Lee, Mehmet C. Vuran, and Shantidev Mohanty. "NeXt Generation/Dynamic Spectrum Access/Cognitive Radio Wireless Networks: A Survey." *Computer Networks,* 50(13), pp. 2127–2159, September 2006.

29. J. M. Peha. "Sharing Spectrum through Spectrum Policy Reform and Cognitive Radio." *Proceedings of the IEEE,* 97(4), pp. 708–719, April 2009.

30. Wikipedia. Field Strength, March 2010.

31. Hajime Yamada. "Developments in Television Band Frequency Sharing Technology." *Science & Technology Trends Quarterly Review,* No. 31, April 2009.

32. Yiyan Wu, E. Pliszka, B. Caron, P. Bouchard, and G. Chouinard. "Comparison of Terrestrial DTV Transmission Systems: The ATSC 8-VSB, the DVB-T COFDM, and the ISDB-T BST-OFDM." *IEEE Transactions on Broadcasting,* 46(2), pp. 101–113, June 2000.

33. D. Gurney, G. Buchwald, L. Ecklund, S. L. Kuffner, and J. Grosspietsch. "Geo-Location Database Techniques for Incumbent Protection in the TV White Space." In *DySPAN'08: IEEE Symposium on New Frontiers in Dynamic Spectrum Access Networks,* pp. 1–9, October 2008.

34. Sang Yun Lee, Sung Hee An, and Yang Moon Yoon. "Area Spectrum Efficiency of TV White Space Wireless System with Transmit Power Control." In *ICACT'10: The 12th International Conference on Advanced Communication Technology,* vol. 2, pp. 1061–1066, 2010.

35. Andrea Goldsmith. *Wireless Communications.* Cambridge University Press, 2005.

36. Joseph Wynn Mwangoka, Jun-Feng Jiang, and Zhi-Gang Cao. "Variable Persistence Sensing Schemes for Opportunistic Spectrum Access." *Chinese Journal of Electronics,* 4th Quarter, pp. 737–742, 2008.

37. M. B. H. Weiss, S. Delaere, and W. H. Lehr. "Sensing as a Service: An Exploration into Practical Implementations of DSA." In *DySPAN'10: IEEE International Symposium on New Frontiers in Dynamic Spectrum Access Networks,* pp. 1–8, April 2010.

38. FCC, *In the Matter of Unlicensed Operation in the TV Broadcast Bands,* ET Docket No. 04-186, Notice of Proposed Rulemaking, FCC OET, May 2004.

Chapter 14

Elements of Efficient TV White Space Allocation Part II: Business Models

Joseph W. Mwangoka, Paulo Marques, and Jonathan Rodriguez

Contents

14.1 Introduction

A business model is a planning tool that is essential for every company, such as a broadband Internet wireless access provider. It describes how a company creates value to customers, owners, and other stakeholders. It is also important in case the services are provided in cross-company collaboration in complex value nets, like in the TV white space scenario where different entities are involved in exploiting the vacant spectrum resource. In this chapter, we have followed the component-based business modeling approach as used in [1] for conveying the economic potential in the TV white spaces. In TV white space information provision and exploitation business models, the key question is related to how the firm will make profit from the business. In this chapter, two basic business models for the economic efficient functioning of the TV white spaces will be discussed. They are:

- **B1** *TV White Space Information Provision:* This model takes into account the possibility to provide TV white space information to interested users. The users, however, are not necessarily willing to pay for the information obtained. Thus the success of this kind of business model is related to the solutions regarding the earnings logic for that company which acts as a TV white space information service provider. All the other business model components, such as the role needed for organizing the revenue flow, have to be done accordingly. The main users or customers of this model are the wireless service providers in the TV white spaces presented in the **B2** business model.
- **B2** *Wireless Services Provision in the TV White Spaces:* This model describes possible wireless businesses to exploit the TV white spaces. They are the main customers of the TV white space information provision business model above (**B1**). The key in this modeling task is to adapt present wireless services to the TV white spaces and consider what kind of technological challenges might the model's service promises, user segments, processes, etc. encounter.

Each role player portrays a different aspect of value in the business model [1]. The *end user* sees value as the utility or benefit gained by using the product or service offered. The *customer,* one who actually pays for the service, is the source of revenue and depicts value in monetary terms. The *value proposition* answers the question of: what customer problem or need will the business attempt to address? Value proposition includes a description of all relevant value elements and drivers identified as important in a TV white space-related business model. Moreover, there are roles without which the proposition could not be done or would somehow change—these are termed as *value creators.* The roles which are needed for the proposition of the service but do not change the value proposition are grouped as *costs* for the service. An interested reader is referred to [16] for further details on the modeling approach.

In the next two sections each business model will be dealt with separately. The roles needed for the business model will be given by describing the value framework interrelations showing the flow of services and payment between them. To that end, the chapter is organized as follows: Section 14.2 presents a TV white space information provision business model, and Section 14.3 presents a wireless services provision business model in the TV white spaces. The two business models have the potential to make the usage of the TV white space functional and efficient. Finally, Section 14.4 concludes the chapter and spells out future research.

14.2 TV White Spaces Information Provision

In order to establish a reliable communication link, a device operating in the TV white spaces must be able to get information on their availability. The spectrum access discovery is utilized by the user in order to discover the most optimal band relative to its utility. One of the important challenges on establishing a link between the devices or between the device and the spectrum information repository is on how to obtain the spectrum availability information. There are a number of approaches that can be used to acquire information on the availability of the TV white spaces. In our context, acquisition is not necessarily restricted to obtaining the number of available white spaces, but can be extended to gathering information as to when (time), how (protocol), for how much (cost), etc. they can be obtained.

The spectrum context broker business model is to provide reliable information on the TV white spaces to would-be wireless access service providers. To understand this model, it is important to grasp the importance of TV white space information and its availability. Generally, information is costly to produce but cheap to reproduce [2]. For example, softwares that cost hundreds of thousands of dollars to produce can be copied to a flash-disk by a click of a mouse. Similarly, a multimillion-dollar sports event—an Olympics game or a soccer match—can be enjoyed by an additional fan with almost no extra cost. The same applies for TV white space availability information. However, producing (and using) this particular type of information is faced with the challenges underlying the dynamics of radio frequency (RF) environments in spatial and temporal domains. Therefore, designing a model that will ensure consistent reliable information on the TV white space availability is key to success, because it will attract many customers to use this information, hence generating the necessary revenue. This will effectively maximize the economic efficiency of the TV white spaces. More details on the model are given in the next section.

14.2.1 The Business Model

Figure 14.1 shows the roles and relationships in the spectrum context provision business model. The main components of the model are: the TV white space users who benefit from the model; the source of revenue, which is through advertisement or direct payment by the user; the value proposition made by spectrum context broker; the value creators which provide spectrum context; the payment system provider;

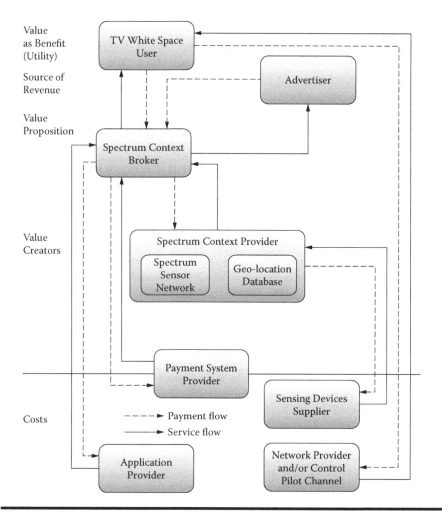

Figure 14.1   Roles and relationships in the spectrum context provision business model (Adapted from [1]).

and the costs which come from the application provider, the network provider, and the spectrum sensing devices supplier (or manufacturer). The value proposition in **B1** is given by the spectrum context broker. The general process of the function of the model is: (1) information collection from geo-location databases and (or) spectrum sensor networks by the spectrum context provider; (2) information processing and repository by the context broker; and (3) information dissemination through the control pilot channel or on-demand retrieval through the Internet. The following roles and related activities have been identified, as shown in Figure 14.1, to develop the set of capabilities necessary to build a value proposal that delivers benefit and reliability to TV white space availability information customers.

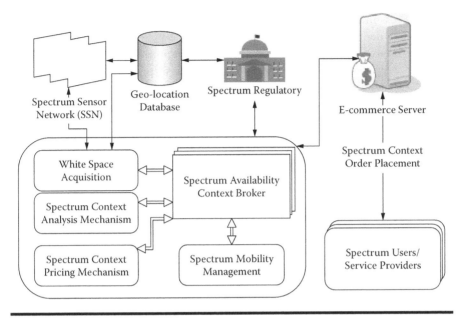

Figure 14.2   Spectrum availability context broker for TV white spaces.

- TV white space user: potentially any spectrum consumer interested in offering wireless services through the TV white spaces. In this chapter we consider the service provider of model **B2** (presented in Section 14.2) to be the main customer.
- Spectrum context broker: puts things together for the TV white space user; packages spectrum context information; prioritizes white space access; matches context to the user profiles and delivers them to the users. Obviously, the spectrum price has to be cheaper or lower than that of cleared bands, and equally provide reliability in order to guarantee quality of service (QoS) for the client.

    As shown in Figure 14.2, the main components of a spectrum context broker are white space acquisition, spectrum context analysis, spectrum context pricing, and spectrum mobility management. The description of each functional block is given below.

    - *White space acquisition:* The collection of raw spectral emission data for a given frequency band, location, and time is accomplished by the *white space acquisition module.* This module may collect spectral emission data from a geo-location database, or from a specialized spectrum sensor network. This could be partly done by employing approaches presented in the acquisition models chapter. This is the first stage for spectrum availability context provision.
    - *Spectrum context analysis:* Once the spectral emission data is obtained, it is then analyzed based on preset criteria. The criteria may be statistical

reasoning to extract spectral emitter's habit, type of service offered, traffic density, etc. Moreover, such information can be processed to obtain opportunities maps to indicate the geographical distribution of the TV white spaces. The information is then classified into different categories depending on reliability and intended users, as well as checked for compliance with regulatory policies.

− *Spectrum context pricing:* Revenue collection is an important function of the context broker which is realized by the context pricing module. Context price is set based on the match between quality of information and user willingness to pay.

− *Spectrum mobility management:* One of the key issues in heterogeneous TV white space environments is spectrum handover in cases where the end user moves from one geographic area to another. The spectrum mobility management module of the broker is to ensure seamless connection by reserving spectrum resources along the path of context changes.

  Moreover, mobility management could be a source of revenue in itself. Assume the following scenario: a cognitive device using the Broker services to acquire spectrum is moving from one location to another while streaming favorite rich media. Upon reaching the white space boundaries, the mobility management module proactively reserves spectrum bands for the service and informs the cognitive network service provider (as well as the end user). The service provider then triggers spectrum handover procedures while maintaining the streaming services quality. In this scenario, the Broker may act as a connectivity monitor to recommend the availability of TV white spaces resources for its client whenever there are chances of losing connection due to spectrum unavailability [3], hence expanding its service portfolio to gain more revenue.

■ Advertiser: invests directly or indirectly in the recommender service to promote its products or services. The advertiser could be reaching the TV white space users who search for spectrum availability information from the Internet.

■ Spectrum context provider: any provider of raw white space availability information, online or offline-based, directly or indirectly, free or paid, etc. Specifically, spectrum context information could originate from a spectrum sensor network or geo-location database. The use of geo-location databases for secondary spectrum usage is widely received by regulators, mainly because it sidesteps the shortcoming of autonomous spectrum sensing. This could come in the form of:

  − *Free database:* This could be an Internet based model where spectrum users search for spectrum availability in their locality for free. The database owner may generate revenue through advertisement. However, monitoring spectrum usage may be hard, and there is no guarantee that there will be no interference from similar spectrum users.

- *Paid database:* This may be an alternative model where spectrum availability is sold to users, preferably after being processed in the context broker.
- *National database:* This could be a government agency-sponsored spectrum database used for future planning and research on spectrum policy changes, protocols for accessing the database, information collection, frequency access, and updating in a reactive or proactive way.

  The information collected from the geo-location database is acquired by the spectrum context broker, which processes it and provides intelligent information as well as context mobility management services to end TV white space users.

- Payment system provider: provides the facility that, from the spectrum context broker side, allows bill delivery and examination (either repeatedly or only once) from the TV white space users to pay them. In the TV white space context information provision service, the payment system provider plays an important role because the TV white space user pays every time it wants to acquire reliable TV white space availability information, and hence its functioning should be reliable to avoid being a single point of failure for the model.
- Sensing devices supplier: the technology company focused on designing and implementing the core components related to spectrum, in this context TV white space usage monitoring. Given the advanced nature of the business, it may be a highly specialized entity with strong innovation and research orientation, with ability to provide spectrum analytical solutions to both regulators, network planners and potential investors.
- Network provider and/or control pilot channel: offers connectivity and accessibility of the TV white space context information to potential customers through the Internet and/or specialized broadcast channels.
- Application provider: provides the application necessary to offer the service.

The TV white space availability information provision service could bring value to its customers especially to wireless services operators demanding reliability in the spectrum resource to offer QoS guaranteed services. Moreover, it facilitates the economic efficiency of the TV white spaces in a number of ways, such as:

- Provides customized TV white space provision services such as long-term lease, a scheduled lease, and a short-term lease, or spot markets.
- Ensures quality of information (QoI) of the available TV white spaces. This further reduces the investment risk and ensures noninterfering TV white space usage.
- Reduces the cost of devices by handling the spectrum sensing on behalf of the end user.
- Increases the energy efficiency of mobile devices by providing an option to switch off the sensing mode and transfer the complexity to the core network.

■ Saves investor's time in acquiring reliable spectrum resources information.
■ Prioritizes allocation of spectrum during emergency situations. In this case, the centralized database and the broker give higher priority to reserve spectrum resources for emergency by temporarily rescinding nonemergency utilization of the TV white spaces in the specific areas of need, thus giving public safety the highest priority.

The model provides a basic idea on the provision of TV white space availability information. In the following section, we extend this discussion by presenting cognitive pilot channel (CPC), which is a popular candidate for providing TV white space availability information to potential users. The section presents some of the motivations for the CPC, its mode of operation, as well as the underlying challenges.

14.2.2 Cognitive Pilot Channel

The existence of a special channel to provide information on available spectrum bands, network services, regulatory policies, etc., could enable effective exploitation of the TV white spaces. In general dynamic spectrum usage, the concept of the cognitive pilot channel (CPC)—*a channel which conveys the elements of necessary information to facilitate the operation of cognitive radio systems*—has been proposed [4,5]. The CPC provides information on spectrum availability and on which radio accesses can be obtained within the locality of a given device. Depending on the source of information, the CPC may also contain information on the wireless access service providers, radio access technology types, regulatory policies, etc. Figure 14.3 shows the basic principle of the CPC, where there is a direct link between the information transmitting station and the TV white space user.

The white space devices make use of the information transmitted in the CPC in order to orient themselves with the immediate environment by selecting an appropriate radio access technology (RAT), by downloading software modules for reconfigurability purposes in case of need, and by obtaining the TV white spaces

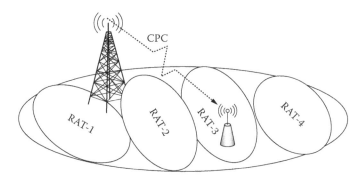

Figure 14.3   The CPC in a Heterogeneous RAT Environment.

to establish an ad hoc network, wireless Internet access, or similar procedures. In particular the CPC supports:

1. The start-up phase in an environment where the white space terminal does not yet know the available RATs and corresponding used frequencies.
2. The selection of the proper network depending on the specific conditions (such as desired services, RAT availability, interference conditions, service cost, etc.). This provides support to joint radio resource management, hence enabling a more efficient and effective use of the radio resources.
3. The reconfiguration of the white space device by providing information which allows the terminal to identify the most convenient RAT to operate with and to download (or activate), in case needed, the necessary software modules to reconfigure the terminal capabilities.
4. The context awareness of the white space device by helping the terminal identify the specific frequencies, operators, and access technologies in a given region without the need to perform long time- and battery-consuming spectrum scanning procedures.
5. The dynamic changes for the white space service provider in the network deployment by informing the terminals of the availability of new RATs/ frequencies.
6. It helps the spectrum regulator to improve the spectrum utilization by enabling the exploitation of available TV white spaces in a given area.

The basic CPC operation is organized into two main phases. One phase is the *start-up*, and the other one is the *ongoing* phase. When a white space device is turned "ON," its first task is to determine its geographical location by means of some positioning methods. After determining its location, the device searches for the CPC. If it detects the CPC, it extracts information corresponding to spectrum and services available in its vicinity. Compared to devices that acquire spectrum through sensing, the device which utilizes the CPC does not need to scan the whole spectrum for the availability of the TV white spaces, because it obtains that information from the CPC.

During the operation stage, that is, when the device is communicating, there is a possibility for the availability of spectrum resources to change. The device needs to obtain an updated version of spectrum availability information from time to time. The *ongoing* phase of the CPC serves to make the communication link more reliable by updating its information in periodic intervals.

Moreover, there are different deployment approaches for CPC, namely (1) out-band CPC, (2) in-band CPC, and (3) combined CPC, as shown in Figure 14.4. These CPC operation modes depend on the physical resource being used. The details are as follows:

■ *Out-band CPC:* In the *out-band* mode, the CPC is conceived as a radio channel outside the component radio access technologies; the CPC either uses a

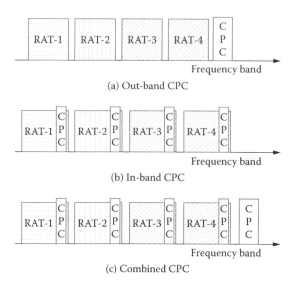

Figure 14.4 Different possibilities for the implementation of the CPC system.

new radio interface, or alternatively uses an adaptation of legacy technology with appropriate characteristics. The main objective of the out-band CPC is to initiate the start-up phase by providing the radio environment information. The concept of out-band CPC implementation is illustrated in Figure 14.4(a). The key advantage of this approach is that any CPC compliant terminal can retrieve the information of the CPC no matter what access technology it operates in. On the other hand, it is necessary to have worldwide common frequency for out-band CPC implementation or to use an adaptation of legacy technology with appropriate characteristics.

■ *In-band CPC:* In the *in-band* mode, the CPC is conceived as a logical channel within the technologies of the heterogeneous radio environment, and hence the information is transmitted by using specific channels of existing radio access technologies. But in this case, the scanning procedure is still involved to acquire knowledge about the RAT where CPC is located. In Figure 14.4(b), in-band CPC implementation is illustrated.

■ *Combined CPC:* In the *combined* mode, both the *out-band* and the *in-band* modes are implemented, but in an alternative manner. The *out-band* CPC is used to provide the *start-up* information, and afterward the *in-band* CPC is used to support the *ongoing* phase. In the *start-up* phase, the mobile terminal listens to the *out-band* CPC in order to obtain the information about basic parameters of the radio environments (such as available networks and their type of services and operating frequency band), then the terminal selects and connects to a network. Once the connection is established, the terminal stops listening to the *out-band* CPC and starts to receive the *in-band* CPC within the registered network, where much more detailed context information and policies can be retrieved. The combined CPC is illustrated in Figure 14.4(c).

Furthermore, the implementation of CPC is split on the basis of how the CPC information is delivered, that is, broadcast CPC and on-demand CPC. In broadcast CPC implementation, all the CPC information is broadcast continuously through a "down-link" broadcast CPC channel, so that it can be received at any time by the white space terminal within the coverage area.

In case of the on-demand CPC, the information is transmitted only when it is required. Its implementation is performed in three different stages: the first is Request; the second is Acknowledgment; and the third is Transmission. In the Request stage, the white space terminal sends a request packet enclosing its geographical location and other preference data by a CPC "uplink" channel. The CPC information provider's base station will Acknowledge receiving the request by sending an acknowledge packet with the identifier of the white space whose request has been received. In the last stage, the CPC information is transmitted through a "down-link" on-demand CPC. The implementation of on-demand CPC is more efficient in terms of power and bandwidth occupation compared to the broadcast CPC. Figure 14.5 gives an illustration of the on-demand CPC approach.

14.2.3 Challenges

There are several challenges in realizing network-centric spectrum context information. These include:

■ Negotiation mechanisms and protocols between context broker and end user have to be developed.

■ In order to ensure the quality of information, especially in case of spectrum handover, there will be a need for an intelligent spectrum sensing network, with enough coverage to satisfy the user. This can be done by miniaturizing conventional spectrum analyzers and networking them to enable cooperative white space discovery or using other mechanisms for coverage, reliability, and power saving. In this way additional accuracy to the information from the national TV white space database will be provided—because the coverage of DTV stations is not radial.

■ Development of learning mechanisms to enable the prediction of spectrum usage habits by the incumbents (DTV, PMSE) as well as other secondary users.

■ Conflict resolution mechanism in case of interference caused by users of spectrum context information from rival context brokers. Otherwise, coexistence and cooperation mechanisms would be more advantageous for an interference-free spectrum usage space.

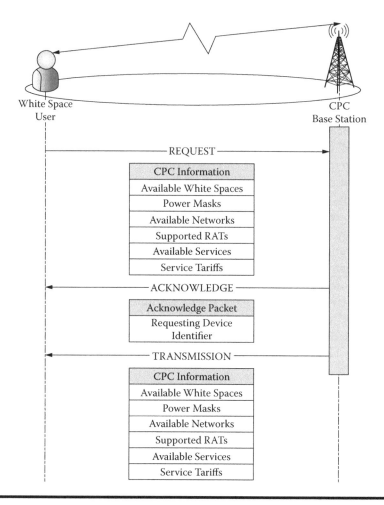

Figure 14.5 The three stages of on-demand CPC with examples of the content of the packets exchanged between the white space user and the CPC base station.

14.3 Wireless Services Provision in the TV White Spaces

The TV white spaces have created an opportunity for technological development to drive the innovation of new wireless services. The opportunity generates incentives for changes in the current business models. For example, the exploitation of the TV white spaces to deliver longer range WiFi-like Internet connectivity services may provide a more substantial value proposition to the customers than operations in shorter-range Industrial, Scientific, and Medical (ISM) band. Moreover, the ability to provide long-range wireless access technology has more market potential in other areas, including the following [7–10]:

1. Business
   - Provide coverage to a large office or business complex or campus
   - Establish point-to-point link between large skyscrapers or other office buildings
   - Bring Internet to remote construction sites or research labs
   - Provide cheap Internet in road, sea, and air transport systems
   - Provide community or campus ubiquitous services such as broadcasts, alerts, security, etc.
2. Cellular networks
   - Increase the capacity of present cellular systems
   - Increase the coverage area of cellular systems
   - Reduce the costs for deployment and maintenance of new services
3. Residential systems
   - Bring Internet to a place where regular cable/DSL cannot reach such as remote areas, etc.
   - Bring Internet to a boat in a lake or sea
   - Share a residential area broadband Internet network
   - Provide wireless networking for home entertainment systems
4. Public safety
   - Bring quick, agile, and cheap Internet connections in areas devastated by disaster
   - Provide additional support in coordination of relief efforts
5. Hospital communications
   - Access to patient records
   - A rapid voice over broadband wireless communication system

Provision of access to wireless services in these segments depends on the availability of TV white spaces resources in a given area. Changes in the availability of the TV white spaces may impact the type of services that can be offered. For example, delay-sensitive real-time applications will require stable and guaranteed availability of the resource; otherwise the resulting poor quality of experience will lead to dissatisfied customers, and hence low revenue performance. Therefore, the level of guarantee on the spectrum resource availability will dictate the type of application to invest in. Below we give the business model in detail.

14.3.1 The Business Model

Figure 14.6 shows the roles in the provision of wireless services through the TV white spaces business model (**B2**). This model could be suitable for the extension of existing wireless services to the TV white spaces, such as Wi-Fi and long-term evolution (LTE) systems; or the creation of new services, such as home entertainment systems and community networks. They provide users with cheap Internet connectivity, capacity increase for current systems, or ad hoc networking through

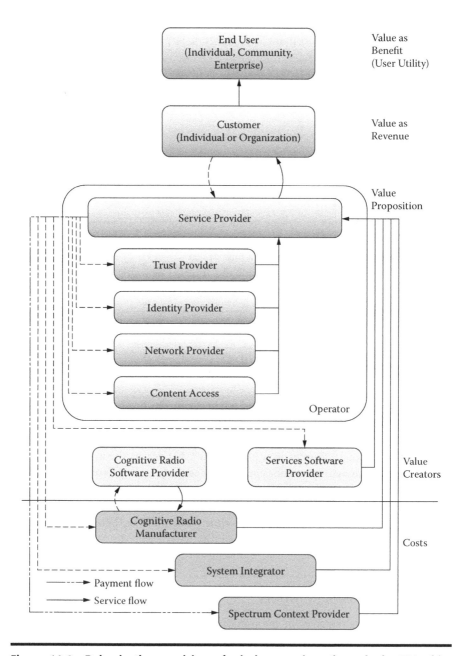

Figure 14.6 Roles in the provision of wireless services through the TV white spaces business model (Adapted from [1]).

the TV white spaces. Obtaining reliable information on the TV white space availability and deploying the network without causing interference to the incumbents, such as DTV and wireless microphone systems, are the key technical challenges. The business model presented in Section 14.1 (**B1**) ensures that the information provided to their customer will not lead to an interference situation.

The value proposition in (**B2**) is given by the service provider. The following role and related activities have been identified, as shown in Figure 14.6, to develop the set of capabilities necessary to build a value proposal that delivers benefits and high value to wireless services customers in the TV white spaces:

- End user: potentially any individual or enterprise with an enabled device.
- Customer: End users if they are the ones paying for using the service.
- Operator: Provide connectivity services including:
  - Trust services
  - Identity services
  - Network connectivity
  - Content access, etc.
- Cognitive radio software provider: provides the software solutions that run in the end user devices.
- Service software provider: provides solutions for the operator.
- Cognitive radio manufacturer.
- System integrator.
- Spectrum context provider (**B1**): provides reliable information on the TV white spaces within the vicinity of the operator's service area.

The model could be for a service provider, such as a mobile operator using the TV white spaces to gain more capacity, or it could be an enterprise in-house wireless services provision system, etc. The service provider in this model could bring value to its customers in wireless services provision through a diversity of applications. These wireless services segments over the TV white spaces could offer applications such as:

- Emails and Web browsing: these are based on best effort, and there are no delay constraints for these applications. These could be easily supported in the TV white spaces.
- Video surveillance: this could be for security purposes in public, business, and residential areas, or for scientific discovery, or monitoring industrial processes, etc. This application requires high bandwidth for video streaming, but no delay constraint due to its buffering capabilities.
- Real-time voice communications: This is an attractive application citing VoIP, for example. With guaranteed TV white space accessibility, the delay brought by spectrum acquisition is alleviated.

- Broadcast services: this could be another potential service based on mobile broadband applications in the TV white spaces. They could be offered through the IP protocol and hence subject to best effort services constraints.
- On-demand streaming media: handheld devices operating in the TV white spaces should be able to receive on-demand video and audio streams. This is a viable model for entertainment content providers, but could also be for other purposes such as education.
- Real-time gaming: This application has real-time delay constraints. Its functioning will depend on the reliability of the information provided on the TV white spaces.
- Access to critical information during disasters or for public safety is an important application. These could be street maps, driving directions or other useful information such as traffic reports, weather forecasts, etc.

The model provides a basic idea on the wireless services provisioning business model in the TV white spaces. In the following section, we extend this discussion by presenting WiFi, which is a popular candidate for operation in the TV white spaces. The section presents some of the motivations for extending WiFi in the TV white spaces, as well as the underlying challenges.

14.3.2 WiFi Extension to the TV White Spaces

Conventional WiFi is operating in the unlicensed 2.4 GHz ISM band. Although its range is short, it has been a very successful technology in delivering cost-effective wireless Internet access in homes, offices, and public areas. The enormous success of WiFi has inspired many researchers and engineers to try to extend its range, to cover a wider area, reaching more users, while increasing revenue to the investor [11–13]. Different ways have been or are being used to stretch the reach of current WiFi systems, some of which include multiple input, multiple output (MIMO) technologies, power increase or receiver sensitivity boosting, using high gain antenna, etc.

Nevertheless, increasing the range of WiFi in the 2.4 GHz ISM band is faced with numerous setbacks. First, using WiFi over long distances requires line-of-sight between the endpoints; therefore its performance is subject to the negative impact of surrounding environment on microwave signals. Second, the ISM band is crowded with a lot of devices which cause interference to WiFi signals. The first limitation can be overcome by taking advantage of terrain elevation, or by using towers to overcome obstacles and provide Fresnel zone clearance. The second limitation is less effective in sparsely populated areas, and may be mitigated by shifting to the less crowded 5 GHz band. Table 14.1 gives the summary of the current methods and solutions used to extend the range of conventional WiFi networks. Nonetheless, these approaches still have to live with the poor radio frequency (RF) propagation characteristics of GHz bands. Therefore, there is still a need to address this problem from a different perspective.

Table 14.1   Challenges and Solutions of Conventional Methods for Extending the Range of WiFi

Challenge	Conventional Solution
Requirement for line of sight between endpoints	Take advantage of terrain elevation
	Avoid areas with obstacles
	Use high towers to provide Fresnel clearance
Vulnerability to interference in the unlicensed band	Operate in rural areas
	Migrate to less-crowded 5 GHz band
Power budget limitation	Use high-gain directional antennas
Timing limitation	Modify the medium access control (MAC) mechanism
Cost and usability	Use cheap antennas
	Employ affordable and easy to use technology

Having seen the success of the traditional WiFi and some of the setbacks, it can be seen that there is still a need for a simpler and cheaper way to stretch the range of wireless Internet connectivity. Compared to the higher frequency ISM bands, the lower frequency TV white spaces have inherent RF propagation properties, such as longer range, better penetration, lower interference, and the possibility to build small antennas for handheld devices. These characteristics make them extremely desirable for long-range high-speed wireless Internet connectivity at a cheaper cost.

Long-range WiFi* over the TV white spaces use case is motivated by the inherent RF propagation characteristics presented above. Moreover, WiFi is a mature, well-understood technology that is inexpensive and easily available. In fact, there are several wireless card vendors considering pushing some version of WiFi to the IEEE standards body for white space networking [15]. So far, there are IEEE P802.11af standardization initiatives to amend the original standard for operation in the TV white spaces. Therefore, high-speed and long-range WiFi-like systems can be built in the TV white spaces by using off-the-shelf components, and coupling them with cognitive capabilities to allow a smarter way to access the spectrum while causing negligible interference to incumbent devices.

---

* In the IEEE TV white space standardization project P802.11af, long-range WiFi is called White-Fi; whereas in the FCC 10-174 second memorandum opinion and order released in September 2010, it is called Super WiFi [14].

14.3.3 Challenges

The TV white spaces are located in the UHF/VHF segment of the radio spectrum. They provide an excellent balance between RF characteristics and antenna size. These features in the TV white spaces open up a lot of opportunities for the provision of new wireless networks services. These could range from the extension of current cellular services to the deployment of new systems such as home and community networks. However, these attractive features are overshadowed by technical, regulatory, and business model challenges. Specifically, several challenges have to be addressed in accomplishing the business model of providing wireless services over the TV white spaces. These include:

- *Device complexity:* In case regulators require mandatory autonomous sensing for protecting the wireless microphones, that means adding sensing capability in the devices, which in turn leads to increased device complexity, cost, and power expenditure. The way around this is for regulators to dedicate specific bands for the operation of the wireless microphones. Otherwise, sensing thresholds should be reduced or eliminated as some regulators are considering. This will be viable if all wireless microphones operations are required to be registered in the geo-location database, and hence their areas of operation can be easily tracked to avoid unintended interference.

  If the information provision business model is functional and is able to provide reliable TV white space availability information, then the need for autonomous sensing could be completely eliminated. This will lead to cheaper and energy efficient devices, hence convenience to the user and profitability to the solution provider.

- *Mobility management:* Seamless connectivity across heterogeneous cognitive radio networks in TV white spaces is an important challenge to achieve ubiquity. The mobility solution should support both vertical and horizontal handovers. Moreover, in order to be able to support most real-time applications, a handover latency of less than 50 ms should be supported. This could be achieved through proactive context-aware enabled mobility management [3].

- *Incumbent protection requirements:* The requirements to protect incumbents in both the co-channel and adjacent channels by the regulators may limit the system performance expectation. These requirements to protect incumbents, like DVB-T and wireless microphones, may also lead to expensive devices (such as sensing devices and the use of very selective filters) and hence unfeasible business models.

- *Fragmentation of the TV white spaces:* The white spaces are fragmented in nature; that is, they are not available in a contiguous manner. In order to efficiently exploit them, transmission technologies with the ability to pool fragmented bands have to be devised.

14.4 Conclusion and Future Challenges

In order to make the usage of the TV white spaces economically functional, this chapter presents two business models. The first is the TV white space information provision business model, where the spectrum context broker sells or provides reliable spectrum context to potential TV white space users. The model combines all the roles related to spectrum context collection, processing, and dissemination. The second is wireless service provision in the TV white spaces, where an operator acquires reliable spectrum context information from the broker to provide wireless services. The model, through technological reconfiguration, exploits the TV white spaces, creating value to customers, and maximizes the economic efficiency of the spectrum resource. In fact, the two models could be integrated into one model by one operator or a consortium of TV white space users for maximal utilization of interrelated resources.

However, several challenges remain for the efficient running of TV white space ecosystem for the future. These include, for example: (i) negotiation mechanisms and protocols between TV white space context broker and end user have to be developed for efficiency in the transactions; otherwise it may become a single point of failure for the ecosystem; (ii) development of mechanisms to ensure the quality of information for TV white spaces availability, which could be achieved through the (iii) development of learning mechanisms to enable the prediction of TV white space usage habits and hence maximize the impact of matching the supply and demand sides; (iv) conflict resolution mechanisms in case of interference caused by users of spectrum context information from rival context brokers; (v) development of coexistence and cooperation mechanisms, which would be more advantageous for the TV white space customers through the sharing of resources for collecting, processing, and retrieving of the TV white space usage information. When these challenges are addressed through the collaboration of regulators, engineers, investors, and end users, this will not only lead to efficient usage of the TV white spaces, but also provide a platform for even higher economic value for society.

14.5 Acknowledgments

The research leading to these results has received funding from the European Community's Seventh Framework Programme [FP7/2007-2013] under grant agreement No. 248560 [COGEU]. We thank the anonymous reviewers for their detailed comments. We also acknowledge Cláudia Barbosa for line editing the final manuscript.

References

1. Ulla Killström, Heli Virola, Luca Galli, Olli Immonen, Olli Pitkänen, and Björn Kijl. IST-2004-511607 MobiLife D10(D1.5): Business models for new mobile applications and services. January 2006.
2. Carl Shapiro and Hal R. Varian. *Information Rules: A Strategic Guide to the Network Economy*. Boston, MA: Harvard Business School Press, 1998.
3. Joseph W. Mwangoka, Paulo Marques, and Jonathan Rodriguez. "Cognitive Mobility Management in Heterogeneous Networks." In *MobiWAC'10: Proceedings of the 8th ACM International Symposium on Mobility Management and Wireless Access*, pp. 37–44, October 2010.
4. Pieter Ballon and Simon Delaere. "Flexible Spectrum and Future Business Models for the Mobile Industry." *Telematics and Informatics*, 26(3), pp. 249–258, August 2009.
5. Zoran Damljanović. "Mobility Management Strategies in Heterogeneous Cognitive Radio Networks." *Journal of Network and Systems Management*, 18(1), pp. 4–22, 2010.
6. Jordi Perez-Romero, Oriol Sallent, R. Agusti, and L. Giupponi. "A Novel on-Demand Cognitive Pilot Channel Enabling Dynamic Spectrum Allocation." In *DySPAN'07: IEEE International Symposium on New Frontiers in Dynamic Spectrum Access Networks*, pp. 46–54, April 2007.
7. Hajime Yamada. "Developments in Television Band Frequency Sharing Technology." *Science & Technology Trends Quarterly Review*, No. 31, April 2009.
8. R. Ahuja, R. Corke, and A. Bok. Cognitive radio system using IEEE 802.11a over UHF TVWS. In *DySPAN'08: IEEE Symposium on New Frontiers in Dynamic Spectrum Access Networks*, pp. 1–9, October 2008.
9. Wikipedia. Long-Range WiFi, October 2010.
10. S. Kawade and M. Nekovee. "Can Cognitive Radio Access to TV White Spaces Support Future Home Networks?" In *DySPAN'10: IEEE International Symposium on New Frontiers in Dynamic Spectrum Access Networks*, pp. 1–8, April 2010.
11. Bhaskaran Raman and Kameswari Chebrolu. "Experiences in Using WiFi for Rural Internet in India." *IEEE Communications Magazine*, 45(1), pp. 104–110, January 2007.
12. Kameswari Chebrolu, Bhaskaran Raman, and Sayandeep Sen. "Long-Distance 802.11b Links: Performance Measurements and Experience." In *ACM MobiCom'06: Proceedings of the 12th Annual International Conference on Mobile Computing and Networking*, pp. 74–85, 2006.
13. Rabin Patra, Sergiu Nedevschi, Sonesh Surana, Anmol Sheth, Lakshminarayanan Subramanian, and Eric Brewer. "WiLDNet: Design and Implementation of High Performance WiFi Based Long Distance Networks." In *The 4th USENIX Symposium on Networked Systems Design & Implementation*, pp. 87–100, 2007.
14. Federal Communication Commission. FCC 10-174: Second memorandum opinion and order. Online: http://www.fcc.gov, September 2010.
15. Paramvir Bahl, Ranveer Chandra, Thomas Moscibroda, Rohan Murty, and Matt Welsh. White space networking with Wi-Fi like connectivity. *ACM SIGCOMM Computer Communication Review*, 39(4), pp. 27–38, October 2009.
16. Ulla Killström, Bernd Mrohs, Luca Galli, Dario Melpignano, Olli Immonen, Björn Kijl, Christian Räck, Stephan Steglich, and Heli Virola. Service architecture enabling advanced mobile applications from business model perspective. WWRF-15, Wireless World Research Forum, Paris, December 2005.

Chapter 15

TV White Space Use Cases

Chin-Sean Sum, Gabriel Porto Villardi,
Richard Paine, Alex Reznik, and Mark Cummings

Contents

15.1 Introduction

In the worldwide quest for more radio spectrum, the use of Television White Space (TVWS) is representative of the more general case of using existing assigned spectrum that is locally unoccupied, often referred to as "White Space." Around the world, there are variations in how TV white space is allocated and in the rules for the use of such spectrum. There are, however, some commonalities in the generalized use of TV white space that warrant a generalized set of use cases to pursue the quest. This chapter explores the use of several application use cases. These are, in some respects similar to other wireless use cases. Then, some coexistence use cases for TV white space being used by different Air Interface Standards (AISs) and/or in crowded or heavily used geographical areas are discussed. This second category of use cases is unique (at this time) to TV white space.

15.1.1 Assumptions

Assumption 1: TV signals are generally broadcast from major population centers, and the strength and utility of the reuse of TVWS wireless broadcast signals is gained at increasing distances from those population centers.

Assumption 2: The use of TVWS radio spectrum is available in unoccupied TV channel spectrum.

Assumption 3: The use of TVWS radio spectrum can be available and assigned to geographical areas following rules of the geographical regulatory agency having jurisdiction.

Assumption 4: The TVWS radio spectrum may vary from regulatory authority to regulatory authority and therefore will need a mechanism to determine what channels can be used in a specific geographic area. Although different regulators around the world are taking different approaches, it appears that all will use some kind of database server accessed over the Internet. These servers can determine regulatory jurisdiction for a specific area.

15.1.2 Definitions

TVWS: TVWS refers to spectrum allocated to a broadcasting service, however, not used locally. It has gained attention after the Federal Communications Commission (FCC) established rules allowing unlicensed devices to use the aforementioned empty spectrum as long as requirements, aimed at minimizing interference with protected users, are met. In this chapter, discussions are based on, but not limited to the FCC regulations on TV white space.

Smart Grid: A smart grid refers to adding communications and machine intelligence to conventional electric distribution systems. It is generally implemented using two-way digital wireless technology. It can (depending on local regulatory decisions) provide real-time (or close to real-time) usage information and, in some cases, control energy-consuming devices at industrial locations and consumers' homes to save energy, reduce cost, and increase reliability. It overlays the electricity distribution grid with an information and net metering system that can optimize the use of energy. Such a modernized electricity network is being promoted by many governments as a way of addressing energy independence, global warming, and emergency resilience issues. Smart meters may be part of a smart grid, but alone do not constitute a smart grid. A smart grid may include an intelligent monitoring system that keeps track of all electricity flowing in the system. It may also incorporate the use of superconductive transmission lines for less power loss, as well as the capability of easing the integration of renewable electricity such as solar and wind. When power is least expensive the user can allow the smart grid to turn on selected home appliances such as washing machines or factory processes that can run at arbitrary hours. At peak times it might be able to turn off selected appliances to reduce demand.

Cognitive Radio: Cognitive radio is a paradigm for wireless communication in which either a network or a wireless node changes its transmission or reception parameters to communicate efficiently, thus avoiding interference with licensed or unlicensed users. This alteration of parameters is based on the active monitoring of several factors in the external and internal radio environment, such as radio frequency spectrum, user behavior, and network state.

15.2 What Are the Applications for TV White Space?

The extension of spectrum occupancy to TV white space has opened up a new dimension for a variety of potential applications. The merit of TVWS occupancy is essentially twofold: (a) providing desirable characteristics to facilitate innovative applications not fully supported by existing technologies; (b) offering resource expansion to existing applications for enhanced performance. One set of desirable characteristics is a function of the fact that TV white space allows for unlicensed

Table 15.1 Potential Applications and Descriptions

APP No.	APP Name	Description
1	Large Area Connectivity	High-data-rate backbone for fixed stations
2	Utility Grid Networks	Connectivity for complexity-constraint fixed stations
3	Transportation and Logistics	Logistics-control for mobile stations
4	Mobile Connectivity	Seamless connectivity for mobile stations
5	High-speed Vehicle Broadband Access	High-data-rate backbone for high-speed mobile stations
6	Office and Home Networks	High-data-rate short-range indoor connectivity
7	Emergency and Public Safety	Mission-critical highly-reliable connectivity

operation in the UHF/VHF bands, which brings airwave characteristics such as more penetration of radio obstacles, positive Doppler effects, and greater distances. Before the advent of TV white space, these bands were unavailable to unlicensed users. TV white space also opens up the possibility of using AISs in unlicensed bands that were previously only usable in licensed bands.

A list of seven potential applications (APPs), expected to utilize the physical advantages offered by the TV white space is presented and discussed. Table 15.1 shows the APPs and their respective descriptions.

This list is meant to be a simple set of examples that defines the key parameters. It is not all-inclusive. For example, it does not include such applications as man-ufacturing floor networks, processing plant networks, hospital networks, mobile health networks, etc.

Communication systems operating in TV white space can offer wider network coverage as compared to wireless communication systems operating in the GHz bands, and therefore are suitable for supporting wireless networks over a large area. In suburban and rural areas, the availability of vacant TV channels can be quite high, thus facilitating high-data-rate communications and seamless coverage. The high-penetration capability of TVWS frequencies is also a desired behavior to enable effec-tive coverage in dense cities and rural terrains. In Table 15.1 and Table 15.2, the APPs listed are those anticipated to be capable of taking advantage of TV white space.

APP1 and APP5 establish high-data-rate backbones to support a collective group of end users in one or multiple sub-networks. The main difference between the two is that APP1 targets fixed stations such as buildings, while APP5 targets mobile

Table 15.2   Application Characteristics and Requirements for the APPs

APP	Bandwidth	Range	Portability/ Mobility	Security	Reliability	Latency Tolerance	Supported Users
1	High	High	Low	High	High	High	Medium
2	Low	High	Low	High	Low	High	High
3	Low	Medium	Medium	High	High	High	Medium
4	Medium	High	Medium	Medium	Medium	Medium	High
5	High	Medium	High	Low	Medium	Medium	Low
6	High	Low	Low	Low	Medium	Low	Low
7	Medium	Medium	Medium	High	High	Low	Medium

stations such as autos, trains, and buses. End users within the sub-network supported by the backbones can share the available bandwidth. As compared to APP1 (Large Area Connectivity), APP2 (Utility Grids) also supports large area coverage with fixed end nodes, but with a lower data rate for complexity-constraint devices, such as wireless meters in the utility grids. APP2 is suitable for cost-efficient battery-powered end nodes requiring a minimum amount of service and maintenance. Another application that operates in low data rate is APP3 (Transportation and Logistics), where the end nodes, such as delivery trucks, are mobile. In APP3, the main operations are identification, tracking, and registration of the moving nodes, which do not require large bandwidth resources. APP4 (Mobile Connectivity) aims to establish seamless connectivity between end users and the base station directly. End users of APP4 are mainly PDAs/smartphones, tablets, and laptop computers, which are essential to be connected all the time, but require moderate bandwidth less than that required in, for example, APP6 (Office and Home Networks). If users of APP4 move into a building or a vehicle, handover can be conducted to local intermediate hubs so that they can be connected as per APP1, APP5, or APP6 for higher communication speed. APP6 primarily supports short-range connectivity targeting mostly high quality multimedia streaming and broadband Internet access. The main coverage of interest is within a building or spaces near the building compounds. APP7 (Emergency and Public Safety) addresses the mission-critical connectivity such as networks deployed in rescue operations and public safety networks. Compared to other APPs, the connectivity in APP7 has to achieve security with higher seamlessness, robustness, and reliability.

Each APP has a set of unique characteristics to address different needs for respective applications. In turn, these application characteristics determine the requirements that drive the direction of successive system design processes. Among others, the application characteristics and requirements for the APPs are bandwidth, range, mobility, security, reliability, latency tolerance, and supported users. Table 15.2 shows how the APPs with respective characteristics are related to the requirements.

The required bandwidth to support the APPs is one of the fundamental radio resource characteristics. APP1 and APP5 aim to construct the network backbone and thus require high bandwidth. APP6 is specified to support high quality multimedia streaming and thus also requires high bandwidth. On the other hand, low-data-rate applications such as APP2 for wireless utility meters and APP3 for transportation logistics require only low bandwidth. In order to obtain a wider bandwidth, technologies such as contiguous and noncontiguous channel bonding may be applied.

The required operating range is set to high for networks targeting a large area such as in APP1, APP2, and APP4. For applications with high likelihood of intermediate hubs or relay stations being deployed in between the main concentrator and end users such as in APP3, APP5, and APP7, the required range is medium. The home and office coverage is set to low range in APP6.

APPs where most of the nodes are either fixed or portable, APP1, APP2, and APP6, are not required to support any mobility. APP5, dealing with high-speed trains and highway buses, is required to support high mobility, while APP3, APP4, and APP7, dealing with mostly human and lower-to-moderate-speed vehicles, are required to support medium mobility. In order to support mobility, the system design should take into consideration the mechanism to overcome the Doppler effect.

Security requirement is set to high in APPs involving logistics, money transaction, or emergency situations such as in APP1, APP2, APP3, and APP7. Security requirements for the more consumer-electronics-related APPs, such as multimedia streaming and broadband Internet access, are set to either low or medium.

Reliability indicating the network tolerance to possible outages is a mandatory requirement for mission-critical applications such as APP7. Similarly, reliability is also essential in APP1, which is the core-backbone connectivity for a large area, and in APP3, which handles the logistics of transportation and delivery systems. On the other hand, in APP2, reliability can be relaxed except in times of utility emergency response.

The latency tolerance requirement translates to the tolerance against delay in the connectivity. APPs handling time-sensitive data such as video transmission have high requirements in latency tolerance. APP6 and APP7 are typical examples of applications that need to be operated in low latency. For applications such as in APP4 and APP5, comprising a mixture of data, voice, and video, the requirement in latency tolerance is medium. Algorithms such as dynamic frequency switching and fast channel transfer may be applied for applications with low latency tolerance.

The number of supported users is another important requirement in the APPs. APP2 is required to support up to thousands of wireless utility meters, and APP4 is required to support a high number of mobile users, therefore requiring a high number of supported users. In a home network, typically, the required number of users is low, as shown in APP6.

15.2.1 APP1: Large Area Connectivity

The first potential APP for wireless systems operating in TV white space is Large Area Connectivity. The large area connectivity constructs the wireless backbone and intermediate backhaul links interconnecting multiple hubs, where the hubs are further connected to the nodes (hence a sub-network), forming a wireless large area network, or a wireless metropolitan area network (MAN). The MAN can be connected to external entities through a high-speed external backbone. A metropolitan area typically ranges from several hundred meters to tens of kilometers, covering several buildings to an entire city. In the rural areas, instead of buildings, the trees, hills, and rivers are the main geographical structures that construct the terrain. Within the large area all the hubs are interconnected by using air interfaces operating in the TV white space. These hubs are further connected to lower hierarchies of nodes by using the same air interface.

Currently, there are various technologies employed to facilitate connectivity within the MAN. Among others, there are cable, optical fiber, radio-wave, and free space optics. Cable and optical fiber are widely used for interconnecting sub-networks. An example is the Metro Ethernet [1] that establishes links among multiple sub-networks (e.g., local area networks or LANs) using cable or optical fiber to form a MAN spanning up to a coverage area of several kilometers. Radio-wave is also a common medium used to establish wireless links within a large area. The Worldwide Interoperability for Microwave Access (WiMAX) [2] is a protocol that employs radio-wave in the 2–11 GHz and 10–66 GHz bands. In the Wireless Local Area Network (WLAN) a.k.a. Wireless Fidelity (WiFi), range extension on the specification is carried out in the 2.4 GHz, 5.15 GHz, and 3.6 GHz bands to cover an area of several kilometers [3]. Exploiting the flexibility provided by radio-waves, High Altitude Long Operation (HALO) networks [4] are used to provide MAN connectivity to end users from a high elevation angle, therefore taking advantage of unobstructed line of sight channels by utilizing aircraft circulating in the skies, high above the terrestrial wireless networks. Additionally, free space optical point-to-point links are also used to connect between sub-networks.

All the existing technologies have respective advantages and disadvantages. Transmission using cables and optical fiber is subjected to lower loss as compared to wireless transmission, but requires higher installation and maintenance cost and time, especially as the coverage area of interest increases. Wireless transmission to facilitate MAN coverage is currently operating mostly in spectrum bands up to several GHz and is therefore subjected to propagation and absorption loss as compared to transmission in the lower spectrum bands such as the very high and ultra high frequencies (VHF/UHF). High Altitude Long Operation (HALO) networks have the advantage over obstructions, but are subjected to challenges such as maintaining uninterrupted connectivity to the intended area under the presence of strong wind gusts in the atmosphere and stratosphere, let alone the general concern regarding crashes of large unmanned aeronautical vehicles. Free space optical links provide relatively simple deployment; their stability and quality are highly dependent on atmospheric factors such as rains, fog, and dust.

To facilitate connectivity for MAN deployment, the TV frequencies in the VHF/UHF bands offer several encouraging advantages as compared to the current technologies. Compared to the currently employed radio frequencies, the VHF/UHF bands are a valuable networking tool for the same reasons they are desirable for TV broadcasting services, which are the longer reaching and higher obstacle-penetrating capabilities. Compared to the wired connectivity, the TV frequencies offer a more cost-effective solution, particularly for coverage of larger areas. The cost and time for installation and maintenance can be reduced significantly. Compared to free space optical links, the TV frequencies are less dependent on atmospheric factors such as the weather. The main setback of the TV white space is the possible unavailability of usable spectrum due to occupancy by incumbents.

The possible physical topologies for the large area connectivity can be star, tree, mesh, or any combinations of the three. In a typical deployment scenario, all the end nodes in a sub-network are connected to a hub or concentrator via air interface operating in the TV white space. The hubs are interconnected via the same air interface to each other, forming a backbone network. Alternatively, sub-networks can also be connected to the backbone via backhaul connectivity. Additionally, the connectivity between the end nodes to respective hubs may also utilize unlicensed white space frequency bands or different air interfaces, such as Bluetooth or WiFi, that use unlicensed or lightly licensed frequencies.

Possible application examples of this use case are discussed below.

■ Rural area connectivity where the end users are wireless service subscribers that are connected to the Internet through several relays and a base station.
■ Campus area connectivity where the end users (e.g., student laptop computers) are connected to hubs in the building. Buildings in the campus area are then interconnected, forming a campus area network.
■ Business enterprise space where the nodes (e.g., company workstations) are connected to hubs. Multiple hubs in the enterprise form a network that is connected to the enterprise central control and management main frame.
■ Municipal and rural area coverage where end users (e.g., domestic-use computers) are connected to separate household hubs. These household hubs are connected to the backhaul link and then to the main hub in the neighborhood, forming a neighborhood area network. Multiple neighborhood area networks form a MAN.
■ Industrial site connectivity where the end nodes are the machines of an industrial site. These nodes are connected to the intermediate hubs, which are ultimately linked to the main control entity of the industrial plant.
■ Military premises connectivity where the end users are radios in military premises and military personnel connected to the hub. The hubs are then interconnected and linked to the backbone of the military command center.

Figure 15.1 illustrates the specific use case of a university campus wireless backbone by means of air interface operating in the TV white space. Fixed directional VHF/UHF antennas professionally installed at the top of faculty buildings provide the necessary TVWS connectivity to support the traffic flow generated within the campus.

Each faculty building has its own sub-networks composed of end nodes (e.g., student laptops, desktops, and laboratory servers), which are connected to a hub (e.g., an access point). These hubs are then connected to the fixed antenna at the top of the building. Upon connecting to its sub-network, an engineering student, for example, could access the library database archive searching for relevant research publications or directly exchange files with an architecture student by having

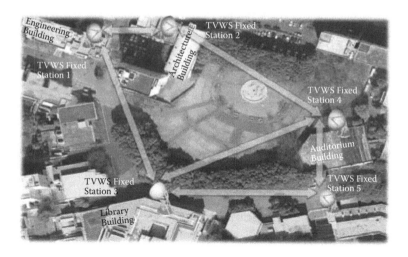

Figure 15.1 Use case of university campus with TVWS wireless backbone.

his/her traffic flow through the TVWS connectivity. If the traffic destination is an external network (e.g., the Internet or another university campus), the traffic is routed through one of the gateways, which connects the MAN to the high-speed external backbone.

The highly penetrating and wide range characteristics of the VHF/UHF bands make it possible to use less transmit power than would be required with microwave links, to interconnect faculty buildings (i.e., to cover the same campus area). Needless to say, the installation of such campus wireless backbone would be less time consuming and more economically viable than if the same application is established by employing fiber optics. Additionally, a multitude of other possible applications, such as wireless backbone for business enterprises providing connectivity between the headquarters to its branches, or connecting industrial sites' machinery, could also be envisioned.

15.2.2 APP2: Utility Grid Networks

The second potential APP for wireless system operating in the TV white space is the Utility Grid Network. Contributing to the efforts of building the ecologically friendly Smart Grid, utility services such as electricity, water, gas, and sewer should strive to comply with the growing demands of constructing a more efficient utility networking system. In order to increase efficiency in utility networks, information and communication technology has to be incorporated into every aspect of utility metering, monitoring, load-control, and demand response. In this potential use case, the utility grid network covers the end-to-end connectivity from the utility provider to utility consumers. At one end of the network is the utility provider, consisting of the systems for monitoring, plus command and control of utility systems. The main system is connected to a transceiver station serving as the origin of the smart utility network. At the other end are the utility consumers with smart metering devices connected to the utility transceiver station via air interface operating in the TV white space. In between the utility provider and the consumers there may be several hierarchies of nodes serving as relays in the information exchange across both ends, also via air interface operating in the TV white space.

Currently, various technologies are used to connect household utility meters to the utility providers. In the existing utility services, touch-based meter reading, power-line cable, optical fiber, and telephone line are some of the technologies used for network connectivity. The touch-based meter reading is one of the conventional methods for metering and billing. A meter reader visits the site with a device that automatically collects the reading from a meter if placed in close proximity. In certain places, there are even touch intensive methods of manual recording by human meter readers. Besides, power-line cable and optical fiber are also common methods to connect the utility provider to the consumers [5]. Another connectivity attempt applied in utility systems is via telephone line [6]. Alternatively, radio-wave connectivity is recently becoming an emerging means of connecting wireless sensors to the control entity [7].

The conventional touch-based method of employing meter readers to conduct on-site tough-and-go metering is expensive and resource consuming. This method will be phased out gradually as new technologies emerge. Human error is inevitable in this method. With power-line cable and optical fiber connectivity, although offering highly reliable networks, the installation and maintenance costs are significantly higher than those of wireless solutions, especially in the long run. Furthermore, to cover wider areas such as the sparsely populated suburban and rural areas, the disadvantages of cable and optical fiber connectivity become exponentially pronounced. Alternatively, radio-wave connectivity is a more promising solution toward efficient utility networks. Various efforts are mobilized to have a unified direction toward employing this technology in utility network deployment, specifically in [8], or an enhanced extension of [7].

For establishing connectivity in a utility grid network, radio-wave seems to be the more suitable candidate technology. Among other choices, the TV VHF/UHF frequencies offer several inspiring advantages. As compared to the power-line cable and optical fiber solutions, network connectivity employing air interface operating in the TV white space offers larger coverage with lower cost of installation and maintenance. In a dense urban area, the high penetrating characteristics make the TV frequencies a suitable candidate for covering multiple households separated by walls, for example, in highly populated apartments. In the sparsely populated suburban and rural areas, the long reaching characteristic makes the TV frequencies suitable for connecting households that are far away from the heart of the network.

Figure 15.2 illustrates the concept of utility grid networks. Smart electricity, gas, and water meters are located at the premises of end users and connected to

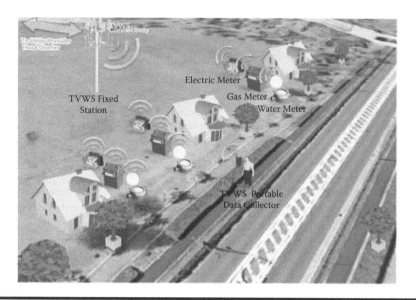

Figure 15.2   Use case of utility grid networks operating in the TV white space.

intermediate hubs and then to the main computing systems of utility providers. The smart meters, intermediate hubs, and mainframes are equipped with transceivers capable of operating in the TV white space in order to establish connectivity.

Through the TVWS connectivity, electricity, gas, and water consumption data of a specific household can be transferred automatically to the utility provider in a precise, fast, and cheap manner. Contrary to the most current common approach, meter readers will not be employed in utility data collection, freeing the personnel for other tasks, and thus adding new values to human resources. Under special circumstances, however, personnel could be deployed to manually obtain the utility data in a specific region, for example, in the case of faulty intermediate hub operation, or in a specific household, in the case of a faulty end user device operation. From the perspective of the utility providers, other advantages of utility grid networks are the ability to connect/disconnect service to the end-users remotely, the flexibility to adjust load-balancing according to local and timely demands, and the capability to respond to emergency situations more effectively. In addition, the utility grid networks application has a remarkable resilience to TVWS outage. In the event of instantaneous channel unavailability, utility data could be transferred seconds, or even minutes later, because delay is not particularly a major issue.

15.2.3 APP3: Transportation and Logistics

The third potential APP for wireless system operating in the TV white space is Transportation and Logistics. In order to increase the efficiency of transportation systems and logistics, where optimizing resource management is an essential factor, advanced networking systems should be established between nodes (e.g., vehicles, cargo) and hubs (e.g., data collectors, relays) for identification, tracking, management, transaction, and telemetry in transportation and logistics. In such advanced networks, the hubs are connected to the main concentrator (e.g., system mainframes) where the control and management entity is located. The connectivity from nodes to hubs, and hubs to the main concentrator is established via air interface operating in the TV white space.

In the arena of transportation and logistics, an existing and still evolving enabling technology is radio identification (RFID) [9]. RFID is used in various applications including public transportation management, shipping/freight distribution systems, virtual payment systems, location-based services, and medical-related services. Currently, the frequency bands used for RFID vary depending on national regulatory bodies and institutions. A fraction of the frequency bands employed by RFID fall within the range of VHF/UHF bands [10,11]. Besides the already-existing enabling features offered by the current RFID technologies, air interface connectivity via the TV white space could also offer several other complimentary advantages. First, the available TV white space is spectrally wider than the spectrum bands accessible by the VHF/UHF RFID, thus providing bandwidth expansion. Second, by using the lower end of the TV frequencies, larger coverage area and higher resilience to mobility (i.e., less impact due to Doppler spread) can be achieved. Together, the hybrid characteristics offered by TV frequencies are able to suit a wider range of application demands. In a nutshell, TV white space is a suitable technology to complement the current effort by RFID to serve as the enabling technology for transportation and logistics. It should be noted that it is not the intention to replace the RFID technology with TVWS communications, but to provide complementary frequency additions to RFID limitations such as operating range and wider bandwidth.

The possible physical topologies for transportation and logistics can be star, tree, mesh, or any combinations. In a typical deployment scenario, all nodes are connected to a hub in a star topology via air interface operating in the TV white space. The hubs may be interconnected with each other in a mesh network via the same air interface. The hubs are then further connected to the main concentrator, either directly or through several hierarchies of relays.

Possible applications for this use case are discussed in the following:

■ Public transportation information system where the nodes (e.g., monitor displays) in the public transports or stations are fed with real-time information such as arrival/departure time and possible delay from the network hubs operated by transport service providers. Additionally, the nodes may also feature multimedia contents such as tourist information, publicity of local events, weather forecasts, traffic condition, and news.

■ Transportation virtual payment system where private vehicle users can conduct payment for highway tolls, parking tickets, and even fines from violation tickets. The same payment system can be applied for public transport payments such as bus, rail, subway, ferries, and even airplanes.

■ Baggage management where the nodes are placed in the baggage and hubs are deployed to perform identification, tracking, and handling of the baggage. The information is then sent to the main frame for collective processing.

■ Freight distribution logistics where the delivery status can be updated in real-time through the connectivity between distributing vehicles (e.g., delivery trucks, ships, trains) and hubs (e.g., access points for monitoring) deployed by the freight service provider in the areas of interest.

■ Shipping container management in the harbor where the nodes are placed in the containers, which are connected to the hubs in the harbor control center for effective tracking and handling.

Figure 15.3 illustrates a specific implementation of transportation and logistics employing air interface operating in the TV white space. This example outlines the networking system facilitating logistics in a transportation system that handles freight and shipping services. The logistic-control center is located at the central office of the transportation company. The control center is connected to the data collector network. A data collector network is formed by interconnected hubs (e.g., access points) acting as data collectors deployed in the area of interest. These data collectors are further connected to the mobile nodes (e.g., delivery trucks) that are distributed within the same area. Through the TVWS connectivity between data collectors and delivery trucks, information can be exchanged, therefore, giving the ability to the central office to add various functions to the transportation service. These advanced functions include real-time location tracking of cargo and parcels,

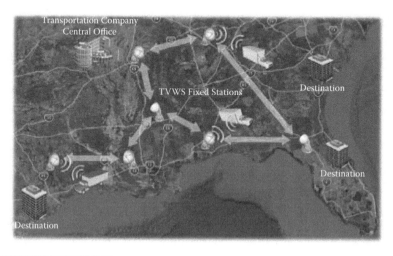

Figure 15.3   Use case of transportation and logistics operating in the TV white space.

protection against cargo theft through surveillance of its trucks during delivery, estimating the best route for delivery based on real-time traffic information collected by its trucks, making predictions on delivery time and others, and so on. All aforementioned added functions contribute to reduced delivery time and costs, therefore leading to increased customer satisfaction. Furthermore, the ability to deploy the connectivity using the VHF/UHF bands complements existing similar services with a wider accessible bandwidth as well as higher resilience to vehicular mobility at the lower end of the TV frequencies. In other words, the use of TV white space to facilitate connectivity brings positive added values to the existing enabling technologies in the current networking systems for transportation and logistics applications.

15.2.4 APP4: Mobile Connectivity

The fourth potential APP for wireless systems operating in the TV white space is Mobile Connectivity. A mobile connectivity network facilitates connectivity for devices that are primarily mobile, portable, or nomadic. The network consists of a main concentrator (e.g., base station) and surrounding mobile nodes (e.g., laptops, tablets, PDAs/smartphones, or transceivers in ships) that are connected via air interfaces operating in the TV white space. The targeted area of coverage spans from several kilometers up to several tens of kilometers. Intermediate relay stations (e.g., access points) may be deployed between the main concentrator and the nodes for range extension. In regions where installation of relay stations requires a higher level of complexity (e.g., the sea), nodes may form ad hoc–based networks to achieve larger coverage. The concentrator is connected to other external networks through a high-speed backbone, possibly with a different means of connectivity.

There are basically two categories of existing systems that support the mobile connectivity networks, mainly targeting applications such as voice telephone, mobile Internet access, video calls, and mobile TV. The first category is the cellular technologies (i.e., mobile phone networks). Earlier cellular systems provide mobile telephony (e.g., GSM) [12], and the later amendments (e.g., GPRS, EDGE, CDMAone, etc.) extend to data packet transmission. Besides, there are also further enhancements that support high-speed mobile broadband access, such as the 3G (e.g., UMTS, TD-SCDMA, W-CDMA, CDMA2000, etc.) [13] and the "beyond 3G" standard (e.g., Long Term Evolution (LTE)) [14]. The LTE-Advanced is envisioned to be the successor 4G standard. The second category is the set of mobile wireless broadband technologies networks. Wireless broadband supports mobile broadband access, primarily formed by the WiMAX family (e.g., Mobile WiMAX) [2,15] and Mobile Broadband Wireless Access (MBWA) [16]. The other wireless broadband technology that enables mobile wireless broadband is WiFi. WiFi now has handoff and mobility capabilities that allow handoff and mobility from WiFi Access Point (AP) to WiFi AP and also WiFi to cellular mobility. It generally supports offloading of high-bandwidth applications from cellular to WiFi, but is

Figure 15.4   Use case of TV white space land mobile connectivity.

becoming more capable of standalone mobile connectivity. Generally, these technologies support a wide range of services from voice to data to multimedia.

The main strength able to be offered by the TVWS connectivity is complementary to the existing mobile broadband access applications, particularly in the aforementioned second category of wireless broadband technologies. The TV white space is capable of offering bandwidth extension in the order of tens, if not hundreds of MHz to ease the traffic of current lower bandwidth technologies. Specific use cases are given in the following discussions to emphasize the potentials of the TVWS connectivity.

Figure 15.4 illustrates one of the envisioned application genres of mobile connectivity using the TV white space, namely, Land Mobile Connectivity. In this use case, PDA/smartphone, tablet, and laptop/notebook computer users have wireless broadband access to data communications through air interface operating in the TV white space. The end users may be stationary at a location, or moving in a vehicle. Common applications are, for example, voice telephony, Internet access, and Web browsing. Compared to connectivity in the GHz band, the VHF/UHF spectrum characteristics offer encouraging advantages to both operators and end users. The longer range and higher penetrating characteristics of the VHF/UHF frequencies translate into a reduced number of base stations deployed over a specific service area, as well as less attenuated signals providing last mile wireless broadband services to end users with higher resilience to service outage. In addition, the higher robustness to mobility inherent to the VHF/UHF bands also results in higher QoS to mobile users using moderate bandwidth-demanding applications, such as video streaming for cellular networks.

Figure 15.5 illustrates another application genre in Maritime Connectivity using air interface operating in TV white space. Base stations located at the shoreline and

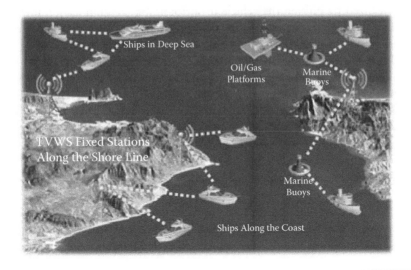

Figure 15.5   Use case of TVWS maritime mobile connectivity.

the ships/offshore structures are connected to allow information flow from land to sea and vice versa. The control center at the shoreline can be connected to conduct data communication, command, and control to the ships/offshore structures. Additionally, by combining the attractive long-range propagation characteristic of the VHF/UHF bands and the implementation of ad hoc mesh networking systems, range extension can be further achieved. Nodes far away from the shoreline can be connected through ad hoc networks formed locally. Alternatively, relay stations built on marine buoys can be deployed for intermediate connectivity. In other words, distant mobile nodes (e.g., deep sea cargo ships) and structures (e.g., oil-gas platforms, sea farms) can be connected to the continent through TVWS connectivity either by employing local ad hoc mesh networks or intermediate marine buoy relays strategically located at some points in the ocean. Furthermore, TVWS connectivity adds significant value to the capacity-limited maritime communication services. Expensive satellite services limited to voice and narrowband data communications available in deep-sea platforms and deep-sea ships will be relieved by the deployment of wireless broadband systems. Such change allows high-speed data communications (hence inexpensive services such as voice communication, high-speed Internet access, and real-time video streaming) and improves working environments for personnel dispatched to such remote locations.

15.2.5 APP5: High-Speed Vehicle Broadband Access

The fifth potential APP for wireless systems operating in the TV white space is High-Speed Vehicle Broadband Access. Broadband access in high-speed vehicles can be realized by linking the vehicle to surrounding communication infrastructure via a network. Generally, the high-speed vehicle connectivity is formed by establishing a backhaul between the hubs (e.g., access points along railway tracks or highways) and the high-speed nodes (e.g., autos, trucks, trains, buses) via air interface operating in the TV white space. The hubs are then connected to the main concentrator (e.g., base station) with the same air interface. In the vehicles, the connectivity between the intermediate relay (e.g., access points in the vehicle) and the end users (e.g., PDAs/smartphones, tablets, and laptops) may utilize either the TV white space or other air interfaces (e.g., WiFi, Bluetooth).

There are several currently available technologies identified as suitable candidates to support such a use case scenario. These potential candidate technologies are WiFi (another IEEE 802 AIS), WiMAX, cellular networks (with a variety of AISs), and satellite communications. In WiFi, the Wireless Access in Vehicular Environment (WAVE) [17] can be used for the connectivity between moving vehicles and roadside infrastructure, operating at the licensed intelligent transportation system band of 5.9 GHz. The WiMAX, AIS can be implemented in a wide variety of bands. Currently, it is only used in bands that local regulatory authorities specifically authorize [15]. Cellular networks, particularly the enhanced version for high data rate and low latency such as LTE, are another alternative to support this application. Satellite communication is also a prospective candidate that is able to provide seamless signal coverage.

As compared to the potential candidate technologies for the high-speed vehicle broadband access, connectivity utilizing the TV white space offers several inviting advantages. As compared to air interfaces such as those in WiFi and WiMAX operating mainly in the GHz band, the TV VHF/UHF frequencies are subjected to lower path loss, longer range, and higher penetration capabilities. These characteristics outline a significant advantage in terms of area coverage in difficult terrains and crowded buildings. Most importantly, the TV frequencies display a more superior resilience to mobility as compared to the GHz band, addressing the primary concern in connectivity for vehicles moving at high speed. As compared to satellite communications, air interface in the TV white space can support a higher data rate with lower latency. In short, the TVWS connectivity is capable of offering complementary advantages, if not outperforming other technologies in this use case.

Possible physical topologies for the high-speed vehicle broadband connectivity can be the star or tree architecture. In a typical deployment scenario, the main concentrator such as a base station is connected to hubs along the railway tracks or roadside. These hubs then form the backhaul connectivity to the moving vehicles, utilizing air interface in the TV white space. In the vehicles, intermediate relays are deployed. All the end nodes (e.g., end user laptops, tablets, and PDAs/smartphones) in the high-speed vehicle are connected to the intermediate relays, either through air interface in the TV white space or other options.

Possible application use cases are as follows:

■ Connectivity for high-speed trains (e.g., Japanese Shinkansen, Korean KTX, French TGV, etc.) with travelling speeds up to a typical 350 km/h where the hubs are constructed along the railway tracks, providing broadband access and multimedia entertainment to passengers in the train.
■ Connectivity to long-distance buses where hubs are deployed along the highways, providing the same services to passengers.
■ Connectivity to subways and underground transportations where cellular and other wireless signal coverage are normally weak or absent. In this scenario, air interface in the TV white space is capable of increasing both the coverage area and resolution, providing broadband access service to passengers.

Figure 15.6 provides an illustration of High-Speed Vehicle Broadband Access application utilizing air interface operating in the TV white space. With the increasing demand for higher data rates and constant need for connectivity imposed by the fast-paced society of today, wireless broadband connectivity must be ubiquitous. That is to say, it must be available even in the most technologically challenging scenarios, like the one inherent to a high-speed bullet train moving with speed approaching a typical 350 km/h, or even higher in the future. Speed in such magnitude causes problems to the end user data communications due to inadequate transition time between adjacent cells and significant Doppler spread. In addition, high voltage machinery in railway transportation creates strong electromagnetic exposure. In order to circumvent such harsh scenarios, the connectivity for high-speed vehicle broadband access in the TV white space can be established by deploying, along the railways, an array of fixed stations (e.g., base stations). The

Figure 15.6 **Use case of high-speed vehicle broadband connectivity in TV white space.**

fixed stations are connected to the intermediate portable relays inside the trains. Connecting to the network, a businessperson travelling from Tokyo to the remote city of Kyoto will have a handful of possibilities to spend the commuting time. The broadband connectivity in the trains supports a wide range of applications, from low-bandwidth-demanding applications such as browsing the Internet and e-mailing to bandwidth-hungry applications such as video conferencing and real-time video streaming. To realize this connectivity, the use case application relies on the attractive VHF/UHF characteristic of being resilient to Doppler spread, therefore allowing the system to support high mobility. Additionally, the reduced path loss and high building penetration characteristics of VHF/UHF also allow a reduced number of fixed stations covering the railway, therefore increasing the cost-effectiveness. Coupling with technologies such as fast handover, resource-demanding real-time applications such as video conferencing and video streaming can be supported, even in the fastest trains in the world.

15.2.6 APP6: Office and Home Networks

The sixth potential APP for wireless systems operating in the TV white space is the Office and Home Networks. This use case seeks to establish ubiquitous high-speed connectivity within the area of office and home. The office and home networks typically require a coverage ranging from several centimeters up to several tens of meters. The network consists of one or several interconnected hubs (e.g., access points), which are further connected to multiple nodes (e.g., end user devices), all via air interface operating in the TV white space. The connectivity between the hubs to external networks may be via any alternative radio interface including a hybrid interface operating in both the TV white space and other existing means.

There are a number of existing technologies, both wireless and wired, employed for deploying the office and home networks. Among such wireless technologies, the more common ones are the Bluetooth, the millimeter-wave wireless personal area network (WPAN), and the WiFi. The Bluetooth [18] is a short-range (10 m) cable replacement technology establishing wireless connectivity among devices within the personal space. The millimeter-wave WPAN [19] is a short-range high-speed communication protocol that targets up data speed in the order of Giga-bps to support applications such as uncompressed video streaming and ultra-high-speed data exchange. WiFi [3] (100 m) is another popular technology used to deploy indoor networks. Besides wireless solutions, wired solutions such as Ethernet, power-line cables, and phone wires are also employed for networks in the home environment. Ethernet [1] is a wired protocol connecting nodes either by using CAT5 wire, coaxial cables, or optical fiber. Protocols involving power-line cables [5] and phone wires [20] are also common in providing coverage for a home network.

In the arena of short-range communications, the major advantage able to be harvested from utilizing the TV white space is the expansion of bandwidth. In applications where there are overlaps in functional characteristics between the TVWS connectivity and the existing technologies with a similar purpose, the TVWS system can be viewed as a complementary expansion of available bandwidth that further boosts the communication speed in times of normal operation and relieves the traffic in times of congestion. Spectrum occupancy segmentation (i.e., dedication of specific applications to specific bands in a hybrid wireless connectivity with multiple bands) or medium occupancy segmentation (i.e., dedication of specific applications to specific means of communications in a hybrid wired plus TVWS connectivity) can be conducted, adapting to respective local and timely demands in order to optimize the use of radio resources. These segmentation schemes will be further discussed in the specific examples following. Generally, the TV white space provides a means for office and home networks to reduce the traffic burden in a specific operation band or medium. In this sense, the TVWS connectivity does not compete, but completes the existing technologies by enhancing the performance of the office and home area networks.

Possible physical topologies for the office and home networks are the star and tree architectures. The hubs installed in the houses or office buildings are connected to the end user devices in a star manner. In the case where intermediate hubs are installed, the tree topology is employed.

There are a variety of possible applications for the office and home area networks:

- Personal workspace connectivity where all the peripheral devices such as printers, scanners, and display monitors are connected and controlled by the main work station.
- Office area connectivity where all work stations can be connected to each other and to a central main frame for purposes of data exchange and centralized management.
- Home area networks where the end nodes such as computers, laptops, PDAs/ smartphones, multimedia displays, stereo systems, game consoles, and telephones are connected through the hub to external entities such as the Internet.

Figure 15.7 illustrates a specific scenario of the Office and Home Networking application utilizing an air interface operating in the TV white space. In this example, all the peripherals in the house including computers, PDAs/smartphones, game consoles, and televisions in the living room, bedrooms, and outside the house are connected to a household main hub. The hub is then connected to external systems such as the Internet. The entire home network can be implemented using the TVWS connectivity. Alternatively, it can also be partially coupled with other existing air interfaces (e.g., the 2.4 or 5 GHz WiFi) in a complementary manner. The "dual-band" network can be optimized through the application-based segmentation of spectrum occupancy. For example, when demand outstrips the capacity of the available WiFi spectrum, TVWS spectrum can be used for overflow. Priority-based segmentation can also be used based on the urgency of the data type, such as handling the less-urgent data-backup operations in the TVWS connectivity, while

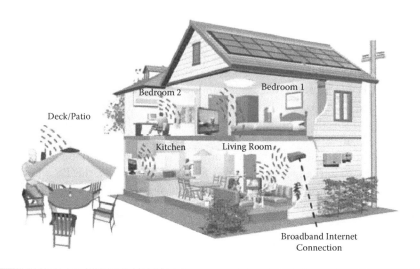

Figure 15.7 Use case of office and home networks in TV white space.

freeing up bandwidth in the 5 GHz WiFi band for other more urgent applications. If the TV white space is coupled with wired solutions such as the Ethernet, then segmentation of medium occupancy can be conducted in the same manner. Similarly, TV white space can be used for congestion relief in cellular networks.

Like any other ubiquitous WiFi today, networking in the TV white space will provide the necessary wireless connectivity to keep the business and the entertainment applications running with high QoS. The large amount of TV white space in the VHF/UHF spectrum allows the radios to take advantage of it. This enables the realization of a broad spectrum of pervasive applications, ranging from the low-bandwidth-demanding to the bandwidth-hungry ones. An important statement about Office and Home Networking based on TVWS occupancy is that it is not intended to be a substitute technology for WiFi today, but adds new spectrum resources, therefore easing the inevitable spectrum crowding faced by congested unlicensed or lightly licensed frequency bands.

15.2.7 APP7: Emergency and Public Safety

The seventh potential APP for wireless system operating in the TV white space is the Emergency and Public Safety network. Emergency and public safety network is a communication network used by emergency response and public safety organizations such as police force, fire, and emergency medical teams responding to accidents, crimes, natural disasters, and other similar events. This network covers connectivity from the command/control entity of the public safety organizations all the way down to the end nodes through several hierarchies of intermediate hubs or relays, via air interface operating in the TV white space. The command/control entity may be the headquarters or commanding offices of the police and fire department. The end nodes can take any form from fixed surveillance cameras/sensors on the streets and other potential disaster areas, to radios in police patrol cars, to mobile radios carried by rescue teams at disaster sites. End nodes have the ability to be interconnected to form mesh networks, to facilitate coverage extension to places where propagation is naturally limited, such as in reinforced concrete buildings and the underground of buildings. Intermediate nodes can be relays or repeaters to extend the range or widen the effective coverage area of the network. Generally, the network should be easily initiated (i.e., in a plug and play manner), with devices simple in design, so as to allow secure, easy operation under stressful situations.

There are several technologies suitable for deploying the emergency and public safety networks. Besides the government-controlled dedicated bands with specific networking systems for the sole purpose of public safety services, there are also other candidate technologies for building public safety wireless mesh networks such as WiFi [3] and WiMAX [2]. The Terrestrial Trunked Radio (TETRA) [21] is a well-established technology for public safety with global presence. It provides voice services with communication security. TETRA release 2 adds higher data rate support in the form of TETRA Enhanced Data Services (TEDS), on the order of hundreds of kbps.

Air interface operating in the TV white space presents a fundamental advantage as compared to spectrum bands in the GHz range: the wider coverage offered by the VHF/UHF wavelengths. Besides, the VHF/UHF bands also display more favorable characteristics in times of emergency response in disaster sites which may demand the highly penetrating feature for critical operations such as a victim search party. For example, the need for communication signals with high object penetration characteristics was significantly pronounced in emergency circumstances in the 9/11 and the London terrorist attacks, when conventional emergency systems had problems maintaining effective communications in the underground of buildings and subways. Additionally, the total available TVWS spectrum is on the order of hundreds of MHz, allowing high-bandwidth-demanding applications, such as disaster-site real-time video streaming.

The typical physical topology for emergency and public safety networks is basically the mesh architecture. All nodes should be interconnected with each other for fast communication during critical times. The nodes also have to be seamlessly and securely connected to the control centers for rapid data retrieval and data upload.

Possible applications for this use case are shown as the following:

■ Public and traffic safety surveillance systems where cameras in targeted locations feed real-time photos/videos of traffic conditions/mishaps, criminal activities, and so on.
■ Emergency surveillance where cameras/sensors are located at potential disaster sites such as shoreline with probable tsunami events, areas around active volcanoes, and areas with possible earthquake occurrence. Most, if not all of

these locations are dispersed outside the dense metropolitan areas and thus require connectivity with wide coverage areas.

■ Mobile real-time video sharing for emergency response purposes where the first responders in disaster sites are required to share videos or telemetry of patient vital signs with qualified medical personnel to increase the efficiency of preserving lives of the victims. In a similar manner, real-time video sharing with the fire department can be useful for the firefighters to better prepare and anticipate the magnitude/type of fire that is taking place. In the field, video streaming from cameras attached to the firefighter helmet can provide the necessary information for the control center to make quick and effective decisions.

■ Fast database access for law enforcement officers where video files of on-site crime recording, witness statements, etc. can be uploaded to the network server for processing and analysis. Likewise, fast database retrieval where criminal records, traffic records, etc. can be downloaded easily to facilitate efficient police investigations. For example, to enable instantaneous criminal suspect identification, the mobile police force has to be equipped with vehicles/person database lookup tools and the capability of taking/transmitting high quality still pictures and fingerprint scans.

■ Navigation and on-site operation assistance where digital maps can be downloaded from the main server to assist public safety organizations to effectively plan and control field operations.

Figure 15.8 illustrates the Emergency and Public Safety Network application utilizing air interface operating in TV white space. The network is composed of end-nodes (e.g., portable devices held by firefighters and paramedics) and the communication hubs installed in the vehicles, forming a local mesh network. This network is in turn linked to the ad hoc command/control unit on-site, and is further connected to the main operational control entity, possibly located in the headquarters. The end-to-end connectivity can be established employing air interface operating in the TVWS connectivity.

As a specific example, a team of firefighters searching for survivors in the underground of a partially collapsed building could communicate with the command/control entity situated near the scene, which in turn provides the team with the building's digital maps to help navigate them through the debris. Such effective and robust communication could take place owing to the high building penetrating characteristics of the VHF/UHF spectrum allied to meshing capabilities with enhanced signal propagation. Once survivors are safely rescued from the debris, video cameras could immediately transmit images of the victims in the ambulance along with vital signs such as heart and respiratory rates to the medical professionals remotely located in the hospital. Such ambitious features are possible due to the vast amount of bandwidth available in the TV white space, enabling the implementation of high-bandwidth-demanding applications.

Figure 15.8 Use case of emergency and public safety networks in TV white space.

15.3 How Do TVWS Operators Coexist?

We have seen some examples of how applications with a need for an Air Interface Standard (AIS) might use TV white space. However, since TV white space is open to all wireless providers and all types of applications, it is likely that there will be multiple white space providers trying to use the same usage block at the same time. In this section of this chapter, we will discuss use cases where there are such requirements for multiple white space AIS technologies operating in the same area.

Some observers argue that AISs currently only deployed in licensed bands will not make use of TV white space. In this argument these observers often point to cellular operators as a significant example. It is instructive to remember that many observers made similar arguments about cellular operators offering WiFi. What happened was that cellular operators began to offer WiFi hotspot services, and some even used VOIP WiFi for femtocell deployment (for example, Sprint in the United States).

Thus, it is likely that the cellular providers may want to use TV white space to overcome congestion in their licensed frequencies and for other purposes. WLAN providers may want to use TV white space to provide more bandwidth in their WLAN Access Point coverage area. TVWS Internet backhaul providers may want to use TV white space to deliver Internet bandwidth to rural customers. The intent of this discussion is to lay the groundwork for development of mechanisms that will allow different AISs and different network operators to operate in TV white space without interfering with each other. The scenarios are those where these different providers, using different AISs, can use TV white space without interfering with each other. One place that these mechanisms are being developed is IEEE 802.19.1.

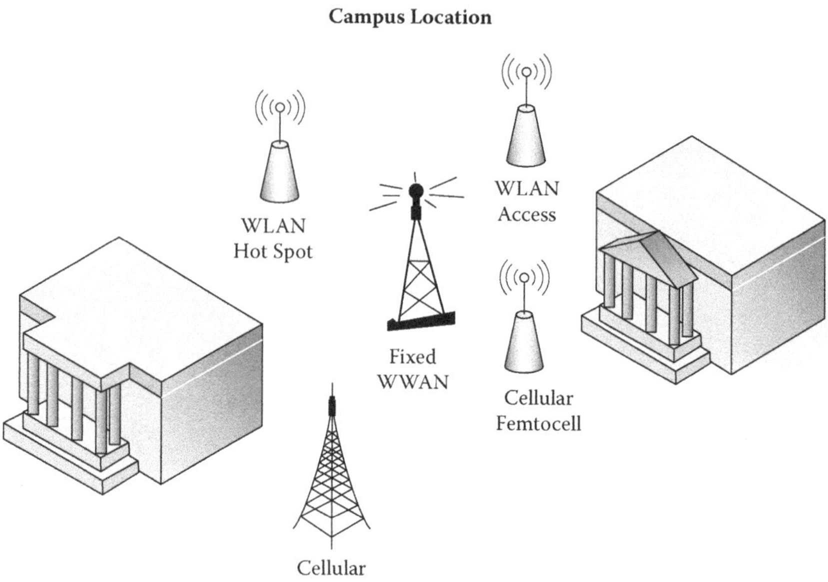

Figure 15.9 Typical campus location with different air interface standards (AISs).

15.3.1 On a Campus (University, Suburban Mall, Office Park)

It is worthwhile to consider what services are provided by each of the white space provider users of frequencies, and therefore the following sections allocate some thought space to these topics. Illustration of a typical campus location is shown in Figure 15.9.

15.3.1.1 Cellular

By the term "cellular provider" we mean any provider delivering cellular (wireless mobility-enabled) services such as cellular telephony and mobile data. The capabilities may be 3G, 4G, or possibly other cellular technologies, most notably those product standards from 3GPP and 3GPP2.

Currently cellular networks support the following services:

■ Voice (via the cellular provider's core network, maybe circuit switched or VoIP based).
■ Data (Internet) access
■ VoIP (via Internet access)
■ Messaging

Figure 15.10 **Femtocell deployment architecture.**

- Emergency services
- Gaming
- Multimedia broadcast

Cellular providers may be using TVWS frequencies in a number of different ways in the same campus area that also contains cellular femtocells, wireless LANs (WLANs), and fixed white space providers, all of whom are using TVWS channels.

Of particular interest in the case of cellular services is the notion of a cellular service's femtocells. To illustrate this, consider Figure 15.10. This shows a femtocell deployment architecture and the links involved.

Of note here are the three links:

- Link 1: Mobile equipment-to-femto-base station link. This is the wireless link of most interest. Currently this is most often a cellular link utilizing licensed cellular spectrum. However, one can envision the cellular operator utilizing other spectra (such as the unlicensed bands and, of most interest to us, the TV white space) to optimize the utilization of their licensed spectrum.
- Link 2: Femto-base station to broadband router link. This is typically a point-to-point wired link.
- Link 3: Broadband router to the cellular network. This link typically uses a broadband provider's network to establish a connection between the femto-base station and the cellular provider's network via a wired cable/DSL/optical fiber broadband service. However, fixed white space links may be the broadband connection acting as a backhaul for the Internet.

15.3.1.2 WLAN

The 802.11-based provider of white space services may be operating in the 100 mW range; but is likely to also use the <40 mW and <50 mW unlicensed part of the white space regulatory spectrum. In a campus, the 802.11-based service is generally provided by an enterprise on an enterprise campus or by the educational institution if it is an educational campus. In the shopping mall example, the 802.11-based service is used sometimes by the mall as an attraction for people to hang out at the mall and at other times by the individual stores to service themselves or themselves and their customers. Internet access is transparent in the 802.11 provider use in that the primary use of 802.11 is for straight Internet access. When a user has straightforward use of the Internet, there is a lot of flexibility in how they might use services, connect to their other providers, and keep their social connections.

The types of services supported by WLAN technologies in the white spaces include:

- Hot spot Internet access
- Open Internet access with restricted QoS and bandwidth making VoIP and streaming multimedia difficult
- Open Internet access with unrestricted QoS and bandwidth making VoIP and streaming multimedia possible
- Proprietary multimedia services (e.g., store-based advertising)
- Gaming

15.3.1.3 Fixed Wide-Area WAN

A fixed wireless wide-area network (WWAN) can refer to a general wide area network coverage including but not limited to the WMAN and WRAN. A fixed WWAN provider use of white space may be similar to the cellular case in providing service directly to end user devices, or it may deliver an access point/femtocell to provide:

- Voice (via the provider's core network, maybe circuit switched or VoIP based)
- Data (Internet) access
- VoIP (via Internet)
- Messaging
- Multimedia (unicast, multicast, or broadcast)
- Emergency services
- Gaming

The 4W white space device could be a candidate for the backhaul broadband service provision, as well as direct connection to end user devices, especially in rural environments. The technologies utilized are likely to be based on 802.22 and fixed 802.16. However, other technologies, such as proprietary 802.11-like solutions, are also known.

15.3.1.4 Control of Operations in the Campus/Mall Use Case

A key feature of the campus/mall use case is the possible presence of a single entity which is either the direct operator of many of the services listed above, or can exercise significant control over individual operators of such services. For example, in the case of a university campus, the university is likely to be the operator of most of the 802.11-based networks, and it can impose restrictions on others (those deployed by students in the dorms or in the labs, for example). Moreover, through established contracts it may gain significant control over the installed cellular femtocells and/ or outsource the operation of some or all of the wireless campus network to a cellular operator.

In the case of a mall, the mall owner is likely to exercise significant control over the 802.11 and femtocell-based networks installed in the mall. Any Fixed WWAN covering the mall would likely be operated either by the mall owner or a wireless service operator under a contract with the mall owner. The same WWAN operator may then be providing coordination services or controls for wireless LANs in the mall. In the case of WWAN, such control could be enabled as part of the service provision model. In many cases (e.g., IEEE 802.22), the Fixed WWAN base station has to be professionally installed. The necessary controlling equipment and software may be provided and configured as part of that installation. However, in the case of WLAN, the control on the utilization of peer-to-peer networks will require methods that can be enabled and used by the network operator without the professional assistance of an installer.

15.3.2 In and around an Apartment Complex

The macro scenario where there are primarily high-density living spaces like apartments is one in which the predominant uses would require small coverage areas, either via WLAN APs and/or femtocells, as shown in Figure 15.11. These 802.11 APs or femtocells would operate under many of the same conditions and deliver similar services as in the campus use case. In the following discussion, only differences between the two use cases are pointed out.

15.3.2.1 Cellular

The presence of a large number of femtocell-using cellular users in a congested apartment scenario implies that coexistence strategies will be required to support such a high density of cells. The use of white space frequencies would enable more channels to support such high-density demand.

15.3.2.2 WLAN

Lots of 802.11 users in a congested apartment scenario imply that the density requires changes in the power levels and channels in order to work together effectively. However, when density reaches critical levels, allocated channels may be sufficient to provide adequate quality of service and could potentially trigger a move into white space. The use of indoor white space frequencies enables an expansion of the available unlicensed frequencies for such use (2.4 and 5.15 GHz). Given the short distance involved, operation of 802.11-based technologies under the TV white space <50 mW and <40 mW rules becomes more likely.

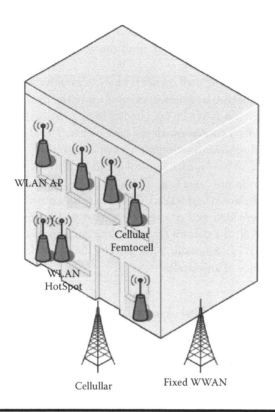

Figure 15.11   Illustration of an apartment complex.

15.3.2.3 Fixed WWAN

In the apartment building case, a Fixed WWAN may be the Internet connection and Internet backhaul via fixed white space (IEEE 802.22) or 802.16 wireless AIS. Non-802 defined technologies may be used as well. Distribution into the apartment building from the single point of the wireless antenna at the apartment complex may be WLAN based, and the WLAN may be utilizing TV band white space.

15.3.2.4 Control of Operations

In the case of an apartment building, each individual apartment dweller may be a fully independent owner/operator of their own small wireless network. Whether this operator relies on an 802.11 or a femtocell-based network extension (or both), each acts independently. Moreover, each apartment dweller may utilize different broadband service providers (in large cities it is not uncommon for 3 or 4 different cable/fiber providers to service the same building, and this may be supplemented by WWAN). Consequently, different ISPs are likely to be used as well. Each apartment dweller may also utilize their own independently selected cellular service provider.

It should be clear that in this case, coordination of white space usage through a single point of control for the building as a whole is not likely. Apartment complex owners, however, may attempt to retain central control if coexistence is not possible.

15.3.3 In the Home

The existing providers of home equipment and services may use <40 mW and <50 mW (sensing) white space devices to extend and multiply their products and services through a new source of frequencies. These will all be addressed separately.

15.3.3.1 Cellular

The cellular providers would like to have white space extensions of the cellular frequencies in the form of additional voice and Internet channels via the cellular 4W regulatory domain in addition to the use of lower-power for white space-based femtocells in the home. The cellular provider use of white space frequencies includes text, voice, Internet, and VOIP. The cellular providers will probably want to provide all their existing services, plus video and some other services that are enabled by either more bandwidth or by more frequencies, including white space frequencies.

15.3.3.2 WLAN

The WLAN providers of home equipment would like to use the white space frequencies to extend and expand the use of currently used unlicensed (2.4 and 5 GHz) frequencies for WLAN. The more bandwidth available from white space frequencies could mean more capable delivery of video or other services over 802.11-based services.

15.3.3.3 Fixed WWAN

The Fixed WWAN providers of home equipment would like to use the white space frequencies to extend, improve, and expand the use of those frequencies for Fixed WWAN. However, other technologies, such as proprietary 802.11-like solutions are also known to operate as WWANs. In considering Fixed WWAN service in the home use case, it is particularly important to differentiate between the outdoor–outdoor and outdoor–indoor antennae setting configuration for TV band white space usage.

The outdoor–outdoor setting corresponds to the likely usage of technologies based on 802.22 and 802.16 standards, as well as non-802 standards, such as DOCSIS. In this case, broadband access, including video service, is provided using up to 4 W effective isotropic radiated power (EIRP) white space channels to service fixed access points. Installation of such equipment may be professionally done.

The outdoor–indoor solution setting corresponds to the use of small indoor equipment communicating with an outdoor base station to provide service into the home. This equipment may be sited by the customer, much like WLAN access points are today. For example, modified 802.11-based solutions can presently reach the kinds of distances discussed in 4 W white space FCC specifications by using 1 W unlicensed frequencies in the 2.4 and 5 GHz bands. It is possible to use TVWS frequencies for wireless backhaul from femtocells and WLAN access points.

15.3.3.4 Control of Operations in the Home Use Case

It is possible that in this use case the home operator becomes a single point of control for a significant number of use case players, including femtocells, 802.11, and smaller non-802 devices (e.g., microphones). However, other white space players (e.g., cellular and white space TV) remain outside of the space of control of the home user. However, proximity of neighbors may result in a need for a much more complex and distributed spectrum management model, more like the case of the apartment complex rather than the campus use case.

15.4 TV White Space and U.S. Broadband Initiative

TV white space will fill a substantial role in the U.S. Broadband Initiative. The need is for the U.S. Federal Communications Commission (FCC) to allow more radio spectrum to be used for broadband access in the limited spectrum available. The FCC has made previous attempts to define and allocate licensed spectrum that is not used efficiently within geographical areas. One such attempt was the use of lightly licensed spectrum in the 3650 MHz band, where there is an authority that manages the light license of the dependent station. Such functionality is used in WLANs via 802.11y. In TV white space, a portion of spectrum may be used in a metropolitan area and not used at all in a rural area. Up until the TVWS initiative, there has not been an efficient way to share that spectrum based on the geographical area of use. The use of a shared database with geographical information, and the ability to reserve a channel through the use of that database, is a major step forward in the extension and expansion of unlicensed or lightly licensed spectra, envisioning use even in currently licensed or government designated spectra and in geographical areas where the spectra cannot presently be used. The TV white space is representative and predictive of this trend, and the expectation is that most spectra will eventually fall under the category of lightly licensed, databased, geographical information for broadband use.

Some argue that TV white space will not be needed if the TV broadcasters are consolidated into a small portion of the existing TV allocation and the remaining spectrum is auctioned off. This is not true for several reasons, including timing, openness to future changes, geographic independence, and relationship to a larger white space spectrum strategy.

The Broadband Initiative, if it is approved, will take some time to achieve that approval. The financial and political forces arrayed are quite substantial. It is difficult to predict exactly how long it would take to achieve approval, but it is likely to be measured in years. In the meantime, TVWS systems could be operating in the public interest. If the Broadband Initiative is approved, even its proponents say that it is likely to take ten years to implement. It is also likely that the implementation process will involve multiple steps, in different locations, over an extended time period. In this case, TV white space could play a key role in the migration process. Given these time periods, it is likely that several generations of TVWS equipment and services could be profitably deployed. Finally, even with full implementation of the Broadband Plan, there is likely to be demand for additional spectrum. What is learned in the TVWS bands can be applied to other bands that are not fully utilized.

15.5 A New Trend: Other Cognitive Spectra

Other cognitive spectra are really the entire radio spectrum. Once a workable concept is proven, then the entire radio spectrum falls into the category of usable if it is not being used by a licensee or a lightly licensed use (Internet-accessed database) within a geographical area.

15.6 Conclusion

TV white space is one of the projects that expands the use of the radio spectrum in a more efficient and effective manner. The applications that it can support are the applications of any radio technology; no wires and, with diligence, no interference to the others in the spectra. This chapter started with a discussion of somewhat generic application use cases applied to TV white space. Then the discussion turned to coexistence use cases. Finally, how TV white space might fit into the Broadband Plan and be applied to other parts of the radio spectrum was discussed. With this as background, it is clear that TV white space can play an important role in providing the capability that the public is demanding for more and more wireless service.

References

1. "IEEE Standard for Information Technology—Telecommunications and Information Exchange between Systems—Local and Metropolitan Area Networks—Specific Requirements. Part 3: Carrier Sense Multiple Access with Collision Detection (CSMA/CD) Access Method and Physical Layer Specifications," IEEE Computer Society, 26 December 2008.

2. "IEEE Standard for Local and Metropolitan Area Networks. Part 16: Air Interface for Broadband Wireless Access Systems," IEEE Computer Society and IEEE Microwave Theory and Techniques Society, 29 May 2009.

3. "IEEE Standard for Information Technology—Telecommunications and Information Exchange between Systems—Local and Metropolitan Area Networks—Specific Requirements. Part 11: Wireless LAN Medium Access Control (MAC) and Physical Layer (PHY) Specifications," IEEE Computer Society, 12 June 2007.

4. M. J. Colella, J. N. Martin, and F. Akyildiz, "The HALO Network," *IEEE Communications Magazine*, Vol. 38, Issue: 6, January 2000, pp. 142–148.

5. "IEEE Draft Standard for Standard for Broadband over Power Line Networks: Medium Access Control and Physical Layer Specifications," IEEE Computer Society, March 2010.

6. "IEEE Standard for Utility Telemetry Service Architecture for Switched Telephone Network," IEEE Standards Coordinating Committee, 21 September 1995.

7. "IEEE Standard for Information Technology—Telecommunications and Information Exchange between Systems—Local and Metropolitan Area Networks—Specific Requirements. Part 15.4: Wireless Medium Access Control (MAC) and Physical Layer (PHY) Specifications for Low-Rate Wireless Personal Area Networks (WPANs)," IEEE Computer Society, 8 September 2006.

8. "IEEE Standard for Information Technology—Telecommunications and Information Exchange between Systems—Local and Metropolitan Area Networks—Specific Requirements. Part 15.4: Wireless Medium Access Control (MAC) and Physical Layer (PHY) Specifications for Low-Rate Wireless Personal Area Networks (WPANs). Amendment 4: Physical Layer Specifications for Low Data Rate Wireless Smart Metering Utility Networks," IEEE Computer Society, March 2010.

9. J. Landt, "The History of RFID," *IEEE Potentials,* Vol. 24, Issue 4, 2005, pp. 8–11.

10. "ISO/IEC 18000-6:2004—Information Technology—Radio Frequency Identification for Item Management—Part 6: Parameters for Air Interface Communications at 860 MHz to 960 MHz," 2004.

11. "ISO/IEC 18000-7:2008—Information Technology—Radio Frequency Identification for Item Management—Part 7: Parameters for Active Air Interface Communications at 433 MHz," 2008.

12. L. Hanzo and R. Steele, *Mobile Radio Communications*, IEEE, edition 1, 1999, pp. 661–775.

13. S. Ni, J. Blogh, and L. Hanzo, *3G, HSPA and FDD versus TDD Networking: Smart Antennas and Adaptive Modulation*, IEEE, Edition 1, 2008, pp. 87–117.

14. A. Ghosh, R. Ratasuk, B. Mondal, N. Mangalvedhe, and T. Thomas, "LTE-Advanced: Next-Generation Wireless Broadband Technology," *IEEE Wireless Communications*, Vol. 17, Issue: 3, 2010, pp. 10–22.

15. "IEEE Standard for Local and Metropolitan Area Networks. Part 16: Air Interface for Fixed and Mobile Broadband Wireless Access Systems. Amendment 2: Physical and Medium Access Control Layers for Combined Fixed and Mobile Operation in Licensed Bands," IEEE Computer Society and IEEE Microwave Theory and Techniques Society, 28 February 2006.

16. "IEEE Standard for Local and Metropolitan Area Networks. Part 20: Air Interface for Mobile Broadband Wireless Access Systems Supporting Vehicular Mobility—Physical and Medium Access Control Layer Specification," IEEE Computer Society, 29 August 2008.

17. "IEEE Standard for Information Technology—Telecommunications and Information Exchange between Systems—Local and Metropolitan Area Networks—Specific Requirements. Part 11: Wireless LAN Medium Access Control (MAC) and Physical Layer (PHY) Specifications. Amendment 6: Wireless Access in Vehicular Environments," IEEE Computer Society, 15 July 2010.

18. "IEEE Standard for Information Technology—Telecommunications and Information Exchange between Systems—Local and Metropolitan Area Networks—Specific Requirements. Part 15.1: Wireless Medium Access Control (MAC) and Physical Layer (PHY) Specifications for Wireless Personal Area Networks (WPANs)," IEEE Computer Society, 14 June 2005.

19. "IEEE Standard for Information Technology—Telecommunications and Information Exchange between Systems—Local and Metropolitan Area Networks—Specific Requirements. Part 15.3: Wireless Medium Access Control (MAC) and Physical Layer (PHY) Specifications for High Rate Wireless Personal Area Networks (WPANs). Amendment 2: Millimeter-Wave-Based Alternative Physical Layer Extension," IEEE Computer Society, 12 October 2009.

20. "Series G: Transmission Systems and Media, Digital Systems and Networks. Access Networks—In Premises Networks. Home Networking Transceivers—Enhanced Physical, Media Access, and Link Layer Specifications," ITU-T Recommendation G.9954, January 2007.

21. "Terrestrial Trunked Radio (TETRA); User Requirement Specification TETRA Release 2; Part 1: General Overview" ETSI Technical Report 102 021-1.

Chapter 16

TV White Space Privacy and Security

Mark Cummings and Preston Marshall

Contents

16.1 Introduction

Security in complex systems such as TV White Space (TVWS) is not a simple, linear scale, where security criteria can be arbitrarily established. It is instead a compromise amongst competing values and objectives, such as the protection of the individual TVWS users' privacy and the right of the TV viewer to receive clear TV service.

Many of these challenges are not unique to TVWS. Any communications device must provide basic security and privacy services for the traffic that is communicated over it, stored on it, or generated by it. Each communications method and standard provides specific methods and tools, such as Wi-Fi's WEP and WPA, the Internet's Transport Layer Security (TLS), and the proposed 802.22 WRLAN. This chapter will address the unique security challenges introduced by TVWS operation, and will not attempt to duplicate the extensive literature on methods for wireless and Internet security. Equally, some attacks are specific to specific architectures, such as incumbent detection in dynamic spectrum access (DSA) systems, or framing of specific system designs, such as those included in WRAN designs. These also are not included due to their system-specific nature [1]. In particular, the lack of internal TVWS communications within the coexistence mechanism of TVWS greatly reduces the attack surface to communications between a single TVWS device and an authorizing database server.

There are a pair of antagonistic parameters that are well accepted in the Information Security field. This traditional pair of antagonistic parameters are labeled Type 1 and Type 2 Errors [5]. At its most basic, Type 1 Error is letting someone (or something) have access to a secured entity who should not have access. It is measured as a percentage of the excluded population let in. Type 2 Error is keeping someone (or something) that should have access out. It is measured as a percentage of the included population excluded. These two parameters are antagonistic because in the current technical environment (some argue that no matter what the technical environment, this will be true), it is not possible to optimize both. As one approaches zero the other approaches 100 percent. Thus a balance has to be struck between Type 1 and Type 2 Errors.

To these two antagonistic parameters, TVWS creates the requirement to add two additional pairs of antagonistic parameters. These three pairs are in some respects independent of each other and in other respects interrelated with each other. In all cases, as with Type 1 and Type 2 Errors, they require that a balance be achieved. What seems the best way to achieve this balance is to determine the cost associated with each Type 1 and each Type 2 Error. That is, the cost of keeping an included party out and letting an excluded party in. Costs have two components. The first is the additional hardware, software, and procedure costs of implementing the security system. The second is a combination of such items as losses from breaches, opportunity cost incurred, loss of productivity, etc. incurred as a result of both Type 1 and Type 2 Errors. This second cost component is often analyzed using a risk analysis technique. The factors that drive each of these two cost components also interact. For example, a decision to decrease Type 1 Error may increase the costs associated with the first component. Once these costs are determined, security systems are tuned to create a combination of costs associated with Type 1 and costs associated with Type 2 Errors such that the total cost is minimized. In this chapter, we will see how this approach applies to TVWS.

The second set of antagonistic pairs of parameters is protection. Protected Users as defined by the FCC include TV receivers, CATV headends, and wireless microphones. For more background information on Protected Users, please see the discussions in Use Cases Chapter [2,3,7]). Protected Users would like 100 percent assurance that they would never be interfered with by a TVWS device. That is, that no TVWS device will emit radio frequency energy in such a way as to create what appears, to a Protected User, to be radio frequency noise that materially impedes the Protected User's normal operation. Of course, the only way to achieve that is to have zero TVWS devices. In fact some suggest that the broadcasters would prefer that there be no white space rules, but rather broadcasters be allowed to lease or sell spectrum. Those with opposing views point out that broadcasters never paid for their spectrum in federal auctions the way recent cellular spectrum users have. Broadcasters reply that broadcast licenses have been sold as "good will" for many decades to the point where it can be assumed that every current broadcaster has paid for spectrum. This set of public policy issues are part of the dynamic behind the political process at play in creating and governing TVWS. As such, they have an impact on privacy and security decision making. On the other hand, potential TVWS equipment vendors would like TVWS to be as inexpensive to develop equipment for, and as easy to use, as possible. Thus they would like no rules, restrictions, or requirements to use TVWS.

The third pair of antagonistic parameters are accountability and privacy. In TVWS these two act in a similar fashion to Type 1 and Type 2 Errors. In order to have "perfect" accountability, all identity and related personal information such as location, time, etc. must be available to everyone all the time. On the other hand, to have perfect privacy all identity and related personal information such as location, time, etc. must never be available to anyone at any time. Here again, if one is maximized, the other is minimized. Here, optimization is also a balancing act that can be accomplished by minimizing the total cost. However, cost may be a bit more difficult to quantify because the information security community has less experience in this area, especially relative to TVWS. The combination of competing industry interests and public policy issues and politics further complicates the analysis. Furthermore, this pair interacts with the other pair (Type 1/Type 2) through the property of identity that can be key in both.

Thus the assurance of privacy and security in any cyber system is a complex and difficult task, at least in terms of current techniques and technology. In a TV White Space (TVWS) device there is the added complication that there are inherent tensions between the driving requirements for security, privacy, and accountability that preclude any unitary approach from meeting the many competing needs. This tension arises from the competing interests in achieving privacy of content and operation, security and integrity of the interference protection mechanisms, and accountability and traceability of any interference that might result from either proper or inappropriate operation.

We can consider that the TVWS system consists of two simultaneous planes of operation. The "*data plane*" [4] provides the same type of service that is associated with existing wireless devices, such as cellular, Wi-Fi, Bluetooth, and other well-established technologies. Where TVWS departs is the "*control plane*" segments that provide interference protection to TV receivers. The principles that underlie these segments in the control plane have no precedent in existing wireless practice or technology, or in fact, in any existing spectrum interference control regime. It is this interference avoidance control plane from which the unique challenges of TVWS arise. It is on privacy and security issues in this control plane that this chapter focuses. There is a reasonable concern about security and privacy in the data plane, but that is not significantly different in TVWS than in other wireless communications systems and is therefore left for consideration in works that focus on those concerns in a general fashion independent of the particular underlying communications technology.

Classical approaches of cryptography and task partitioning do not address these unique considerations. A TVWS device must disclose information which many users might consider private in order to meet the obligations to ensure noninterference. For example, without user and location information, the device cannot perform geographic database verification of operating frequencies in an accountable manner. However, user information of this nature violates principles of privacy that are generally assumed for services such as cellular and Wi-Fi, so these tensions are inherent in the concept of TVWS.

Another consideration is that it is contemplated that TVWS devices will exist within a competitive environment. There might be a competitive advantage to a device that might "skirt" the rules and thus potentially have a higher probability of causing interference to TV receivers and other Protected Users as well as other Unprotected Users (radio devices confirming to the regulator promulgated White Space Rules and also Coexistence Procedures developed by such groups as IEEE 802.19.1). There are two reasons that one can believe that an initially certified device might become a cause of interference.

1. DSA regulation is defined, at least partially, in terms of behavior. Conventional Part 15 devices are specified using nontemporal measures, such as power output, out-of-band emission, etc. These static characteristics are relatively simple to test, compared to the dynamics explicit in any spectrum sharing protocol. Industry certification of protocol conformance often included this, but to the spectrum regulator this is a new form of metric, and it is unlikely that, in a rapidly changing and dynamic market, all devices can get a full

evaluation. In particular, it is unlikely that all of the updates will be able to be fully vetted.

2. The temptation to improve performance to more closely achieve the potential of the hardware will be very high. The regulation of software defined radio (SDR), in general, is a work in progress. As journalists run A to B comparisons between competing products, there will be an inherent drive to push the boundary to the point where the regulatory parameters are exceeded.

Even if the vendor itself did not condone such behavior, there is an entire underground of "hacks," "jailbreaks," and other "enhancement" software that could be loaded into a device to "improve" its performance. Hacks of WiFi routers are widely available on the Internet, and many of these modify regulated characteristics, such as frequency or power. Because this is a world market, vendors build equipment that is capable of operating to the full extent of permissible performance in any of their markets. Restrictions imposed by other regulators therefore are potentially vulnerable to software transfer from other national versions, or modification. On the other hand, there are constraints that flow from technical and economic factors governing the development and manufacture of integrated circuits. Although TVWS is new in some aspects, it shares some characteristics with WiFi. In the area of skirting rules, WiFi devices could be seen in some respects as having similar advantages. The experience in the WiFi sector is thus somewhat instructive in TVWS.

It might be argued that software such as Android proves that a large ecosystem can exist without risk to the interference characteristics of wireless devices. Yet Android operates at the layers above the physical waveform and signaling protocols by design. It is not at all comparable to the regimes contemplated by TVWS, or SDRs in general.

Some can argue that one aspect of TVWS that is unique is that the device manufacturer and user have little incentive to execute the interference avoidance features of the regime, or at least have little incentive to perform these functions well. Others point out the experience of the Citizen Band (CB), where out-of-control usage destroyed a major market. These people argue that large, well-established manufacturers see themselves as having an incentive to make sure that TVWS "works" to preserve the market opportunity. If user and manufacturer self-interest is absent as a controlling force, the security architecture of a TVWS is the sole mechanism to ensure compliance with the spectrum sharing regime.

On the other hand, others point out that TVWS could be the first of an increasing use of the White Space concept in spectrum allocation. Thus, large industry players who would like to see the White Space device market grow and prosper have an incentive to conform to the rules and make TVWS work in all respects, including protection of Protected Users and giving purchasers of TVWS equipment a good user experience. These people recognize that small unscrupulous companies trying to make a quick buck may have an incentive to field "defective" equipment, but they point out that, given the economies of scale at work in the semiconductor

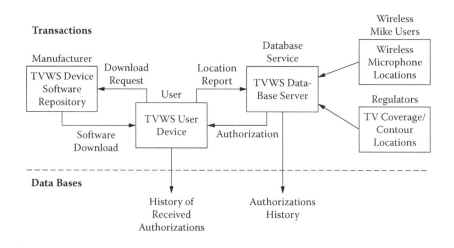

Figure 16.1   Core elements of the TVWS architecture.

sector, it may not be easy or economically attractive to do so. In support of this view, they point to the WiFi experience. The security and privacy concerns are thus not simply the privacy and confidentiality issues that are common throughout the Internet. A much more profound set of challenges must be met to not only ensure privacy and confidentiality, but also accountability for actions, assurance of the integrity of the spectrum sharing regime, and the privacy of the user.

In the following section, we will consider two mechanisms to ensure the security of the TVWS architecture. These involve the transactions between the elements of the architecture and the content of the databases retained by the architecture elements. Figure 16.1 illustrates the critical transactions and databases that will be central to the security strategy.

16.2 Encryption Methods and Security Protocols

Throughout this section, we will refer to several types of encryption methods. A brief summary of these is shown below:

- **Symmetric:** Symmetric encryption is the commonly applied technique of using some key material to both encrypt and decrypt a message. This method is the most convenient to process (lowest processing resource requirement). It is a reasonable technique between two users. It has the disadvantage that any device that can decrypt a message also can encrypt it. It is thus not possible to authenticate the sender of a symmetrically encrypted message (nonrepudiation), because any device that holds the key to decode it can also encrypt using it. These systems are useful in a peer-to-peer arrangement when all nodes are equally trusted, have no "competitive" incentives, and can protect the key material. Thus this process creates confidentiality, but not a trusted identity (assuming more than two parties hold the key). This confidentiality is within the pool that share the key. Any compromise of the key by any member makes the confidentiality of all members nonexistent. An example of this is Wi-Fi Protected Access (WPA) encryption, which protects communications and access to only users who hold the key, but provides no confidentiality between these users, because all users have the same key and thus can read each others' packets as if they were in the clear.

■ **Asymmetric:** Asymmetric encryption uses different key material to encrypt and decrypt messages.* The decryption key is referred to as a "public key," and is typically widely distributed. The encryption key is the "private key" and is held by the authorized generators of messages. The public key cannot encrypt messages, so a properly decrypted message is proof that the sender held the private key, and thus the action of sending the message cannot be denied, or repudiated, because it cannot be replicated by any other sender. This process is often not intended to create confidentiality, but it constitutes a "signing" action. However, the process can be used to form a symmetric key, which can be used for traffic encryption, as described above. The two parties interact through several messages to generate this key; however the key itself is never transmitted between the nodes, and knowledge of the transmissions is not adequate to determine the key that was generated. This avoids the need to preposition or exchange key material other than the original public credentials. Also, it segregates confidential sessions, because the key formed interacting with one certificate holder is unique from that formed when interacting with another certificate holder. Again, using a WiFi example, this service would not provide for the ability to restrict users of Wi-Fi network, but it would provide confidentiality for each user's traffic.

Although these constructions are the basis for the security architectures developed in this chapter, the details of each are outside the scope of this chapter. For more details, please consult standard texts in each area.

Hashed Message Authentication Codes (HMAC) implement the authentication feature. They create a hash from message content and the private key. A recipient can compare this hash with the message and the private key and verify both the identity of the sender, and the integrity of the message (since any change in the message, or an incorrect private key will create a hash that will not match the message). For example, if a message authorizes a transaction, the same message cannot be modified without a change in the HMAC code, and thus it is impossible for anyone but the original signer to modify it. If the message includes temporal

---

* The specifics of the number theory that enables the construction of these two key materials is beyond the scope of this book.

information (such as a sequence or time validity), the message cannot be later retransmitted without detection.

Another example is the Transport Layer Security (TLS) [6] and Secure Socket Layer (SSL) protocol. SSL, now incorporated into TLS [6], is commonly used in current Web-based e-commerce applications. Instead of providing the end-points with a unique key set, the SSL protocol has a mechanism for the end-points to manufacture a session key. Many people in the industry feel that the process of manufacturing the key set provides an adequate level of security for most applications. Once the key set is manufactured, it can be used to protect information flowing over the link. If there is a trusted certificate authority available, SSL can also be used to verify the identity of a server such as a TVWS Database Server. However, unless there is a certificate distribution mechanism for each TVWS device, it is not possible to verify the identity of the end-point (TVWS device). The Internet HTTPS protocol uses TLS to validate Internet hosts and provide some security for connections.

This protocol can establish the identity of at least one end-point through asymmetric public/private credentials to assure trust in the end-point of the connection. Using HMAC as an example, a client process or node can access the bank server with any credential, but only one possible credential can enable a trusted connection to the server, presumably the bank's server. Two-way authentication can also be performed, such as in a corporate virtual private network (VPN). In this case, only a specific authorized remote computer(s) is allowed to form a connection, and it must prove so by asymmetric identification. A common symmetric key* is formed through nodes message exchanges of values derived from the public and private keys, and this symmetric key can then be used to ensure traffic confidentiality.

The Internet HTTPS protocol uses TLS to validate Internet hosts and form secure connections. HIP (Host Identification Protocol) uses a combination of asymmetric encryption and an identity server to both assure identity and protect identity privacy. These standards are widely implemented and applied in the Internet world, so they are robust, and also have extensive research into possible vulnerabilities (of which numerous examples have been identified and, presumably, repaired).

16.2.1 Accountability

A principle of radio regulation is that transmitters are accountable for their radio frequency (RF) emissions. This is particularly true in a service such as TVWS that operates in a noninterference status, which requires that they cease operation if it is determined that they are causing interference to Protected Users. The process for assuring that a device operates properly thus implies that the identity of an interfering node be available for interference resolution. Similarly, in a geo-location-based

---

* Even though these exchanges can be intercepted, without knowledge of the two private keys, the symmetric key that is formed cannot be determined.

system, the device would want to have a nonrepudiatable* message authorizing that operation was permitted, so that the device could demonstrate that it had followed the required policies and protocols appropriately. The accountability is therefore absolute because both the requesting and authorizing parties to the geo-location transaction have high assurance that the node authorizing and requesting have the identity which they assert.

Although the emphasis in most TVWS discussions has been on ensuring that the devices honor the restrictions provided by the geo-location database, a potential concern is also that the database may be too restrictive. That accountability for Protected Users being too restrictive in their service contours, thereby unnecessarily restricting the operation of TVWS devices should also be considered that is that the area designated as not available for use by white space devices may be larger than actually needed to protect protected users.. However, this is primarily a function of the rules and procedures developed by the regulator. Thus, it is the regulator's responsibility over time to tune the rules governing contours to optimize the performance of the system. It is recognized that regulators are subject to difficult political pressures, but there does not appear to be another way to resolve this issue.

16.3 Privacy Challenges

Now we confront the antagonistic parameters of accountability and privacy. The TVWS rules are based on a set of fundamental premises. One is that all TVWS devices must contact a TVWS database provider to get a list of potentially usable channels in their location. In doing so, some are required to reveal the device's identity and its owner's identity. For many of the others, identity can be deduced. In all cases, they have to reveal their location. Thus, privacy is one of the areas where there is a fundamental challenge between the spectrum sharing regime and the accepted principles of confidentiality. This problem is not unique to TVWS devices; cellular technology also has the same challenge with the pervasive opportunity for disclosure of location and identity information. However, cellular operators are subject to a number of statutory controls over disclosure of customer information. At the time this book was authored, there were no similar controls over information dissemination and retention for operators of TVWS devices or TVWS database operators. Even the statutory controls have been circumvented by illicit "pretexting,"† or by location-based services that are inappropriately accessed by third parties.

Certainly a statutory obligation for confidentiality is desirable, even if it is not completely adequate. The rest of this discussion will focus on the technical

---

* Nonrepudiation means that the holder of the message can "prove" that only a specific node (or class of nodes) could have generated the message.
† That is when a third party imitates the account holder and obtains unreleasable information under a pretext of a disclosure that is permitted to the account holder.

protection. This can be supplemental to statutory protection, or, in the absence of statutory action, a primary mechanism.

16.3.1 Confidentiality of Identity

For device types that do not have to reveal their user's identity or for which the user has entered false identity information, there are several ways that identity can be revealed. One way that identity can be revealed is through purchase information. Another way is through warranty information. A third way is through patterns of use. For example, a device which operates at the geo-location of a specific house for a significant portion of its usage can be ascribed to a user residing in that house. Then other information sources can be used to associate an individual with that particular house.

Identity need not be immediately associated with an individual's name, address, or other personal data. It is enough that it be associated with a pattern of behavior (on and off times, location, access patterns, etc.) to establish a violation of personal information. Even fixed devices can portray information about personal habits which could be exploited by an observer.* Yet, the identity must also be traceable back to the original source of the transmission in case of interference.

16.3.2 Confidentiality of Location

Location information is one of the mechanisms by which an observer could associate specific nodes to identity of the users. It can also compromise personal behavior in the same manner as cell phone location services. This is an area where some people are particularly sensitive. Some people feel that the ability of an individual, organization, or government to know where a person is at all times is a particularly egregious invasion of personal privacy and a path to the loss of important universal human rights. On the other hand, it is important to note that this varies greatly by country and culture. Some also state that it varies by age, where young people are less concerned. Here again, opinions differ. Some feel that younger people are willing to have their location known, but they want to control when, how, and to whom that information is available.

Location and identification information provides increasing correlation of behavior as the database retention period is increased. It is important to understand how this information is needed to resolve interference, validate the operation of the spectrum sharing regime, and ensure accountability.

Exposure of location information without respect to identity may also represent a risk. Currently, some systems use the concentration of cell phones, or their speed moving through a particular area to determine the presence or absence of traffic jams on freeways. The same approach could be used to predict the appearance of

---

* For example, house burglars used travel information obtained from social networking to identify houses that were likely to be unoccupied.

flash mobs in political protest situations. Such spontaneous protest groups have played a key role in democratization in some countries, and the ability of undemocratic governments to use this information to prevent such protests could be seen by some as a loss of basic human rights.

16.4 Security, Integrity Challenges

It is likely that the implementation of a TVWS device will be heavily software-based. Previous generations of consumer equipment were much more limited in their capability to cause interference. WiFi equipment was only capable of emitting on frequencies that would not interfere with licensed services, and the cost advantages of very fixed-behavior integrated circuits precluded extensive software control. Cellular devices offered more levels of software control, but the basic operation of the RF section remained tightly in the control of the cellular operators, even when devices were "hijacked" or "rooted"! With TVWS, the consumer is buying and controlling a device whose interference avoidance behavior is solely in software, and where there is no operator to control all of the software that is loaded onto the device.

This has significant impact, because it has been hard for device manufacturers to ensure that the software validated by the factory and certified by the regulator is the only software that can be run on a device. Without this assurance, it is possible that TVWS devices will have their interference protection features amended to reflect more aggressive spectrum usage. Simply put, the device software can be a threat to the spectrum sharing regime. Most consumer equipment is relatively constrained in its capability to interfere due to limitations in frequency coverage, power, fixed filter, and other constraints. Wireless LAN technology is also typically in a commons frequency band where other devices must accept interference, and often are tolerant of interference due to their unprotected status. A TVWS device, on the other hand, may be capable of operating on a wide range of frequencies. It may be allowed to emit higher power than typical WiFi devices, and most importantly, it is operating in bands where significant interference could be caused to Protected Users.

Additionally, there may be advantages to TVWS users to defeat some of the interference protection mechanisms to enhance access to spectrum. Therefore TVWS devices may become a target of modified software that seeks to defeat the protection and accountability features, much as some consumer WiFi and base station hubs have been modified to open up nationally imposed spectrum or power limits, but with even more disruptive effects.

It is worthwhile considering how these features could be used to defeat the types of attacks that are common in today's Internet. Access to the TVWS database will be via the public Internet, so the same attack and exploitation methods that are common on the Internet can be anticipated to be applied to attack these systems. Some common attacks and their mitigation through these techniques are shown below:

■ **Man in the Middle:** In a man-in-the-middle attack, an unauthorized node relays communications between two nodes that are mutually authorized. In TVWS, such a node might want to modify authorizations, collect usage information from users, or route requests to a TVWS database service for which the node was not an authorized user. The asymmetric encryption feature enables both users to validate that the messages were both initially created by the stated endpoints, and were not modified by any intervening node. It also is a mechanism to form a content protection key that assures that the node monitoring the traffic cannot read the traffic. IPSEC also includes the IP address within the signed message, so that any attempt to modify and relay the message through an intervening node is detectable.

■ **Domain Name Server (DNS) Poisoning:** Severe vulnerabilities have been demonstrated in many of the implementations of the DNS service. A false domain name server entry can cause a node to initially access a man-in-the-middle node, but when that traffic is then sent to a proper node for processing, the inclusion of a different IP address on the signed returned message will be apparent. Alternatively, the node can masquerade as an authorized TVWS database server, but this would be immediately apparent by its lack of an appropriate credential.

■ **Denial of Service:** A denial-of-service attack overwhelms the resources (communications or processing) of a node by generating more requests than it can receive or process. There is little that can be done to resolve attacks that are so intensive that they overcome the bandwidth to the server. However, the use of asymmetric encryption enables the processing node to detect the identity, and thus the validity, of the requesting node. Processing of the request can be more rapidly terminated, thus minimizing the resource expenditure and reducing the effectiveness of the attack.

■ **Malware Exploits:** A fully configured Internet device has over 64,000 individual ports open to attack by malformed packets, open vulnerabilities, and other attack vectors. Asymmetric encryption enables the device to reject third party messages that are from any unanticipated source, which can immediately reject most attack messages. Limiting the number of ports that are accessible is important to avoid attacks on the underlying operating system, which will most likely be a commercial product with numerous vulnerabilities on other ports and invocable applications. These must be inaccessible from the Internet side of the device. The TVWS authorization must be operated in a "sandbox" that is isolated from the over services on the device.

16.5 Directions toward Solutions

In the following discussions there are a series of discussion items that are not easy to describe in a single word, but come up over and over again. To support these discussions, these items are labeled with a phrase and then given a label. The label consists of number or letter and a single capitalized word. The single capitalized word is only a label and should not be confused with the meaning of the whole phrase that it points to.

From the preceding section of this chapter we can see that there are three major areas of concern:

Privacy and security of communications between the TVWS device and the database service

Privacy and security of software download to the TVWS device

Privacy of user identity, location, and activity information communicated to the database service

Although they are interrelated for ease of discussion, each of these will be considered in a separate section below.

16.5.1 Privacy and Security of Communications between the TVWS Device and the Database Service

The TVWS device must be certified by the regulatory agency for the country it will be used in. Upon certification, the specific model certified is given an identification number. In use, the TVWS device must send messages to a database server that include this number along with location information and in some cases user identification information. It appears that there are three levels of security possible:

A.CLEAR)—Send everything in the clear.

B.WEBSEC)—Use some combination of web commerce security tools such as SIP, SSL, HTTPS, etc.

C.ASYM)—Use some form of asymmetric encryption with private keys distributed in a separate secure process to both TVWS devices and TVWS Database Service Providers.

A.CLEAR) has the advantage of being very easy to implement. However, it provides no privacy protection for the end user and no accountability for the device. Anybody who wants to capture end user information can simply listen and catch the messages as they go by. Similarly, an approved regulatory device ID can be captured by an unapproved device and used to impersonate the approved device. On the other hand, it is the cheapest and easiest alternative to implement.

B.WEBSEC) provides some protection for end user privacy. It can be argued that since this is being widely used for e-commerce in conjunction with Web services today, it is acceptable. Others argue that it is not sufficient for existing e-commerce applications. One might ask, if it is not sufficient for e-commerce applications today, why is it being used? Those that argue it is insufficient, answer that there are a series of other controls that are used in conjunction with these techniques. For example, credit card statements are sent to individuals on a monthly basis and provide a mechanism for individuals to contest unauthorized transactions. Also, credit card companies agree to take the financial burden in case of some types of security breaches. The credit card companies see the losses from inadequate security as a cost of doing business and are worried about consumers losing confidence in the system if they add too obvious security technology. Industrial security techniques go well beyond these consumer techniques. These people argue that, therefore, such e-commerce techniques are not adequate for TVWS accountability (due to the focus on protection of content via encryption, but weak or nonexistent validation of the server and user node identity). Accountability protection is a problem because the regulatory ID is available in portions of the system for portions of time in such a way that it can be copied. This approach is attractive to existing Web service providers because it is a technology that they are very familiar with. From their perspective, it has little cost to implement. Device manufacturers see it similarly.

There is also the potential threat of a rogue Database Service Provider, for mischievous reasons, some currently unknown financial benefit, or for political terrorism reasons to pose as a regulatory agency-authorized and correctly operating Database Service Provider. Some of the tools in this category do allow a service provider to provide a digital certificate that indicates an authority derived from a trusted source. These certificates provide some assurance and some accountability. However, it is important that the integrity of the process of trust be maintained. It is not enough to have certificates offered by a Database Service Provider, if the basis for determining the trust of their issuing authority is unknown! Therefore, the trusted process must be present from the initial development and load of the base software, as well as any subsequent updates.

C.ASYM) can provide both good accountability and privacy protection. Asymmetric encryption can be used to protect the manufacturer ID and device information while in transit from the TVWS device and the Database Service Provider. It can also be used to protect the device and user identity information in transit to the Database Service Provider. It can also be used to provide more reliable exchange of certificates to indicate that the Database Service Provider is a trusted source, and it can provide very good accountability trails that can even differentiate between different service points of the same Database Service Provider. A cost analysis needs to look at actual costs and perceived costs.

The actual costs to the device vendor are relatively small. The processing and storage required to implement the asymmetric encryption is well within the capabilities of what is already likely to be there for other reasons. To reduce development time, it is possible that IP will be licensed, but the cost of the licensed IP, if any, should be very low. The vendor will have to set up processes and procedures to manage and secure the keys involved. This may require some relatively small additional labor expense.

For vendors who have not had previous experience with asymmetric encryption (asymmetric encryption of this type is not common today in consumer electronic products), the perceived cost may be somewhat higher just because of anxiety about something new. Because it is new, some vendors may overestimate the costs involved and so resist this approach. However, even with overestimation, the impact is seen as being over once the product is shipped.

For Web service companies considering being the Database Service Provider, C.ASYM) creates processing overhead added to processing each incoming message. In the context of today's technology, the increase in processing required is small. If there are a large number of messages generated by a very successful uptake of TVWS devices, the increase in processing is still modest. However, if a precedent is set, the demand could materialize for widespread use of asymmetric encryption to protect users of Web-based services. That could create an increase in processing requirements, and that prospect scares some Web service providers. Others suggest that protecting users of Web-based services is long overdue, and these fears should not be allowed to stop the provision of needed security. So, here as for the vendors, it is perceived cost rather than actual cost that is high. There is a caveat, though. That is, that the regulators around the world harmonize their approaches.

For regulators, the issue is global circulation of TVWS devices. If every country has its own certification process and gives out its own manufacturer, equipment, and encryption key, the burden on device vendors and Database Service Providers becomes greatly magnified. In A.CLEAR) and B.WEBSEC) the messages between the TVWS devices and the database just become much longer. In C.ASYM) the TVWS device now has to have a list of encryption keys and countries. Since the device has to know its location when it sends a message to a database, this by itself is not too great a burden. However, without coordination and harmonization, multiple approval processes can become burdensome. What is needed is a system similar to that used with certain cell phone standards today, where several certification labs in several parts of the world perform certification testing that is accepted by all administrations. Using this kind of certification process, the TVWS device vendor only has to have a new device certified once and for C.ASYM) only gets one asymmetric key. This makes the whole operation much less costly for all involved. As long as the number of Database Service Providers remains relatively low and the various country, region, and international regulatory bodies are well coordinated, the above approach is likely to be adequate. If, however, coordination is not good and the number of Service Providers becomes large, HIP may become necessary.

16.5.2 Privacy and Security of Software Download to the TVWS Device

Now, let us look at the question of software download and then come back to how to choose between the available alternatives. Here again there are several choices:

W.NONE)—Provide no protection against injecting new software, which makes a previously certified TVWS device now able to operate without obeying the TVWS rules.

X.FIRST)—Require that TVWS devices only operate using software contained in internal ROM that cannot be changed or overridden.

Y.ID)—Use the certification process described in B.) above to make sure that new software comes from a trusted source.

Z.IDMAC)—Use an asymmetric key system to make sure that new software comes from a trusted source and that it has not been changed en route.

W.NONE) relies on a perception by the overwhelming majority of people that it is not worth the trouble to try to modify the operating software of TVWS devices in order to circumvent the rules and that the damage done by the very few who try is minimal. This is the general approach taken by the Citizen's Band (CB) industry. In that case, it turned out that the assumption was wrong. Large numbers of people sought to gain advantage by circumventing the power limitations in the Citizen Band Rules and thereby created so much interference for the rest of the users that the general public stopped using CBs. Thus CB went from a successful mass market industry with large numbers of consumer electronic companies producing product for a large group of users who used their CBs frequently to, at best, a tiny niche market that all the consumer electronic companies had exited.

X.FIRST) provides a good security solution, but it has a serious Type 2 problem. There is no way for vendors to fix software bugs in the field, nor is there a way for TVWS to evolve using software downloads to update devices in the field. The experience with cellular handsets was that this kind of restriction was very expensive for the vendors. Possibly of more import is the fact that TVWS is a relatively new approach to spectrum allocation. As such, it is likely to need to grow and evolve based on experience. Thus a software update security approach which prevents fielded devices from evolving is an extremely "expensive" solution.

Y. ID) opens the door to field updates while providing some protection against rogue Software Update Servers. However, it is weak in preventing software updates from being corrupted in transmission. It is also weak in accountability of which of several access/service points of a Software Update Service a particular software update came from. This approach is one that has been developed and widely used in updating software for individual PCs and cellular phones. In that context it operates within an institutional and incentive structure which does not exist in TVWS. In the cellular case, there is a very powerful network operator that controls all the nodes in the network. In the PC case, there is an individual owner who has a big incentive to protect his/her own data, operation, etc. Even so, there are stories about breachers in these areas every day. Those industries are continuing to evolve their security systems, trying to balance the cost of protection with the cost of breaches. In this evolutionary process, there are strong voices that maintain that approach Y.) is not sufficient for the constrained context of PCs and cell phones.

In the TVWS context, there is no operator to maintain strong central control. Also the TVWS users' incentives are different. Corrupted software could lead to the end user feeling that they personally were getting better performance initially. By the time that enough TVWS devices have been corrupted for the end user to notice performance degradation, things may have gotten so far out of hand that a collapse similar to the CB collapse may not be preventable. This is particularly true if both the response to regulatory rules and the response to collaboration standards such as those being developed by IEEE 802.19.1 are affected. Of course the Protected Users are likely to experience degradation quickly and use the Black List feature and regulatory policing to remove the corrupted device from service. For small numbers of corrupted devices scattered in time, this is a potential solution. For large numbers of devices corrupted in relatively small blocks of time, this may not be a fully effective solution.

Z.IDMAC) provides a potential comprehensive solution. It can be used to certify that the software comes from a trusted source. Through the use of HMAC or full encryption of the code, it can be protected from corruption in transit. Full accountability of the access point/service source can be provided. The costs are similar to those described in C.) above.

Deciding between the various alternatives is challenging because it combines tension between powerful interests, highly technical matters that are not well understood by all parties, and conflicts between actual and perceived costs. As advocates for privacy and security, the authors would recommend that C.ASYM) and Z.IDMAC) be chosen. The political process is more likely to look for a compromise and lean toward B.WEBSEC) and Y.ID). Therefore the pragmatic choice is to recommend that B.WEBSEC) and Y.ID) be implemented in such a fashion that they can evolve into C.ASYM) and Z.IDMAC) as experience more clearly dictates the need.

16.5.3 Privacy of User Identity, Location, and Activity Information Communicated to the Database Service

Now we come to questions of security and privacy surrounding the personal information that ends up at the Database Service Provider. This can be broken down into three objectives:

1. CORUPT)—Protecting the Database Service Provider's system from being corrupted
2. DISC)—Protecting end user privacy from unintended disclosure
3. HR)—Protecting end user privacy from intended disclosure which is in contravention of basic human rights

We will start the discussion with 3.HR). The question here is what kinds of information about individual activity should not be available. Various countries have various approaches to this question, and there is also United Nations policy in this area as well. TVWS is likely to be first implemented in the United States. Therefore, in this discussion we will use the legal and public policy situation surrounding privacy in the United States as the context. In that U.S. legal and public policy context, it is generally agreed that a citizen has the basic right to assume that his everyday activities can be undertaken with an expectation of privacy unless a court has provided law enforcement authorities with a warrant. The privacy problem that TVWS creates is that, as discussed above, TVWS Database Service Providers are collecting information that can remove that expectation of privacy through revelation to nongovernmental organizations/individuals or revelation to governmental organizations without proper judicial authority. This is the conflict between the antagonistic parameters of accountability and privacy.

This problem can best be addressed by a combination of strict policies about the distribution of personal information supported by policies regarding "*forgetery*." "*Forgetery*" is a word coined to be obviously related to "memory" in a systems sense. That is, just as memory must be systematically maintained, the process of disposing of information must be systematically maintained. When computing systems were first developed, memory technology was so fragile that, if very special efforts were not taken, information would disappear. Now, memory technology and general digitization has progressed to the point where very large amounts of personal data are captured and without special efforts at protection can be aggregated in ways that defeat previously existing guarantees of privacy. Policies about release of personal information have proved important and helpful, but experience has shown that, if personal information is kept, it tends to leak out in spite of the policies. Therefore, selective destroying or "*forgetery*" of personal information must be an important part of privacy protection.

In the policy area, TVWS Database Service Providers should be required not to reveal any information collected to any private individual or organization (including other commercial or government services provided by the same company operating the Database Service) without the express written authorization of the affected individual. Furthermore, an "opt out" authorization should not be accepted. An informed "opt in" authorization should be required. However, this is insufficient by itself, because of the difficulty of maintaining this protection if large amounts of personal information are "laying around." If active steps are not taken to make sure that sensitive data is destroyed, employees get the unspoken message that protecting it is not important, and information starts to leak out. Therefore, "*forgetery*" must be employed. For accountability reasons, logs of activity on TVWS Database Servers must be kept. A specific time limit needs to be set, after which all such information is destroyed. A good starting point for consideration of that time limit is in the range of 48 to 72 hours. This requires that, if there has been an interference event with a Protected User, the system (including the government portion of the system) must act within three days. That does not seem to be an unreasonable expectation.

In 2.DISC) the risk is that personal activity information stored in the TVWS Database Service is obtained without authorization and in spite of steps by the Service Provider to prevent it. Internal losses can be protected against by encrypting the data as it is stored in the Service Provider and using accepted division of duty techniques to manage the encryption keys. Additional steps may need to be taken to prevent access to the information while it is being processed, if that processing is occurring in a multi-application environment such as a Cloud Computing facility. The major external vulnerability is the link between the regulator and the TVWS Database Service Provider. This link can be protected by making it a point-to-point link and not part of the public Internet and then using heavy encryption with keys that are changed frequently on a varied schedule. Symmetric keys and double/triple encryption may be sufficient here.

1. CORUPT) is similar in some respects to 2. DISC) because a major route of attack could be the link with the regulator. However, there are other ways that a Service Provider can be corrupted. This is a constantly evolving arena. Commercial and government Web service providers in general are finding that, as they develop techniques that close off one avenue of attack, other avenues are discovered. The TVWS Database Service Providers should monitor this "spy/counter spy" dynamic and implement each protection as it becomes available.

16.6 Conclusion

We have seen that TVWS privacy and security are characterized by a series of pairs of antagonistic parameters. Some of these pairs have been with the industry for a long time, and some of them are new, generated by unique aspects of TVWS. We have also seen that the privacy and security solution space is complicated by the political interactions of large commercial players with significant financial considerations. Given that political process, it is likely that somewhat less than fully optimal security provisions will be initially implemented and that the focus of security professionals in this space should be to insure that there is an evolutionary path available to those stronger security techniques as it becomes clear that they are needed. Finally, there are a set of privacy threats that are inherent in the TVWS Database Server, and these must be protected by a clear and relatively short deadline for information destruction combined with strong internal security techniques and policies.

TVWS specifically and White Space in general can play an extremely important role in providing the public with the spectrum necessary to meet society's growing demand for highly productive wireless communications. Mandating adequate security and privacy protection can make sure that the White Space technique is available to meet this growing need.

References

1. Kaigui Bian and Jung-Min Park. "Security Vulnerabilities in IEEE 802.22," in WICON '08 Proceedings of the 4th Annual International Conference on Wireless Internet, 2008.
2. Federal Communications Commission Office of Engineering and Technology. Initial Evaluation of the Performance of Prototype TV-Band White Space Devices. July 2007.
3. Federal Communications Commission Office of Engineering and Technology. Evaluation of the Performance of Prototype TV-Band White Space Devices Phase II. OET Report: FCC/OET 08-TR-1005, October 15, 2008.
4. H. Khosravi and T. Anderson. "Requirements for Separation of IP Control and Forwarding," RFC 3654. The Internet Society, November 2003.
5. William Stallings. *Cryptography and Network Security: Principles and Practice.* New York: Prentice Hall, 2006.
6. The Internet Society. The Transport Layer Security (TLS) Protocol Version 1.2, RFC 5246, August 2008.
7. Federal Communications Commission. Noticed of Proposed Rulemaking in the Matter of Unlicensed Operation in the TV Broadcast Bands. Docket 04-186, May 13, 2004.

Index